BT EXACT COMMUNICATIONS TECHNOLOGY SERIES 3

Voice over IP
systems and solutions

Other volumes in this series:

Volume 1 **Carrier-scale IP networks: designing and operating Internet networks** P. Willis (Editor)
Volume 2 **Future mobile networks: 3G and beyond** A. Clapton (Editor)

Voice over IP
systems and solutions

Edited by

Richard Swale

The Institution of Electrical Engineers

Published by: The Institution of Electrical Engineers, London, United Kingdom

The Institution of Electrical Engineers,
Michael Faraday House,
Six Hills Way, Stevenage,
Herts. SG1 2AY, United Kingdom

British Library Cataloguing in Publication Data

Voice over IP.
(BT communications technology series; no. 3)
1. Internet telephony
I. Swale, R. II. Institution of Electrical Engineers
004.6

ISBN 0 85296 024 7

Typeset by Mendez Ltd, Ipswich
Printed in the UK by T J International, Padstow, Cornwall

ιχθυς

CONTENTS

PREFACE

The landscape of the information technology and communications industry is periodically marked by the advent of new technologies that rise to challenge either the structure of the industry or the nature of the market-place. With the benefit of hindsight, a few technologies, such as personal computers and cellular mobile radio, can be considered special since they have been able to achieve both. This is because they are not only distinct technologies in their own right — and so have opened up new opportunities for applications that have ensured their success — but because they have also presented challenges to established practices.

McAuley's axiom states that: "If a system is of sufficient complexity, it will be built before it is designed, implemented before it is tested and outdated before it is debugged." Although 'Voice over IP' (VoIP) has been in existence for many years, and much longer than many appreciate, only relatively recently has it evolved from a simple capability for transporting voice communications into a much more powerful technology capable of changing the way voice applications are constructed, delivered, marketed and sold. It is in these areas that VoIP can still be considered an emerging technology and it is in these areas that VoIP must cross the technology acceptance chasm. There is every indication that this acceptance has already started to happen and the question that has been in the forefront of most people's minds has therefore migrated from "Will it work?" to "What will the industry look like when it is delivered at scale?"

While VoIP has traditionally been associated with the public Internet, that is only one manifestation of the technology. Another is its deployment in managed IP networks, including enterprise networks, where voice and real-time traffic can be isolated from not only the public Internet but also such vagaries as the absence of end-to-end quality of service, concern over network integrity, and a potential lack of security. Indeed these idiosyncrasies point towards the evolution of a new generation of converged networks that are neither the public switched telephone network that we know today nor the public Internet in its current form. Rather these networks may exhibit the paradoxical characteristics of being both and yet neither at the same time, in the sense that best-effort end-to-end applications may be supported concurrently with those requiring more careful and serious treatment with an appropriate service surround. The importance of this latter point should not be underestimated, since it is this service surround that must provide the vehicle for generating the realistic revenues required to justify the necessary investment, thus avoiding the 'free lunch' that has brought so many 'dot-com' companies to a grinding halt.

The industry as a whole — from customer to service provider, from equipment manufacturer to regulator — must continue to face, and overcome, the many and varied challenges presented by the development and adoption of VoIP. And yet, from certain, uninformed, perspectives, such concerns may appear inconsequential. In reality, however, the underlying issues are incredibly complex, since the separation of application from transport, that is implicit within VoIP technologies, challenges the very foundations of voice — and data — network design and blurs the boundaries between fixed, mobile and multimedia. Yet while providing challenges to existing networks and practices, VoIP also affords new openings for those who are able to both envisage the opportunities and also be sufficiently fleet of foot to make their way through the turmoil that has become this vibrant industry to seize them.

This book considers some of these issues and opportunities. However, as with any broad subject area, it is always a question of what to leave out rather than what to include. In this case there is only sufficient space to present a selection of the views of those who are working on the various facets of VoIP. I therefore take this opportunity to thank my many colleagues, from across both BTexact Technologies and the wider industry, who have been able to contribute to this book by sharing their knowledge and experiences. I hope you find it to be both informative and enjoyable to read.

Richard Swale
VoIP Technologist, BTexact Technologies
richard.swale@bt.com

ACKNOWLEDGEMENTS

I imagine every book has a tale to tell concerning its production — and this one is no exception. Beyond the authors and contributors themselves, there have been a large number of people involved in creating this book who have helped or contributed in a variety of ways. To them I owe my thanks. In particular I would like to acknowledge Ian Dufour and the editorial board of the BT Technology Journal for accepting my initial proposals for a book that curiously was re-engineered into the April 2001 edition of the BT Technology Journal. My thanks also go to Eric Allard who, as my manager through most of the production of this book, provided me with so much encouragement right from the first days of my involvement with the VoIP industry and afforded me the flexibility and freedom to contribute to its development in so many ways.

I would also like to thank Andrew Jell, Pat Hicks, Dave Clough and Karen Garnett of Mendez Ltd for preparing the illustrations, typesetting the book and helping me proof read it, and Roland Harwood of the IEE publishing department for his frequent encouragement to both Andrew and myself to finish it. I am also indebted to my parents who supported me through some difficult times during the writing of this book and in particular to my father, David Swale, for his critical eye.

Finally I would like to thank Bethany and Isaac Swale for providing the continued motivation for completing this book.

Richard Swale

CONTRIBUTORS

T Baden, VoIP Network Solutions, BTexact Technologies, Adastral Park

R H Brennan, Principal, Telxxis, San Mateo, CA, USA

A B Catchpole, IP Telephony Solutions, BTexact Technologies, Adastral Park

Q G Collier, System Switching and Intelligence, BTexact Technologies, London

S J T Condie, VoIP Solutions Designer, BTexact Technologies, Adastral Park

G Hare, Softswitch Designer, BTexact Technologies, Adastral Park

J R Harrison, Signalling Protocols, BTexact Technologies, Brentwood

A J Heron, VoIP Technology Evaluation, BTexact Technologies, Adastral Park

K D King, Signalling and Performance, BTexact Technologies, Adastral Park

R R Knight, VoIP Standards, BTexact Technologies, Adastral Park

P A Mart, Marconi plc, London

C J Middleton, Connectivity, BT Ignite Solutions, London

S E Norreys, Signalling and Protocols, BTexact Technologies, London

J C Parr, Telspec plc, Rochester, Kent

R J B Reynolds, PsyTechnics, Ipswich

A W Rix, PsyTechnics, Ipswich

B Rosen, Marconi Networks, Warrendale, PA, USA

Dr G Rufa, Professor of Physics and Mathematics, Mannheim University of Applied Sciences, Germany

P G A Sijben, Lucent Technologies, Huizen, Netherlands

A Stephens, VoIP Technology Evaluation, BTexact Technologies, Adastral Park

R P Swale, VoIP Technologist, BTexact Technologies, Adastral Park

Dr D J Thorne, Advanced Communications Networks, BTexact Technologies, Adastral Park

G Travers, Technology Architecture Standardisation, BTexact Technologies, London

A H Warren, VoIP Technology Evaluation, BTexact Technologies, Adastral Park

C D Wilmot, VoIP Multimedia, BTexact Technologies, Adastral Park

Dr D R Wisely, Mobility Architectures, BTexact Technologies, Adastral Park

1

INTRODUCTION TO VoIP

R P Swale

1.1 Introduction

Voice over IP (VoIP) has been associated with eye-catching headlines in the media — specifically with the expectation of 'free' telephone calls over the Internet. Not surprisingly, voice on the Internet has been forecast to have a substantial effect on traditional long-distance and international markets and there is plenty of evidence [1] to support this claim as a reality. Yet voice over IP, IP telephony and voice over the Internet are among the many terms that have emerged since the mid-1990s to describe various technologies and applications for transporting and setting up voice and multimedia calls over an Internet protocol (IP) based packet data network. The very fact that such a range of terms abound continues to cause much confusion within the industry for all kinds of stakeholders — including end users, service providers, equipment manufacturers and policy makers. This has been compounded by the wide-ranging — and sometimes subsequently proven to be wildly inaccurate — claims about what the technology can offer, the relationship it has with existing services, and the likely future impact upon those services. All of which makes it rather difficult to separate fact from fiction, and 'gold-rush fever' from sensible business opportunity and practice. So what is VoIP and what does it really mean for the communications industry? This book aims to go some way to answering this question by examining the full spectrum of what VoIP could be, should be and currently is.

1.1.1 Taxonomy

Some commentators refer to VoIP as 'voice over the Internet'; truly speaking this refers to the specific use of VoIP on the global public Internet. It is particularly important to appreciate this subtlety since there are many IP-based data networks deployed that are not part of the true Internet *per se*. For example, there are many data networks that are not directly connected to the public Internet, such as managed networks and private intranets. In fact this class of data network even includes those found embedded within the operation of existing public telephone networks.

Therefore, in the absence of uniform terminology within the industry, it is useful to classify the various kinds of technology that fall within the umbrella of providing voice communications over IP networks. In considering such a classification, it is important to highlight that the very nature of IP as a data transport technology means that any voice solution can in theory be readily extended to support full multimedia capabilities. This is because an IP network has no inherent understanding of voice *per se* — it is 'media neutral'. Video, voice and data packets are no different from the IP network's point of view. However, in practice, the end-to-end transmission characteristics required for voice, video and data applications are very different, as will be apparent from the discussion of voice transport over IP networks given in Chapters 2 and 3. The ability to control or manage the network capacity between a sending and receiving point is therefore crucial to achieving the reliable transport of such media. However, as discussed in Willis [2], these mechanisms are still evolving for IP networks. So dedicated or carefully engineered networks are most likely to be the first that are able to deliver stable and sustainable transmission performance. It is in these specific areas that the true multimedia flavours of VoIP are likely to take off most quickly due to the size of these networks and the ability to control network traffic — essential requirements, as discussed below in section 1.4.

From a theoretical point of view, voice communications may be considered to be a member of a family of real-time conversational communications applications. This taxonomy recognises the wide range of differences in capability between different network applications found in practice. From this perspective there is the need to highlight the difference between these applications and draw distinctions between those that aim to mimic the full range of facilities — the 'telephony service' as discussed in Chapters 6 and 8 — available from the public telephone network, and those that do not. In doing so there is also the need to consider whether the application aims to address the associated end-to-end quality-of-service issues or whether it is simply a 'best-effort' application. These considerations are reflected in Fig 1.1, where one such taxonomy is proposed. The distinctions used in Fig 1.1 are maintained in the remainder of this book. However, when contemplating this, or any other similar, taxonomy, it is important to fully appreciate the potentially wider implications — particularly when considering issues such as speed to market and the impact upon, and of, regulation.

1.1.2 Drivers for VoIP

It is asserted that there are essentially four main stakeholder groups that influence the adoption and long-term viability of any communications technology. These are users/customers, service providers/network operators, equipment manufacturers, and policy makers/regulators. Each group has a distinct set of perspectives that are important to appreciate and contrast, as outlined below.

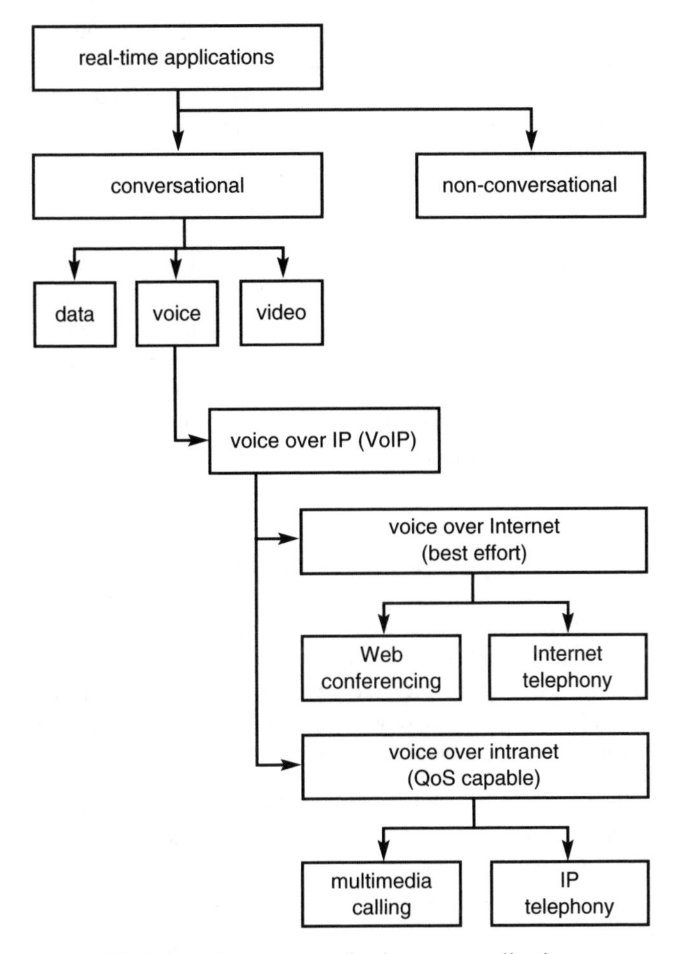

Fig 1.1 A taxonomy of voice as an application.

- Users

  This group has many diverse as well as common needs for communications technologies. While some users are extremely sensitive to end pricing, others are motivated by the potential benefits to be obtained from additional functionality and will gladly trade price for functionality, ease of use and flexibility. Some are intensive users of communications services, others are occasional users, while yet others rely upon services for emergency purposes only.

- Service providers

  From a VoIP perspective, service providers have essentially two main drivers in that they are motivated to supply existing services at lower cost bases where possible and to develop additional lines of business through the development of new applications and technologies. Any technology, such as VoIP that offers the promise of migrating voice, data and multimedia services to a common network platform, and therefore reducing cost while enabling new applications and services, will be of major interest.

- Manufacturers

  Simplistically, manufacturers are motivated by the opportunity to generate revenues from communications equipment production and associated services supporting the operation of the equipment through its lifetime. While some manufacturers address high-volume, low-profit-margin market segments, others concentrate on low-volume, high-margin opportunities. The components required to build the VoIP systems discussed in this book may support business opportunities for both approaches.

- Policy makers

  Policy makers possibly have the most difficult role within the industry — they seek to balance the interests of service providers (and, by implication, manufacturers) and users. From a technology point of view, this needs to be achieved within a framework that recognises both what is available and what is feasible, in a spirit that will foster economic development and prosperity for all. The opportunities to get it wrong — through either inappropriate action or unconscious inaction — are therefore innumerable.

  Possible consequences of such errors include end users being deprived of reasonable choice, the absence of valid business models able to support sustained development by service providers or equipment manufacturers, or the invalidation of an existing business model and the wider socio-economic impacts that can ensue.

The various relationships between these stakeholders are shown in Fig 1.2. This highlights the natural tensions in these relationships that are important to appreciate. In particular, when considering any new technology, including or possibly especially VoIP, the consideration of the balance between the various interests is of prime importance.

It is this balance that ultimately defines the likely success or longevity of a specific communications technology by determining whether sustainable business models truly exist.

Where such a balance exists, it justifies the investment levels that long-term commitment to such a national communications infrastructure requires. Possibly it is the lack of appreciating this point that has so quickly invalidated some of the early claims ascribed to VoIP.

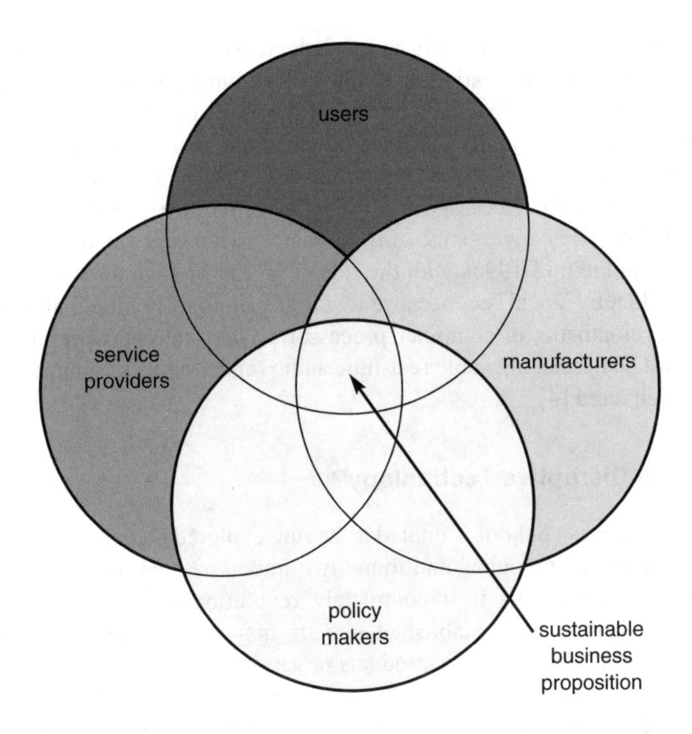

Fig 1.2 Relationships between the stakeholders in the VoIP industry.

Underpinning these concerns is the fundamental issue of whether the technology itself is viable for large-scale deployment from a number of points of view, including operational characteristics, scalability and performance. One of the more effective ways of resolving these concerns with the technology is simply to try it out, as discussed in Chapter 17.

1.1.3 The Convergence of Enabling Technologies

While VoIP has caught the imagination of the communications and computing industries in recent years, it is not widely appreciated that the technology is most definitely not new in concept. In fact, the real origins of VoIP date back to the early 1970s when experiments, driven by requirements from military applications [3], were undertaken on the DARPAnet — the forerunner of today's public Internet. However, for commercial mass-market applications, the computing power required for real-time multimedia simply did not exist outside the relatively expensive mini-computers available at that time. In addition, the voice processing technology necessary for reducing the inefficiencies due to packet overheads inherent within a VoIP system, as discussed in Chapters 2 and 3, was itself an emerging art when these experiments were first published. Fundamentally it has taken the advent of affordable, ubiquitous computing with integral multimedia support in the form of personal computers, and data networking in the form of IP, to make 'voice over packet' a real possibility. In the case of personal computers it has taken the development of audio and video capabilities — driven by the computer games market — to deliver the necessary computing infrastructure to enable VoIP to come to fruition today. The convergence of these factors determined that VoIP would hit the headlines in the mid-1990s with the release of the first commercial VoIP client software product by VocalTec, because it was at that point in time that a number of unrelated developments in computer processor power, data networking and voice processing all coalesced to enable real-time audio (and video) communications to be affordably delivered [4].

1.1.4 A Disruptive Technology?

The Harvard Business School is quoted as having explored the characteristics of so-called 'disruptive technologies' and found two important common features. Firstly, a technology is disruptive if it completely revolutionises the market-place; an implication of this is that established players may quickly see their businesses disappear as the demand for their products or services are completely usurped by a new technology — invariably driven by a new market entrant. Secondly, and most importantly, disruptive technologies tend to appear less than attractive when they start out as they generally under-perform established technologies in a number of areas. However, they tend to have features that are beyond the capabilities of established technologies to deliver. For example, early 3.5 inch hard disk drives had lower bit densities than the established 5.25 inch technology — yet how many people would have notebook PCs if only 5.25 inch hard disk drives were available? Not many! But what really makes a technology truly revolutionary is its ability for rapid improvement in performance and capabilities to levels beyond the reach of existing solutions.

So the question is: 'Is VoIP a disruptive technology for the communications industry?' The answer to this can only be found by considering the evolution capabilities of the technology, taking into account the players, and reflecting on the acid tests of: 'Does it work for real?' and 'What opportunities does this afford?' In essence, this is the thrust behind the work that BTexact Technologies, along with its predecessor organisations at Adastral Park, has been leading [5, 6].

1.1.5 The Telephony Baseline

Before engaging in a more detailed discussion of VoIP, it is useful to establish a frame of reference for any expectations there may be for VoIP by considering some basic characteristics of voice telephony, as it is currently known — otherwise it is possible to fall into the trap of drawing an overly simplistic comparison and missing some key points. For the purposes of this discussion, there are essentially four important characteristics of existing telephone networks and services.

- Everything is a voice call

  Existing voice networks have evolved over many years to form complex networks capable of providing services from one end of the globe to the other. As such they are highly optimised for the switching and transport of 'telephone quality' speech. While this makes these networks less suited to the transport of other 'media', commonplace applications, such as facsimile, have emerged to overcome these deficiencies. There is an expectation therefore that end users can simply hook up to a 'telephone' and use it for a large number of applications beyond simple voice communications. They may, for example, send faxes, log on to computer networks and the like.

- Everything is circuit switched

  One of the important aspects of circuit-switched networks is that they integrate call control and service logic — features like call divert and transfer — with the basic transport mechanism. In other words, the signals requesting routing, connection and disconnection largely follow with the media stream. Where the media stream goes, the control goes, and, along with it, the full essence of what the 'telephony' service is. This actually makes the introduction of some features, such as emergency calling, relatively easy to handle while others are made more complex.

- Everywhere is part of the network

  The telephone network links all industrialised nations across the globe. A voice call is a voice call is a voice call, wherever you are and to whatever location you are calling. Only voice quality and cost are the key differences.

- Every time you want to use it — it's there.

  The telephone network is so pervasive in modern society that our very lives often depend upon it being there. So there is a very high expectation of it always working and always being available.

1.2 How it Works

As a starting point, consider the simple case of two users wishing to communicate using VoIP. Each user has a multimedia-capable personal computer with an appropriate client software application running on it, as discussed further in Chapter 4. Each is connected to a common IP network to provide simple data packet communications. Before a call can take place, the first major issue that needs to be addressed is: 'How can the calling user find the called user and establish contact with them?' This requires some form of signalling that is understood by the calling and called users' clients, as shown in Fig 1.3. It also requires a suitable naming and addressing scheme, as discussed in Chapter 9. Ideally the signalling messages in the VoIP system will be based upon some common format to enable a wide range of terminals, and hence users, to communicate. This mandates the use of industry-accepted standards, as discussed in Chapter 5 — otherwise people will only be able to call each other if they are using the same equipment from the same manufacturer. This is not as strange as it may seem, since this was precisely the case with the first generation of commercial VoIP products and was one factor that inhibited wider adoption. Clearly this is unacceptable for any viable large-scale public communications service. A common, open, industry-accepted standard is therefore desirable from both a commercial and operational point of view as it enables the growth and development of the market-place, as discussed further in Chapter 5.

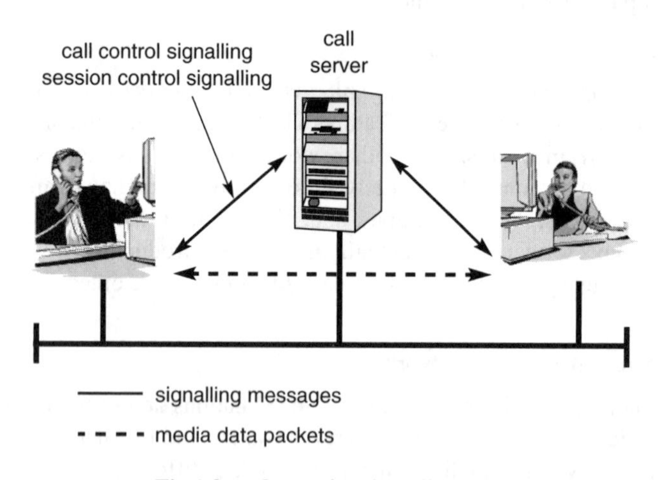

Fig 1.3 Contacting the called user.

To facilitate users establishing communication with each other, it is advantageous if they can share a common meeting point. This enables users to find each other irrespective of whether their IP address changes from one session to another — which is typically the case for dial-up sessions and why it is inappropriate to consider simple IP-based addressing as a naming scheme for end users of VoIP applications, as discussed further in Chapter 9. Figure 1.3 above introduced such a meeting point, which is known generically as a call server. This is a network resident server that enables users to place their current addressing information in a location accessible by potential callers. As a consequence it may help mediate the calls in the network. In doing so, it can also provide many valuable additional functions. These could include intelligent call handling, for when the user turns their terminal off, and assistance in maintaining network performance by supplying a form of admission control. The call server may also offer a crucial point of integration for value-added applications, of the kind described in Chapter 13. These may include features that are more easily provided in the VoIP environment such as 'presence' and 'instant messaging'.

Having established contact with the called user's computer, the next major challenge to be overcome is the decision concerning the nature and format of the media to be involved in the call session. This needs to establish whether the conversation will involve audio only or whether it will also include video or data or any combination of these three. In switched-circuit networks, such as the public telephone network, decisions of this kind concerning the media that will flow over the connection between the calling and called parties are largely taken by implication. More specifically, it is assumed by the PSTN that the call is going to be a 3.1 kHz speech call[1]. However, in the VoIP environment such assumptions are not likely to be valid and so it is necessary to establish, on a call-by-call basis, precisely which types of media are going to be required. In addition, the amount of network bandwidth that will be needed must also be considered if anything other than 'best-effort' quality of service is to be achieved. Ideally all these agreements are reached through mutual negotiation — typically known as a 'capabilities exchange'. This approach in the packet environment offers the additional benefit over switched-circuit methods of allowing new voice and video codec technologies to be developed independently of the network infrastructure. In particular, there should be no impact upon the network as new codecs are deployed, with the possible exception of where changes in end-to-end bandwidth requirements occur.

Having completed the 'capabilities exchange' phase, the media streams — in this case, for simplicity, assumed to be voice only — can be initiated between the two users. This is achieved by establishing a set of media channels between the two clients by sending appropriate signalling messages informing each client where to start sending IP packets. This is sometimes referred to as opening a 'media stream' since there will be a sequence (or stream) of IP packets transmitted. However, the ability to transmit a stream of IP packets is not particularly useful unless there is something of interest to put inside them. For voice communications, the basic principle involves digitally sampling speech signals obtained from a handset or other audio device connected to the client, as shown in Fig 1.4. The digital speech samples may then be subsequently encoded into a predetermined format using a codec before being framed into IP packets and transmitted out over the IP network. As discussed in Chapters 2 and 3, the choice of codec and framing strategy are essential pieces of information transferred as part of the 'capabilities exchange' phase previously described.

During transmission, the traffic loading and other factors in the IP network may cause packets to be lost, delayed or mis-ordered. At the receiving side, the client therefore has to reconstruct the correct sequence of the packets that were transmitted as best it can, recognising the variations that can arise in packet delay and loss. This

---

[1] ISDN does provide a certain degree of flexibility in this area, although it remains comparatively restricted in its capabilities in terms of the media supported.

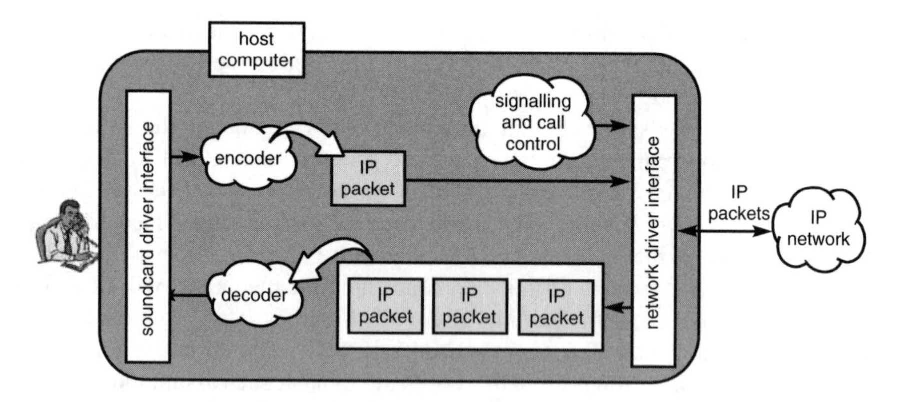

Fig 1.4 How a voice over IP client works.

is achieved by using a special packet store known as a 'jitter buffer' that enables received packets to be correctly ordered before being decoded and played out to the receiving client's headset or loudspeaker. However, this is entirely reliant upon the transmitted media packets containing information that enables the recipient to correctly order any received packets — and identify where any packets have been lost. In practice, this process occurs in both directions simultaneously as each party in the call will typically be both transmitting and receiving speech packets. While ensuring that the speech packets are decoded in the correct order, the action of the jitter buffer can also introduce a substantial amount of delay in the speech signals. As discussed in Chapter 2, this is because each packet received has to queue in the jitter buffer until it reaches the head of the queue. There is therefore a natural trade-off between jitter buffer size and delay, the detail of which is discussed in Chapter 3.

Once the call has finished, the signalling between the clients closes the media streams and indicates that the call has ended. The client then awaits a subsequent call to be either made or received. Although the example has described a single call, the IP environment allows calls to be made and received at any time, so it is quite possible to have many calls that are currently active. Calls may also involve multiple end-points and also any combination of voice, video and data. This is in stark contrast to the PSTN where multiparty calls and multiple concurrent calls are difficult to handle.

It is important to note at this point that the separation of the signalling and media introduces a substantial difference between switched-circuit networks and VoIP networks. Signalling can be much more flexibly routed in the IP network, whereas in the PSTN it is generally confined to close coupling with the telephone circuits with which it is associated. A consequence of the technology is therefore to open a separation between the transport network and the telephony service application. However, while this is theoretically possible, there remain largely ignored issues when this approach is scaled to the dimensions of current public voice networks.

Resolving some of these issues forms the basis of ETSI's TIPHON project that is discussed in detail in Chapter 16.

1.2.1 Connecting with the Wider World

Voice over IP between client devices enables users to enjoy the benefits of interactive, conversational communications in a truly multimedia environment. However, since most people are currently connected to circuit-switched public telephony services, there is the need for people using VoIP client devices to be able to make and receive calls from people served by existing networks and services, as discussed in Chapter 7 and illustrated by example in Chapter 15. This requires the use of a specialised network component comprising the functions of both an IP client device and a conventional telephone with the ability to bridge calls between the two. During early research into VoIP, BT coined the term PSTN-IP-gateway — or PIG for short — for this component, since it provides a gateway function between the PSTN and an IP network.

For a call to be made from an IP device to a conventional telephone, the call control functions in the VoIP network must first locate an appropriate gateway and then route the call from the gateway to its intended destination, as discussed in Chapter 9. This type of structure is reflected in the initial H.323-based architecture developed by ETSI's TIPHON project [7], and is shown in Fig 1.5. Within this architecture:

- H.323 end-point signalling and control flows are supported across reference points A and C;

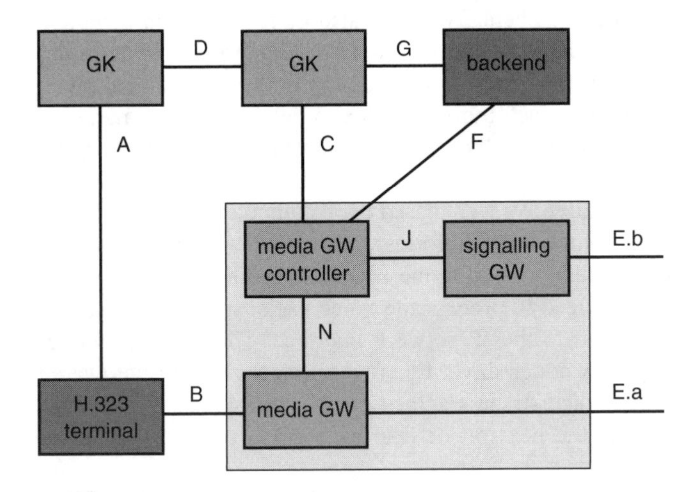

Fig 1.5 ETSI's initial TIPHON H.323 network architecture.

- inter-domain signalling flows (e.g. H.225.0 Annex G) are supported across reference point D;
- media flows are supported across reference point B;
- 'back-end' functions, such as intelligent call routing and service management, are provided by information flows across reference points F and G.

Since signalling and media interfaces in current switched-circuit networks can be implemented as either circuit associated or not, the signalling interface in this architecture can appear as E.a or E.b. This architecture also highlights the decomposed gateway architecture discussed later in this chapter and in more detail in Chapter 7.

A structure analogous to the inter-domain gatekeeper architecture, relating location servers rather than gatekeepers, is envisaged by the TRIP protocol developed within the IETF's IP-Tel working group. Aspects of this are discussed in Chapters 11 and 14. In either approach, however, there is the potential that the choice of gateway may be based upon a wide range of conditions — including the price of PSTN egress and latency in the network. Once the target gateway has been selected, the VoIP call is routed on to the gateway and a second, conventional, PSTN call covering the final leg on to the PSTN telephone is then made from the gateway to the telephone. The reverse sequence of events occurs where the call traverses the opposite direction, with the initial phase locating a VoIP gateway and the call completing its routing on the IP network side.

The consequence of connecting with the PSTN is, of course, that wider service aspects need to be considered, some of which are discussed in Chapters 6 and 9. For example, at a simplistic level, switched-circuit telephones can only use numerical addresses — therefore VoIP-based clients must be addressable using normal, diallable, telephone numbers. There are also more subtle and complex issues, such as how emergency calling features should be supported. For example, what is the true location of the emergency — the gateway or the IP client making the call? Such aspects are therefore much more complex to deliver into the market-place at scale than simply addressing the physical media translation between the circuit and packet environments.

Initially, VoIP gateways were based upon computer/telephony (CT) technologies and PC-based computing platforms[2]. While this approach enabled gateway solutions to be rapidly brought to market, it did exhibit a number of deficiencies — largely due to the need to process the voice packets through the host processor in order to gain access to the IP network interface. The major manufacturers of CT components quickly added direct Ethernet interfaces to their hardware to overcome the need to take voice streams across the PC back-plane. While this provided an evolutionary step, the next major leap came from the direct integration of VoIP

[2] It is interesting to note that BT had experimental gateways working in 1996 that supported multiple E.1 spans and scaled to hundreds of ports.

gateways within access router hardware. These developments led to the port cost of gateways falling from well over \$1k per 64-kbit/s port to around \$300.

In what may ultimately prove to be a sideways step, the decomposed gateway architecture emerged (see reference points B, C, F, N, J and E shown in Fig 1.5), as discussed in Chapter 7. This approach separates the functionality of a VoIP gateway into its three constituent elements:

- signalling gateway (SG) which interfaces signalling of the packet environment (reference point J) with that used on the PSTN interface (reference point E.b);
- media gateway (MG) which provides the actual media stream processing (reference points E.a and B);
- media gateway controller (MGC) which supplies the control and co-ordination of the gateway function (across reference point N).

Reference point E is split between E.a and E.b depending upon the presentation of the SCN signalling as previously discussed. The decomposed gateway architecture (see Fig 1.6) enables very high-density gateways to be developed. Prices for the actual media gateway ports using this approach have reduced to well below \$100 per port. However, this development has been offset by the complexity of the MGC components. These have been compounded by the associated confusion concerning whether the end-to-end control of calls should be integrated within the MGC, in an external call server (such as the gatekeeper shown in Fig 1.6), or shared

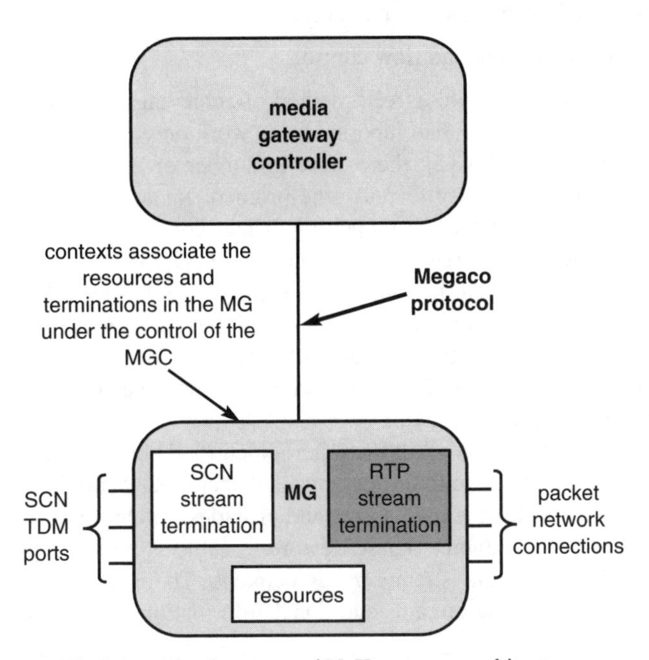

Fig 1.6 The decomposed VoIP gateway architecture.

between the two in a 'soft-switch' architecture. In fact, this turns out to be a rather fundamental issue since there are some wider architectural issues that arise as a consequence of this decision. Specifically these centre on the ability of the solution to be readily extended to support multimedia clients — a point that is revisited later in this chapter.

1.3 Exploring the Technology

1.3.1 Media Transport over IP

Possibly the most fundamental challenge for any VoIP system is the simple transport of voice samples over the packet-based infrastructure of an IP network designed to meet the needs of data applications. Specifically, a complete solution for the transport of real-time media over an IP network needs to be able to appropriately handle:

- timing and synchronisation of, and between, individual media samples;
- the effects of lost packets;
- the effects of delayed packets;
- mis-ordering of packets at the receiver;
- identification of the transported media type;
- transmission monitoring and flow control.

Without a viable, common, agreed method of achieving this, VoIP simply does not exist. Although the IETF had reported some work on voice transport over an IP network in the early 1970s [2], there were a number of issues that needed to be resolved, including the ability to reuse the protocol for a number of applications beyond simple port-to-port voice tunnelling. These concerns were addressed by a new protocol known as the real-time protocol (RTP) [7]. Due to its careful design and inherent flexibility[3], it has emerged as the *de facto* specification for the transport of time-sensitive streamed media over IP networks. More significantly, of all VoIP technology components, excluding IP itself, it is possibly the only common element found in use. Even so, RTP is not without its peculiarities when considering multi-vendor interoperability.

RTP alone cannot address all of the criteria identified above — however, it does enable a complete application to be constructed that can. RTP supplies a wrapper for media data that identifies the media type and provides synchronisation information in the form of a timestamp. These features enable individual packets to be reconstructed into a media stream by a recipient. To enable flow control and management of the media stream, additional information is carried between the

---

[3] RTP is based upon integrated layer processing concepts.

sender and receiver using the associated RTP control protocol (RTCP). Whenever an RTP stream is required, associated out-of-band RTCP flows are established between the sender and receiver. However, some implementations dispense with the RTCP information! RTCP enables the sender and receiver to exchange information concerning the performance of the media transmission and may be of use to higher level application control functions. Typically RTP is mapped on to, and carried in, UDP packets. In such cases, the RTP session is usually associated with an even numbered port and its associated RTCP flows with the next highest odd numbered port.

In the case of a typical two-party VoIP application, it is necessary to transport media in both directions simultaneously. This means that there needs to be four connections between the two terminals — one each for RTP and RTCP in both directions, as shown in Fig 1.7.

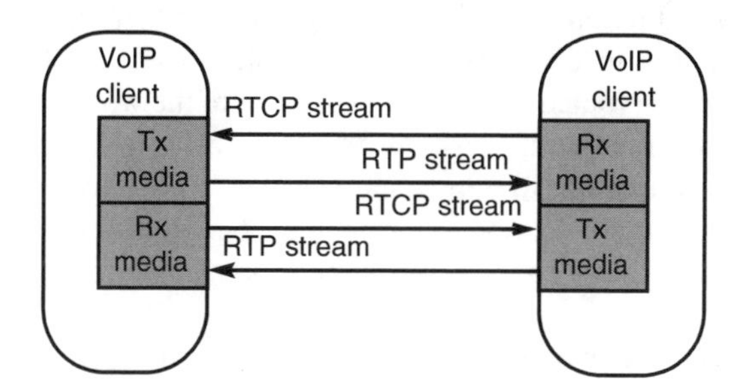

Fig 1.7 Full-duplex real-time communications using RTP.

1.3.2 Voice Transport over IP

Although RTP provides a mechanism offering the possibility for transporting real-time media over an IP network, it is insufficient to consider this as a complete solution for VoIP. This is because the voice quality as perceived by the end user is affected by additional factors including:

- choice of codec technology;
- effects of packet payload framing strategy;
- effects of packet loss.

These and other factors are discussed in greater depth in Chapters 2 and 3, where the need for careful engineering choices of the technology components, including the design of the IP network infrastructure, are highlighted. Economic factors also need to be considered since the very transport of voice over an IP packet infrastructure introduces bandwidth efficiency degradations due to the overhead in transporting RTP, UDP and IP headers, in addition to the actual media data. Put more simply, VoIP is not an efficient method of transporting voice, as discussed in detail in Chapter 3. However, this deficiency needs to be carefully balanced with the advantage of the ease with which differing voice codecs or other media can be accommodated on the same infrastructure simultaneously. Even so, it has taken a considerable period of time [3, 8] to enable IP to stand the remotest possibility as a candidate solution for real-time media transport.

1.4 Call Control

Voice over IP is not only a completely different way of transporting voice, it also offers completely different ways of controlling call sessions. This is because the signalling — or control — messages, that establish, maintain and disconnect the media flows in a VoIP application, are themselves routed across the IP network. As a consequence, there is no compulsion for these messages to pass directly between the calling and called terminal nor is there any need for them to even pass through the same paths as the media packets. In principle, there is nothing to constrain control messages being routed around the other side of the world simply to establish a media flow between adjacent desks in the same office. This aspect of VoIP is probably the most important to appreciate because it is really in this respect that VoIP has the opportunity to revolutionise the communication-services business environment, creating new opportunities and potentially transforming existing networks.

There are essentially two approaches to signalling that may be adopted within VoIP systems. These go by a variety of terms within the industry. Here they are termed direct-routed signalling and server-routed signalling, and are shown in Fig 1.8. The simplest of these is the direct-routed signalling model which is a strict peer-to-peer model — all the control messages required to support the VoIP application are exchanged point to point, directly between the calling and called users.

From a theoretical point of view this effectively represents a complete convergence of the user and control planes in the ITU-T network model. While this scenario enables basic connectivity features to be implemented, it lacks the ability to support more complex call treatment and behaviours. For example, it is not able to handle behaviours when the called user has switched off their terminal — a common scenario that is encountered in mobile networks.

In contrast, the server-routed model provides a rendezvous and mediation point between the parties involved. It can also handle much more complex functionality including call screening, call treatment in the event of unavailability of the called terminal, etc. The downside of this model is the potential complexity of the server technology and the balance between call control residing in the end-point or residing in the server.

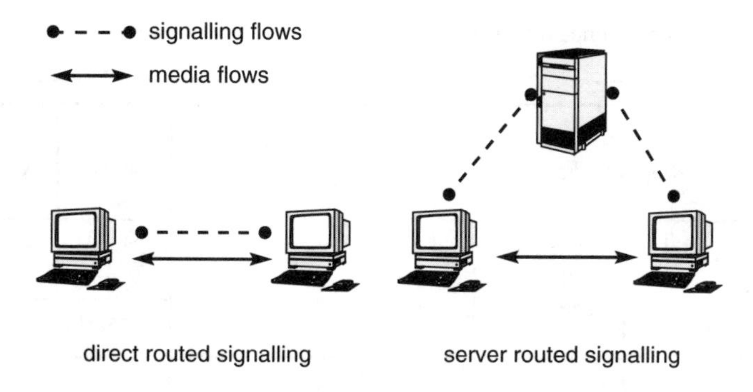

Fig 1.8 Signalling modes in VoIP networks.

Nowhere have the effects of this tension been more apparent than in the evolution of the decomposed gateway architecture shown in Fig 1.6. This is because, for a simple gateway solution, such as would be used for the trunk level switch replacement in the PSTN discussed in Chapter 6, it is easy to omit multimedia client support from the implementation and move server-based call control into the media gateway controller. In doing so the network model effectively degenerates into a direct-routed model rather than a server-routed model. The consequence of this being that it becomes very difficult to subsequently extend the solution to encompass multimedia client devices and has therefore been found to be a serious flaw in such designs.

1.4.1 The Need for Standards

As with any communications technology, VoIP needs well-defined and industry-supported methods for signalling call control information in order to succeed. Without such standards, the ability to communicate between users becomes at best severely restricted or difficult to achieve. Initial implementations of commercial VoIP solutions used proprietary techniques until industry developed a consensus around the use of ITU-T's multimedia conferencing standards as a useful starting point. More specifically it was the development and promotion of the H.323 specification, which is discussed in detail in Chapter 10, that provided the initial focal point within the industry. Derived from related specifications for multimedia conferencing over ISDN, H.323 defines an architectural framework (see Fig 1.9) that encompasses the ability to use it in both direct-routed and server-routed signalling modes. Within this architecture, the server-routed signalling mode is known as 'gatekeeper routed' signalling due to the term used within the H.323 specifications to describe the server component.

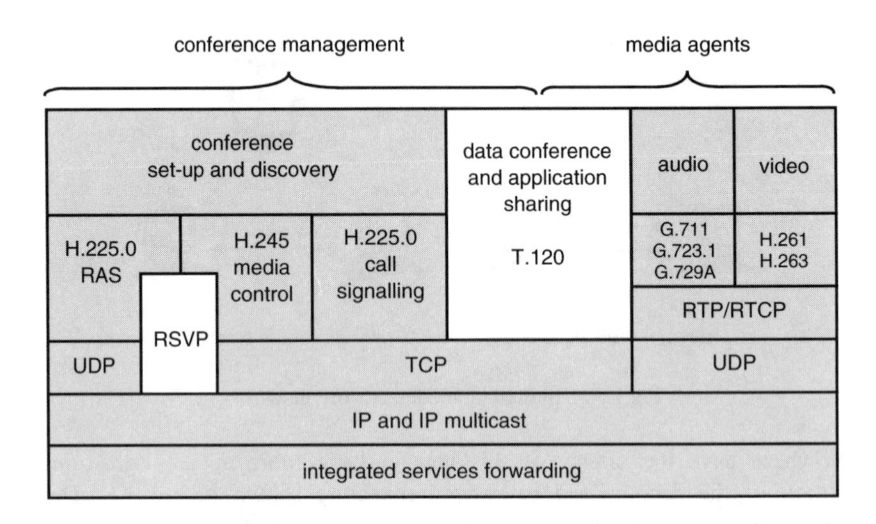

Fig 1.9 Protocol architecture for H.323.

H.323 is an immensely powerful technology, incorporating many features that can be switched on or off depending upon the network deployment context. It is by the careful choice of these options and appropriate design of gatekeeper-based applications to route signalling messages that H.323 networks may be scaled to very large dimensions. A simple call flow for an H.323-based network is shown in Fig 1.10, where the calling client initially engages in a registration sequence to identify it to the network. This type of behaviour is an essential feature of a VoIP system and effectively provides a degree of inherent terminal and user mobility functionality since the client may register from anywhere in the connected IP network.

In contrast to H.323, the IETF has been developing a competing, but potentially complementary, architecture for multiparty, multimedia conferencing on the Internet [9] as is shown in Fig 1.11. The session initiation protocol (SIP) is a component of this architecture and provides the basic session control mechanism used within it. As discussed in Chapter 11, SIP has gained a substantial following within the industry by offering the potential for an easily implemented method of establishing and controlling basic voice calls. From its very lightweight inception, SIP has been developed to address the challenges of being used outside basic point-to-point voice calls and an overly simplistic direct-mode signalling model. As can be seen by contrasting the same simple call flows using H.323 (Fig 1.10) with those using SIP (Fig 1.12), the differences between the technologies [10] are not always as obvious as some may wish them to be, which belies the true roles each should be able to play.

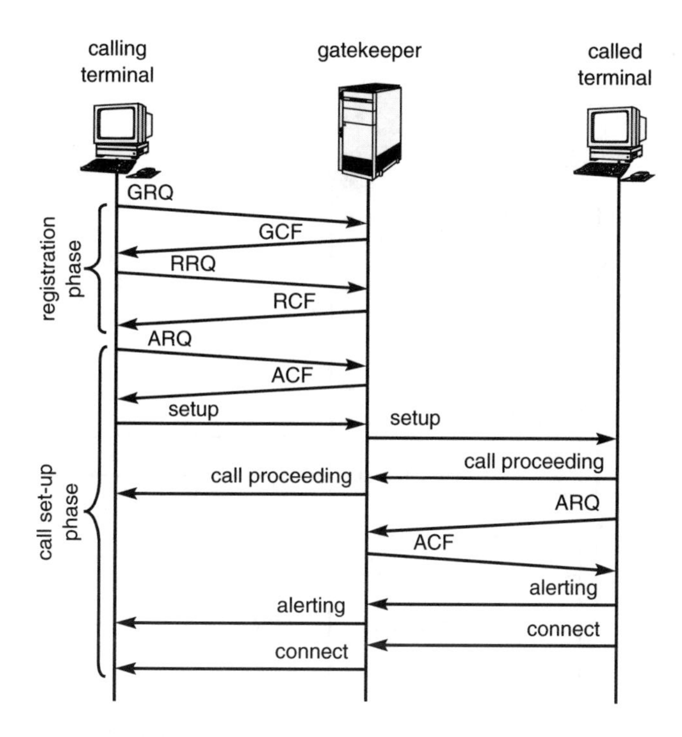

Fig 1.10 Typical registration and set-up in H.323.

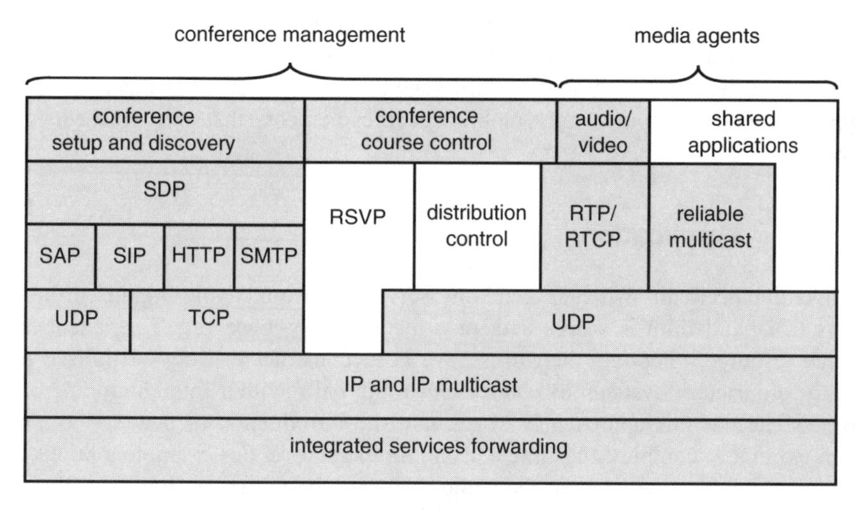

Fig 1.11 IETF multimedia conferencing architecture.

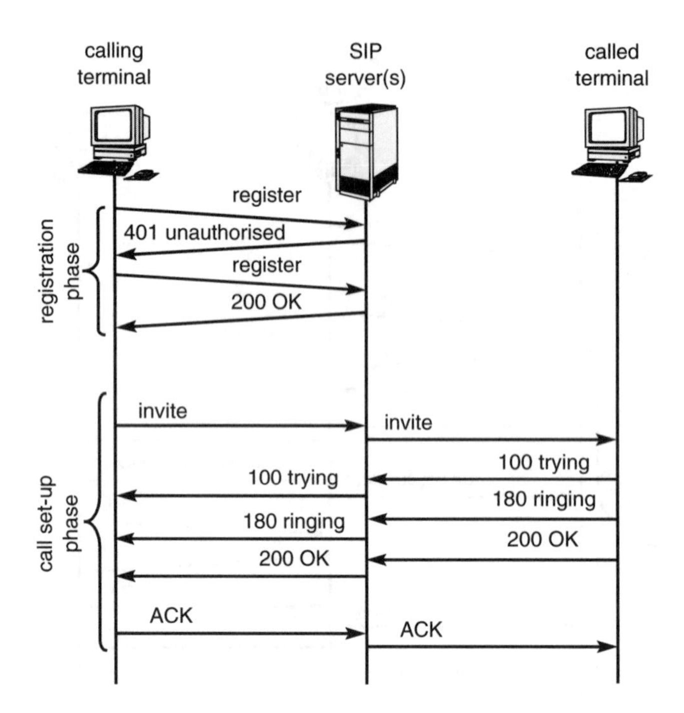

Fig 1.12 Typical registration and call set-up using SIP.

SIP offers an easily implemented, powerful, control environment capable of scaling to very large networks due to its simple message request/response format. This, combined with its relative immaturity compared with H.323, encouraged its adoption in the access segment of third generation networks, since this affords the opportunity to incorporate any mobile-specific elements that were subsequently identified.

1.5 Applications

Conventional circuit-switched telephony services are effectively organised on the basis that bandwidth is scarce and must therefore be conserved. To manage this scarce resource, telephony networks have been constructed using centralised and highly proprietary systems to control and route calls within this environment. A consequence of this approach has been that the introduction of new services has been extremely complex and time consuming to achieve. For example, options for changing the characteristics of the audio delivered are extremely limited and those that are actually possible require the insertion of specialised and expensive equipment such as voice compression systems. Similarly the closed nature of the call processing and service control are inhibitors to the provision of additional services. Yet, in the context of an IP environment, the actual telephony function is potentially just another application running on the network. Simplistically it is entirely possible to readily link applications and media together in an *ad hoc* manner to produce complex solutions highly customised to a specific need — and control these streams from anywhere in the world. Examples of ways in which the technology could potentially be applied include:

- improving internal communications, such as:

  — access to corporate networks from smaller sites over integrated voice and data networks to branch offices,

  — managed virtual gateways and private networks for voice as a replacement for VPNs,

  — intranet-based corporate directories, personal numbering and messaging services for people on the move,

  — directory services and graphics-based conferencing services from the desktop;

- collaborating:

  — extranet-based multimedia VPNs for telephony and data sharing,

  — collaborative tools for data conferencing and videoconferencing;

- improving access to the market:

  — Web-based call centres offer the ability for sales agents to push Web pages to the caller and assist in the completion of selections and payment for services while the customer is on line to the Web site, provided the network quality can be sustained;

- reducing service costs:

  — in certain cases, fax and voice on IP have a lower cost base than conventional TDM approaches for certain voice applications; the potentially lower capital costs associated with gateways make it relatively easy to introduce new points of presence, reducing egress costs to the PSTN.

Further aspects of VoIP applications are considered in Chapter 13. In reality, providing reliable, real-time-streaming, media applications over IP networks remains a challenging engineering problem to solve at scale.

1.6 VoIP in Practice

In 1996 BT's Systems Engineering unit, the predecessor of BTexact Technologies, at Adastral Park, was tasked with examining both the facts and the opportunities afforded by VoIP [4] under a project known as the 'Initial Feasibility Study' (IFS). This initiative was typical of the exploratory work undertaken by large incumbent service providers at that point in the VoIP development story. By the autumn of 1996, the IFS project had produced prototype solutions able to demonstrate all possible basic call scenarios involving VoIP, as discussed in Chapter 13, and exposed a number of both technical and commercial implications. These initial results were subsequently championed by BT's Global Marketing unit under the 'Concert Voice on IP Solutions' (CVIPS) project for the consideration of the development of global voice networks.

1.6.1 'Concert Voice on IP' Trial — Slaying the Dragon

Conventional wisdom within the telecommunications engineering community of the mid-1990s held that VoIP was not — and was unlikely to become — a serious candidate technology for real-time conversational voice-related applications. This view was not unreasonable since early VoIP systems exhibited very poor performance, tending to provide high inherent transmission delays[4], poor call processing support and non-existent interoperability between client software and gateways when procured from more than one supplier. The first major objective of the CVIPS project was therefore to establish the viability of constructing a voice over IP network in all its aspects to prove — or disprove — categorically the viability of the approach. With this aim, an extensive assessment was made of the vendor community early in 1997 to establish what was actually available commercially, without taking the laboratory prototypes produced by the initial feasibility study through manufacturing. On the basis of this study, equipment was chosen for deployment in a network trial. The particular equipment was selected on the basis that it demonstrated the key features sought:

- separate and distinct client, gateway and call control functions;
- PBX-style call control features;
- ability to support PC-to-PC, PC-to-telephone, telephone-to-PC and telephone-to-telephone scenarios;
- ability to generate call detail records;
- acceptable audio quality with low inherent delay appropriate for use in long-haul networks.

In essence these still remain the core features of any VoIP solution. Although the product was not standards compatible (a desirable feature to facilitate a wide customer base for services and open the market for further procurement), it was probably world leading in terms of the above features it offered at the time.

The technical trials were separated into two phases of activity — a laboratory trial and a network trial. The laboratory tests concentrated on subjective tests to examine how the audio quality compared with existing PSTN and VPN services.

---

[4] >500 ms per round trip

These tests were carried out using the subjective testing suite at Adastral Park and employed established procedures for blind tests to compare VoIP and conventional calls. The successful completion of the laboratory tests provided the justification for proceeding to a network trial spanning Sweden, the UK and the USA — linking Telenordia, BT and Concert PBXs via the Concert Internet Plus service. This network became operational during the summer of 1997 and was used for voice quality tests using volunteers. The results of these initial experiments have been translated into a wide range of activities, some of which are described in Chapter 13, and demonstrate the benefits of being in the enviable position of being able to rapidly leverage a diverse range of development resources to rapidly exploit an emerging technology.

1.6.2 VoIP — So What?

Having established the basic principles of engineering a VoIP network solution through the CVIPS project, the main challenge has been to translate this into commercially successful services and applications. Although the trial equipment previously described was made to function, there were a number of issues that would have had a major impact upon its successful commercialisation as a service. These included a lack of:

- operational management process support, specifically remote management and diagnostic facilities[5];
- application development and creation tools;
- stability and reliability of hardware and software components.

A series of spin-off projects have leveraged the CVIPS work to address most of the aspects of VoIP, from simple best-effort Internet calling and PSTN trunk replacement, through to rich multimedia applications on QoS-enabled IP networks. The development of a portfolio of products and services has been concentrated on the establishment of propositions derived from subsequent focused technology trials or specific opportunities, some of which are reported in Chapter 13.

1.6.3 What Does This Mean?

Having considered the various essential aspects of a VoIP solution, it is appropriate to reflect on what this means in practice for the various stakeholders previously considered.

[5] Some of these requirements were indirectly fed into the development of the Megaco protocol by inclusion within the ETSI TIPHON specifications [11].

- User perspective

  Considering the end user's main interests in terms of functionality and cost, it is highly likely that VoIP will ultimately prove to be an important technology. From the point of view of cost, low performance services have emerged and are likely to remain an option, while higher performance, more reliable, services mean that end-to-end IP QoS mechanisms of some description will be sought.

  These costs are likely to be offset, at least in part, by cost reductions arising from savings in terms of no longer having to manage separate networks for voice and data applications. Similarly VoIP is attractive from a functionality point of view, as it is apparent that new applications continue to emerge — the genuine quality, and utility of which, are for ever improving.

- Service provider perspective

  As generic QoS management and control capabilities are still emerging for IP networks, the long-term benefits of a fully converged IP network remain to some degree precisely that — long term. There is little doubt, however, that there are opportunities to make this a reality and much can be done by way of interim steps that build towards the longer term. In doing so, a much richer applications infrastructure can be created that should enable many of the applications envisaged today, as long as the equipment materialises to deliver it in a cost-effective manner. However, it is increasingly a possibility that the margins to be made in certain service areas — such as QoS-enabled IP packet transport — will force consolidation among service providers in the longer term.

- Manufacturer perspective

  The potential for a wide range of network components and associated products continues to grow for the VoIP manufacturing industry. With deployed volumes of existing products very much in the early adopter stage, there are no clear market leaders in many areas for what may ultimately prove to be either a massive global consolidation or fragmentation for the supply of certain major components. If the industry fragments too far, the basic components on which service providers can develop sustainable business models may become prohibitively expensive.

- Policy maker perspective

  The benefits for and impact upon users, service providers and manufacturers continue to raise the stakes in an increasingly complex and globalised environment. Determining a valid path for the industry therefore becomes an increasingly difficult task without seeing the economic fabric of the industry disintegrating.

1.7 Summary

VoIP has been found to be capable of delivering effective voice calls. However, current solutions require careful network engineering at present to provide sustained voice quality. With a number of large players very active in the development of equipment, the performance of the technology continues to improve rapidly and unit costs continue to decline. On a cautionary note this has been driven by a large amount of predatory pricing and confusion between price and cost. There is little doubt that these aspects are not sustainable in the future.

With inherent capabilities for the support of multimedia, VoIP offers a rich fabric upon which more advanced applications can be built. Through the use of the ubiquitous networking offered by suitable IP network solutions, such applications can be readily built without the need for complex network infrastructure upgrades. The current challenges remain in the area of achieving the right functionality from both the IP network and a controlling application layer that recognises the stakeholders identified and described in this chapter.

However, VoIP needs to be set in the wider context of multimedia broadband networks — including third generation wireless — as it becomes less viable to contemplate single-service-type networks in the future. In other words, VoIP enables, and should be considered part of, a new generation network architecture and industry that reflects the separation of multimedia bearer services from true content. In this context, VoIP is only likely to reach its full potential where the networking concerns have been appropriately separated from content and application concerns.

1.7.1 Panacea — is VoIP a Disruptive Technology?

It is apparent that VoIP continues to demonstrate all of the qualifying characteristics of a disruptive technology. However, from the point of view of voice transmission, VoIP is really only a potentially disruptive technology for the reasons discussed further in Chapter 3. More specifically while it uses more bandwidth to transport media due to the inherent packet overhead, the ability to accommodate flexibly a multiplicity of different bandwidth voice services concurrently on the same network represents a major differentiator. The question is whether this actually means anything in practice to the end user. Conversely from the point of view of call control, VoIP clearly offers a more flexible method of handling call sessions. The issue remains, however, as to how this flexibility is linked to QoS control mechanisms in other than purely best-effort services. The greatest area on which VoIP is likely to have an impact, though, is in the area of network applications. Without any doubt, the IP environment is better suited to applications development than a switched-circuit telephone network. The problem will be how to blend the reliability and security of the telephone network with the potential openness of IP networks. As with any disruptive technology, however, there will be winners and losers.

1.7.2 PIG's Ear — What are the Key Challenges?

While the benefits of VoIP technologies may be apparent, there are currently a number of fundamental technical challenges that remain to be solved within the industry. These mainly centre upon the provision of applications in anything other than a purely best-effort environment — specifically the issues of:

- achieving manageable QoS across IP networks;
- resolving complex interactions at the myriad of boundary points in IP networks;
- achieving appropriate network and service management functionality.

There are also issues concerning the IP devices that will be required for multimedia applications in a converged network, more specifically, the production of IP devices with sufficient functionality at an acceptable cost compared with existing solutions.

Fundamentally though, the winners in this game will be those stakeholders who listen to and take into account the perspectives of the other players in this increasingly complex industry — who would want to be anywhere else ....?

References

1 '*Voice over packet networks*', Probe Research Inc (August 2000).

2 Willis, P. J. (Ed): '*Carrier-scale IP networks: designing and operating Internet networks*', Institution of Electrical Engineers, London (2001).

3 IETF: '*Specifications for the network voice protocol (NVP)*', RFC 0741 (December 1977) — http://www.ietf.org

4 Lewis, E.: '*Andreesen's law versus the techno treadmill*', IEEE Internet Computing, **1**(5), p 96 (September-October 1997).

5 Babbage, R., Moffat, I., O'Neill, A. and Sivaraj, S.: '*Internet phone — changing the telephony paradigm?*' in Sim, S. and Davies, J. (Eds): '*The Internet and Beyond*', Chapman & Hall, London, pp 231-254 (1998).

6 Swale, R. P. (Ed): '*Voice over IP*', BT Technol J, **19**(2) (April 2001).

7 ETSI: '*Telecommunications and Internet protocol harmonisation over networks (TIPHON), network architecture and reference configurations*', ETS 101 313 v0.4.2 — http://www.etsi.org/tiphon

8 IETF: '*RTP: A transport protocol for real-time applications*', RFC 1889 (January 1996) — http://www.ietf.org

9 IETF: '*The Internet multimedia conferencing architecture*', draft-ietf-mmusic-confarch-03, work in progress — http://www.ietf.org

10 Cordell, P. J., Courtenay, J. M. and Rudkin, S.: '*Conferencing on the Internet*', in Sheppard, P. J. and Walker, G. R. (Eds): '*Telepresence*', Kluwer Academic Publishers, Dordrecht, pp 74-98 (1999).

11 ETSI: '*Telecommunications and Internet protocol harmonisation over networks (TIPHON), requirements for a protocol at reference point N; media gateway controller to media gateway*', ETS 101 316 v1.1.1 — http://www.etsi.org/tiphon

2

ACHIEVING VoIP VOICE QUALITY

R J B Reynolds and A W Rix

2.1 Introduction

The success of VoIP is strongly influenced by customer opinions of call quality and how this quality compares with that of the public switched telephone network (PSTN).

The unreliable nature of the Internet and initial product offerings, where cost savings were often gained at the expense of quality, led to an image that VoIP is of a worse quality than circuit-switched networks. The 'poor quality' image has, to some extent, accompanied VoIP as it has moved from the domain of the Internet enthusiast to being used in today's carrier-scale networks.

Carriers are eager to squeeze maximum efficiency from their networks but understand that there is a minimum quality level required to achieve customer acceptance. Enterprise customers, now adopting VoIP, are also sensitive to noise, distortions and general impairments in their voice communication solutions.

Yet, tainted by the experience of experiments on the public Internet and earlier generations of equipment, one of the most commonly discussed concerns held by people contemplating VoIP is that it will never deliver the 'appropriate quality'. Significant steps have been taken over the last few years to achieve higher quality systems, and this chapter provides a view on some of the issues and choices that VoIP system designers face. In particular, one issue to be considered is: 'What is the appropriate quality?' As will be shown in this chapter, the achievement of high-quality VoIP solutions is not an impossibility. Rather it is simply an engineering challenge.

This chapter considers the issues behind delivering an 'appropriate' level of voice quality by showing how design choices ultimately affect, and potentially limit, a customer's perception of VoIP quality. It concludes by describing some of the new signal processing techniques that are helping to measure and optimise the performance of VoIP solutions.

2.2 Designing for Quality

It is important to realise that perceived quality is technology independent. The customer generally does not know, nor wants to know, about the underlying technology when using the telephone. Their demands are the same, regardless of whether VoIP or other voice network technologies are used. Most of the standardisation work [1-4] for delivering a desired quality level is therefore applicable to VoIP as well as the PSTN. Although this appears to be a superfluous statement, all too often VoIP solutions fail to meet key requirements such as use of the recommended delay thresholds or inclusion of the correct telephony filtering or level control. This ultimately results in the customer receiving poor service.

This section considers how different components of a VoIP system can have an impact on a customer's perception of quality. The challenge is to design an efficient solution that delivers a 'just-good-enough' quality level for a particular application.

For a VoIP system the three most significant influences on a user's perception of quality are the:

- performance of the VoIP device;
- performance of the underlying network;
- end-to-end delay.

A particular quality level can therefore be achieved by trading one of these degrees of freedom against the others. However, for high-quality systems the scope for trade-offs is greatly reduced. Based upon the degrees of freedom identified in this area by ETSI's TIPHON project [5] (see also chapter 16), Fig 2.1 illustrates how the performance range to achieve a specific QoS class greatly increases as the required QoS level reduces.

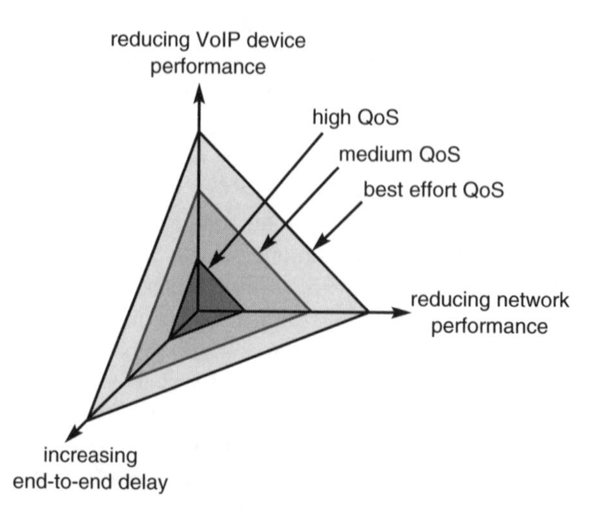

Fig 2.1 Trading performance to achieve desired QoS.

While there are a number of factors relating to the IP network's contribution to the end-user's experience, this chapter concentrates on VoIP device and delay performance. However, the impact of network performance parameters such as packet loss and jitter do receive some consideration.

2.2.1 Core Design Issues

One reason for choosing VoIP is its potential to act as a multiplexing technology to deliver voice and data services more efficiently than the PSTN. Good speech quality is critical in order to deliver a commercially viable VoIP system. This can come at a price, however, since it is possible to create a high-quality system, but usually at the expense of increased network bandwidth requirements (see Chapter 3). Since low-bandwidth allocations typically result in some reduction in speech quality, there is a technical challenge in delivering high-quality speech while achieving high network efficiency. The challenge is to deliver 'just-good-enough' speech quality at a specific bit rate.

There are various ways to trade bandwidth and speech quality to meet a target quality level, but there is no definitive answer on how this is to be achieved. This section describes those aspects of a VoIP system design that have the greatest bearing on the user's perception of speech quality:

- speech coding;
- packetisation efficiency;
- silence suppression;
- error-concealment methods;
- jitter buffer implementation;
- codec tandem performance.

These are considered in turn below.

2.2.1.1 Speech Coding

Most domestic PSTN networks operate with speech sampled at 8 kHz and an 8-bit nonlinear quantisation scheme according to ITU-T G.711 [6]. This encodes at 64 kbit/s and introduces little audible distortion for most types of signal.

In a number of applications, however, a much lower bit rate is desirable, either because capacity is limited (e.g. mobile) or to maximise the amount of traffic that can be carried over a trunk connection. Current ITU-T Recommendations include codecs that compress to as low as 5.3 kbit/s, although quality at this bit rate is well below that of G.711.

To maximise interoperability it is usual for VoIP gateways and clients to offer one or more standard codecs. Table 2.1 lists some common, standardised, voice codecs with their associated bit rates.

Table 2.1 Voice codecs.

Codec	Bit rate (kbit/s)	Coding technique
G.711	64	Pulse code modulation (PCM) [6]
G.726	40 to 16	Adaptive differential PCM (ADPCM) [7]
G.728	16	Low-delay code excited linear prediction (LD-CELP) [8]
G.729	8	Algebraic code-excited linear prediction (ACELP) [9]
G.723.1	6.3/5.3	Multi-pulse max likelihood quantisation (MP-MLQ)/ACELP [10]
GSM FR	13	Regular pulse-excited long-term predictor (RPE-LTP) [11]
GSM EFR	12.2	Algebraic code-excited linear prediction (ACELP) [12]

The choice of codec is to some extent dictated by the bandwidth on offer which determines the maximum bit rate for the codec, and in turn the maximum speech quality that the system will achieve under ideal conditions. It is important to note the fact that just because a codec is standardised does not in itself make it suitable for mass telecommunications applications.

In general, the lower the bit rate, the lower the quality perceived by the listener. However, more modern codec designs are driving up the quality for a given bit rate. This is illustrated in Fig 2.2, which shows the quality of speech passed through some of these standard codecs, measured using PAMS [13] (see section 2.3.2).

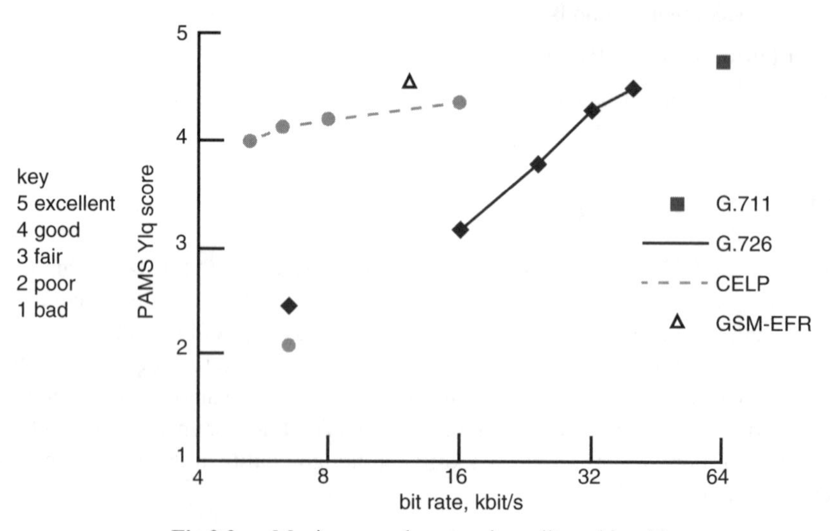

Fig 2.2 Maximum codec speech quality achievable.

The highest quality basic codec is still G.711 at 64 kbit/s. The low-complexity G.726 codec only offers good performance at 32 kbit/s or above, and is comparable with the more recent, though computationally intensive, GSM-EFR at 12.2 kbit/s and G.728 (CELP) at 16 kbit/s.

Given a demand for high quality, it is clear that choosing the lowest bit rate codec will not suit a large proportion of today's VoIP market. Therefore most systems tend to offer G.711 and at least one low bit rate codec. This allows the operator some flexibility to trade between quality and bandwidth — potentially on a per-call basis.

2.2.1.2 Packetisation Efficiency

A further consideration when choosing a codec is how much speech can be placed in an IP packet. This is partly dictated by a codec's frame size. For example, GSM uses a fixed 20-ms frame and therefore packets must be a multiple of 20 ms, whereas G.711 can be any arbitrary length.

The amount of speech placed in a packet has a direct impact on the underlying network efficiency. VoIP is inefficient for small voice packets while large voice packets lead to long delays (see Chapter 3) — which is inappropriate for real-time communications. Typically packets contain between 10 and 30 ms of speech. This provides a practical trade-off between network efficiency and increased delay. Some increase in bandwidth overhead may be justified by the economic benefits of running a single combined voice and data network.

It is important to note that because of IP protocol inefficiencies, reducing codec bit rates below a certain level makes limited sense. This is primarily due to the IP, UDP and RTP [14] headers contributing towards a 40-byte overhead to each packet. At 20-ms packet spacing (a single GSM frame or two G.729 frames), this equates to a 16-kbit/s overhead. With efficiency defined as the reduction in bandwidth compared to a 64-kbit/s PSTN channel, the graph in Fig 2.3 shows the efficiency gain from using lower bit rate codecs (all calculations assume 20 ms of speech per packet).

By comparing only codec efficiency, G.729 shows an 87.5% saving over G.711, but when accounting for the IP headers this saving is reduced to 60%. These calculations do not include header compression [15], which would provide additional savings.

When selecting a codec it is also important to know how well it will code non-speech signals such as background noise. When telephones are used in a noisy environment the ability of a particular codec to encode the background noise can have a significant effect on the perceived quality.

Mobile telephone codecs are designed to cope with high-noise environments and therefore it has been suggested that these are probably the best low bit rate codecs for noisy environments.

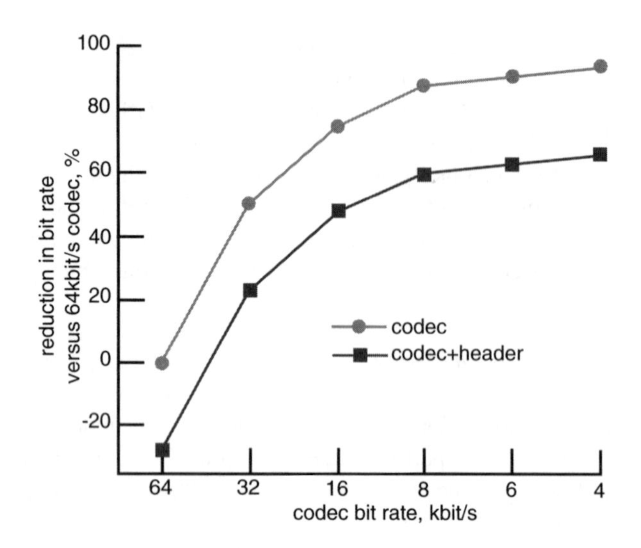

Fig 2.3 Codec efficiency with/without IP headers.

2.2.1.3 Silence Suppression

Silence suppression, or discontinuous transmission (DTX), describes the process by which periods when the user is not speaking are neither coded nor transmitted. This allows the overall system bandwidth requirements to be reduced and can produce an average bandwidth saving of some 40%. DTX is implemented by the use of a voice activity detector (VAD).

The problem with silence suppression is its susceptibility to front and back-end clipping. This describes the situations where the VAD triggers too late or too early, cutting off speech at the beginning or end of a sentence. It has therefore proved difficult to design a VAD that works perfectly in all circumstances, despite extensive research and development effort.

In addition to the problem of deciding what is and is not speech there is the issue of how to generate a signal to fill in the silent periods. This generated signal is called comfort noise. Without comfort noise the telephone system feels dead to the end user, but a mismatch in noise between when the person is and is not talking is distracting and leads to an impression of poor quality.

When choosing to develop a VAD an understanding of the application is critical [16]. For example, a VAD that labels more of the signal as speech, suppressing transmission only when the talker is definitely silent, will deliver higher quality than a more aggressive VAD. This is of course balanced by a slight increase in bandwidth, but there can still be a significant saving over a system without DTX.

2.2.1.4 Error Concealment

VoIP systems can suffer from a degree of packet loss during normal operation and there can also be packets dropped by the jitter buffer [17]. The ability of a VoIP device to conceal packet loss makes a significant difference to its performance.

Certain standardised codecs (e.g. G.729, GSM-EFR) include their own error-concealment methods. However, the use of proprietary error concealment can provide an improvement in quality over these standard methods and is applicable to other codecs.

An alternative to error concealment is the insertion of silence in place of lost packets — this gives a clicking effect which users find annoying. For example, at moderate levels of packet loss a G.711 implementation with no error concealment can lead to a lower quality than a low bit rate codec with error concealment.

At the expense of an increase in delay and bit rate, it is also possible to add redundancy to allow some errors to be corrected or lost packets reconstructed. Techniques available include forward error correction or duplication of frames across multiple packets.

An example of how packet loss affects quality is shown in Fig 2.4. This shows the distribution of quality scores observed in a number of measurements of a G.729 VoIP system at different packet loss rates. The range of scores occurs because a lost packet may coincide with a critical or non-critical part of the voice stream.

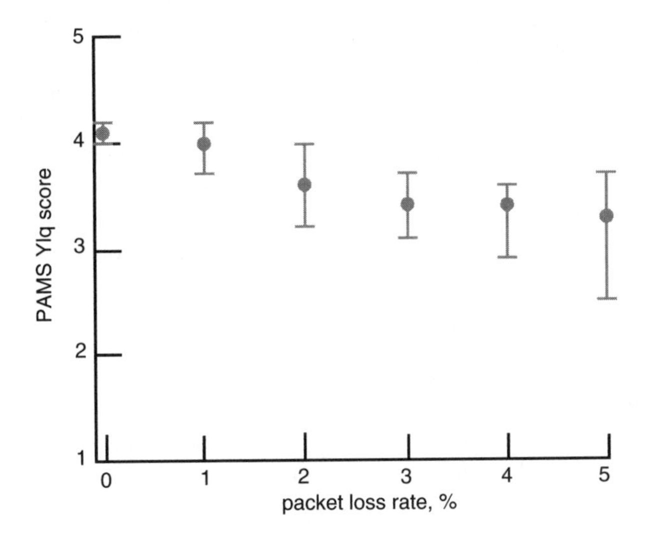

Fig 2.4 The range of speech quality measured and its mean for a given packet loss rate.

2.2.1.5 *Jitter Buffer Implementation*

A jitter buffer is needed to smooth over the distribution of packet delay that is characteristic of IP. It is important to note that a gateway's jitter buffer has an impact on the system's delay and speech quality. A careful balance is needed between adding too much delay, which impairs conversation, and dropping packets, which reduces the speech quality.

Jitter buffers are categorised into two types — static and dynamic. Static buffers use a fixed-length buffer and are the easiest to implement and manage. Any packet that arrives late is simply discarded. The jitter buffer size is normally configurable, which adds some flexibility for tuning a given VoIP system, but generally a static buffer requires a well-managed underlying network to keep jitter within the size of the buffer.

The more sophisticated dynamic jitter buffer has scope to adjust the play-out point in the buffer based on a history of the arriving packet jitter. This suits a network with a more erratic jitter profile, but can also be of benefit on a well-managed network. If the managed network is performing better than specified, e.g. below 10 ms of jitter rather than below the 30 ms planned, the buffer can adjust to reduce the overall delay of the connection. This delay reduction improves the perceived conversational quality experienced by the customers.

An important design consideration for dynamic jitter buffers is when to adjust the play-out point. Although it is sometimes possible to do this during speech without the user noticing, it is far better to make the adjustment during silence where the user will not 'specifically notice' the change.

The notion of not 'specifically noticing' a change in delay is because when asked to identify delay changes during silence the user is generally unable to do so. However, subjective tests show that delay changes are perceptible subconsciously during normal conversation.

This suggests that delay changes should be made in small steps and kept to a minimum during a conversation or the perceived system quality will be reduced.

2.2.1.6 *Codec Tandem Performance*

When designing any voice transmission system it is important to know how well it will work with existing networks. It is not enough to simply establish a connection — there is a need both to ensure adequate speech quality and to minimise delay. Typically where networks join, the speech traffic is passed as 64-kbit/s PCM. If the originating system has coded the speech in another format, a decode is required. This results in a transcoding or tandeming of speech codecs.

Generally the quality of the combination cannot be better than the poorest link, and may be noticeably worse if two or more low bit rate codecs are included. The order is also important. Because these systems distort speech in a nonlinear way,

G.729 followed by EFR will not produce exactly the same quality as EFR followed by G.729.

Delay also increases significantly with tandeming. For example, mobile networks introduce delays of around 100 ms each way. A mobile-to-mobile call with a VoIP trunk could easily exhibit 300-ms one-way delay.

These problems can be avoided by 'tandem-free operation', where the systems negotiate a common codec which is used end-to-end. It is expected that this will become more prevalent, but the potential quality improvement is balanced by a costly engineering challenge.

2.2.2 Delay

ITU-T Recommendation G.114 [1] provides limits for one-way transmission time (delay) on connections with adequately controlled echo (see Table 2.2).

Table 2.2 G.114 limits for one-way transmission time.

One-way transmission time	User acceptance
0 to 150 ms	Acceptable for most users
150 to 400 ms	Acceptable, but has impact
400 ms and above	Unacceptable

ETSI TIPHON defines four QoS classes [5] and ascribes a delay budget for each (see Table 2.3).

Table 2.3 TIPHON QoS classes.

	Best	High	Medium	Best effort
End-to-end delay	< 100 ms	< 100 ms	< 150 ms	< 400 ms

TIPHON systems with delays greater than 400 ms are still classed as best effort, although this is only a target value. Also TIPHON system specifications assume echo is adequately controlled. The graph in Fig 2.5 shows how for a specific network configuration the quality degrades with increasing delay. The graph also shows how this degradation is further affected by the presence of echo.

There are situations where longer delays must be tolerated, such as for access to remote regions, but the general impact of delay on quality does not change. The challenges of achieving this, using modern data networks, are considered in Chapter 3.

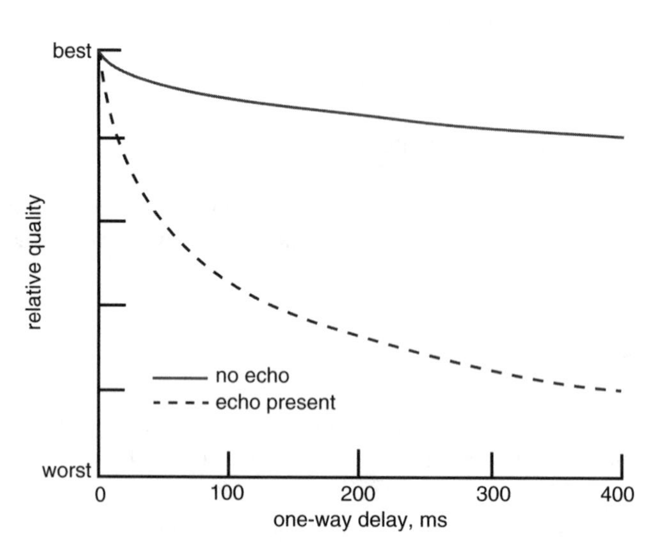

Fig 2.5 Impact of delay on quality, with and without echo [18].

2.2.3 Echo

It is readily understood that the conversational impairment due to echo increases with its level and delay. A large amount of work has been conducted to determine the combined effects of talker echo with delay. Recommendations on its control are summarised in ITU-T G.131 [3].

There are essentially two sources of echo in today's PSTN; these are shown in Fig 2.6.

The delays introduced by packetising the speech and removing network jitter are long enough to make the system susceptible to echo problems. Echo cancellation is therefore likely to be needed in most VoIP systems. This is in contrast to the PSTN where echo cancellation is only necessary on long-haul connections.

Short-delay echoes are rarely distinguished from side-tone unless either the round-trip delay exceeds 30 ms or the echo level is extremely high. For this reason echo cancellation is not required on short-haul PSTN connections, where round-trip delays do not exceed 30 ms. However, round-trip delays of VoIP systems are unlikely to be less than 30 ms, ensuring that some form of echo cancellation is invariably required.

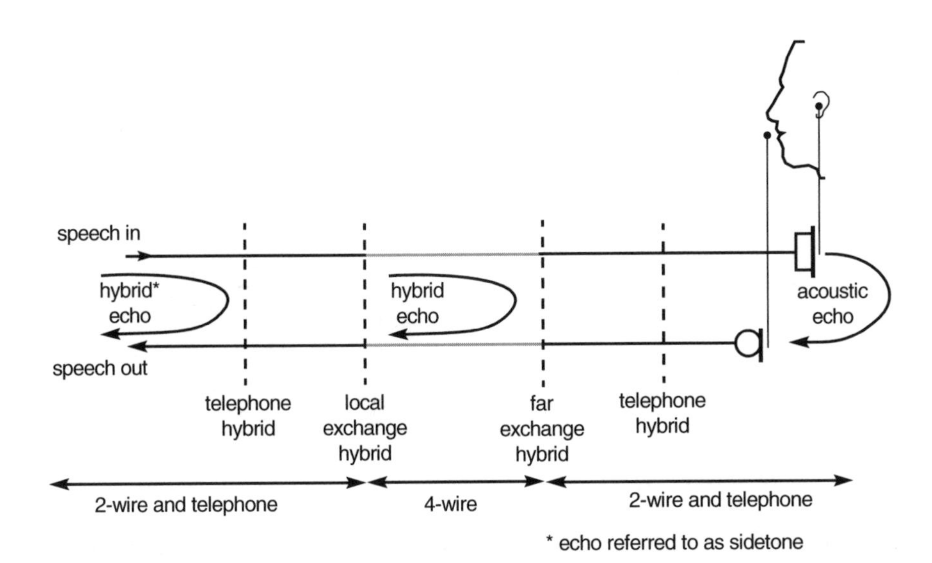

Fig 2.6 Sources of echo in today's PSTN.

If a VoIP system connects to a local PSTN, echo cancellation is probably needed to cancel the local hybrid reflections. If the system does not connect to a local PSTN, echo cancellation should still be included to remove any acoustic echo.

As a final note, in order for a VoIP device to be considered as high quality, the performance of its echo canceller should adhere to ITU-T Recommendation G.168 as a minimum.

2.2.4 Speech Level

Speech level is a key factor affecting ease of communication. The most effective measure of level is active speech level (ASL) with a measurement method defined in ITU-T P.56 [19]. Within the core PSTN, ASLs are expected to be around –18 dBm0 and hence the telephone network is optimised for this level.

The graph in Fig 2.7 illustrates how perceived quality varies with level. Active level control (ALC) systems are starting to be introduced to keep level in the optimal range. Of course it is desirable to have correct levels in the system as ALC may introduce other degradations.

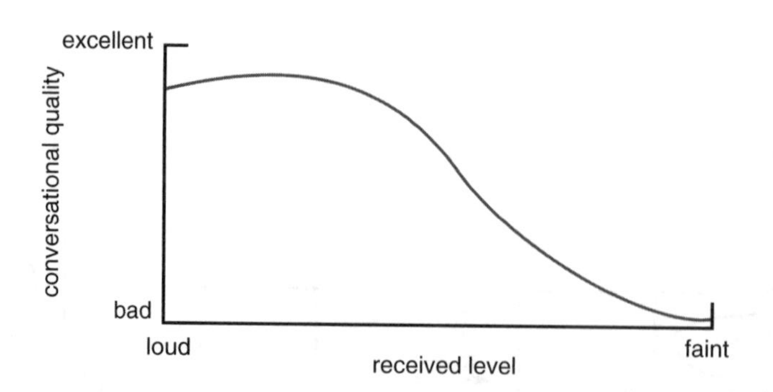

Fig 2.7 Typical impact of level on quality.

2.2.5 Wideband Codecs

Current telephone systems typically allow the transmission of audio frequencies up to about 3.4 kHz. Speech and music contain many components at higher frequencies, which are filtered out and result in an unnatural sound. In contrast, CD audio produces frequencies up to about 20 kHz, the highest that can be detected by people with unimpaired hearing. Also, codecs for wideband telephony and teleconferencing, transmitting audio frequencies up to about 7 kHz, have been available for some years. However, these have not become widespread due to problems with quality and interworking.

New wideband codecs are under development and offer the potential for good quality at similar bit rates to the existing PSTN codecs, while allowing easy interoperation with the PSTN. Typically an existing narrowband codec, such as EFR (at 12.2 kbit/s), is used as the starting point for the wideband codec. This is extended by encoding the higher frequencies separately from the lower band frequencies — usually with relatively little overhead — and multiplexing the two data streams together. The resulting bit-stream at about 16 kbit/s can be decoded into wideband audio, or the higher frequency component can be dropped, without the need for re-coding, for transmission over a legacy connection.

2.3 Testing for Quality

New voice services have traditionally undergone a significant amount of subjective testing to evaluate their performance. Although subjective testing still plays an important role, a proportion of the work necessary has been replaced and supplemented by new objective test methods. This section provides a description of new VoIP test methods, specifically those for measuring speech quality and delay. It should be noted that other test methods, such as for level and echo, are still based on traditional techniques.

2.3.1 Terminology

2.3.1.1 Subjective/Objective Testing

There is no absolute physical definition of speech quality; the only 'baseline' available is the subjective perception of human listeners. There are two methods for testing end-to-end speech quality:

- subjective tests, which seek the average opinion of users [20];
- objective tests, including comparison methods against a known reference signal [13, 17, 18, 21-25].

Subjective tests aim to find the average user perception of a system's speech quality by asking a panel of users a directed question and providing a limited response choice. For example, to determine listening quality users are asked to rate 'the quality of the speech' by voting on the scale shown in Table 2.4.

Table 2.4 Subjective 'Quality of speech' scale.

Opinion	Score
Excellent	5
Good	4
Fair	3
Poor	2
Bad	1

A mean opinion score (MOS) is calculated for a particular condition by averaging the votes of all subjects. The results of subjective tests are influenced by a wide variety of factors. It should be noted that subjective tests take a long time to perform, are costly, and ill suited for investigating large numbers of parameter combinations.

Objective testing techniques measure physical properties of a system. Typically, for telephony systems, this is either achieved by the injection of a test signal or the monitoring of live traffic.

The measurement is repeatable, efficient and, in comparison to subjective testing, fast. It is therefore ideally suited to testing large numbers of parameter combinations.

2.3.1.2 *Intrusive/Non-Intrusive Testing*

Objective measurement techniques can be categorised as intrusive test methods or non-intrusive test methods.

- Intrusive

  The system under test is taken out of service and tested, usually by the injection of a test signal. In terms of a telephone network, this can be limited to a single telephone channel. During this time the system does not carry customer traffic and therefore cannot generate revenue. In some instances testing may, in fact, incur a real cost such as when testing over third-party networks. However, a benefit of intrusive testing is that it can be done when no customers are on a system (during development before live traffic is present) and allows control over external factors.

- Non-intrusive

  Measurements are performed on live customer traffic so that network capacity is not lost during testing. Non-intrusive techniques [25] allow for a larger number of tests, at minimal cost, since traffic sampling is possible. However, these measurement techniques are more difficult to develop and are typically less accurate than intrusive techniques. For instance, what is the effect of a dog barking in the background?

2.3.2 Measuring Speech Quality

With modern digital compression techniques it is no longer possible to determine speech quality by using simple engineering metrics such as signal-to-noise ratio (SNR) and total harmonic distortion (THD). This is in part because low bit rate coding techniques use knowledge of human physiology and perception to hide quantisation distortion in areas of the signal that humans cannot perceive. To overcome this problem with assessing the performance of such codecs, more complex engineering tools have been developed that also account for human physiology and perception. These techniques essentially model the human hearing system and are capable of technology-independent assessment.

There are a few major players developing such techniques. The three most significant techniques, in terms of world recognition and availability, are described below.

- PESQ

  This is the new ITU-T Recommendation P.862 for the end-to-end intrusive assessment of telephony networks. PESQ is the result of collaboration between the authors of this chapter and KPN. PESQ-based tools will become available over the coming months [22].

- PAMS

  This is the authors' intrusive speech quality measure for assessment of end-to-end network quality. PAMS was the first method to provide robust assessment for VoIP, in the latter half of 1998 [13, 17].

- PSQM

  This is the original ITU-T Recommendation for intrusive speech quality assessment (P.861) developed by KPN and adopted in 1996 [21]. However, the scope of the standard is limited to the assessment of codecs and not networks. This standard was withdrawn when P.862 (PESQ) was adopted in February 2001 [22].

All of these predict a measure of listening speech quality, PESQ on a scale (−1 to 4.5), PAMS on a MOS-like scale (1 to 5), and PSQM on a degradation scale (0 to 6.5). PAMS Ylq scores are comparable with the ITU listening quality scale introduced in section 2.3.

When testing with these techniques, as with all measurement tools, some care is required. The main areas in which extra care is required is in the preparation of test signals, namely:

- pre-filtering;
- test signal length;
- appreciation of talker-dependent quality aspects.

2.3.2.1 *Test Signal Pre-Filtering*

Depending on the interface characteristics of the system and test device, a pre-processed narrowband signal [26, 27] is required for testing. Using a wideband test signal results in excess low and high frequency content being transmitted into the system under test. This causes a form of power overload at the codec and generally results in a systematic drop in perceived quality. This can be rectified by processing the test signal through an appropriate filter [26, 27].

Testing performed on a G.729 VoIP test bed demonstrates this effect (see Table 2.5).

Table 2.5 G.729 VoIP test results.

Test file	PAMS Ylq
Non-filtered test signal	3.76
Narrowband filtering test signal	4.3

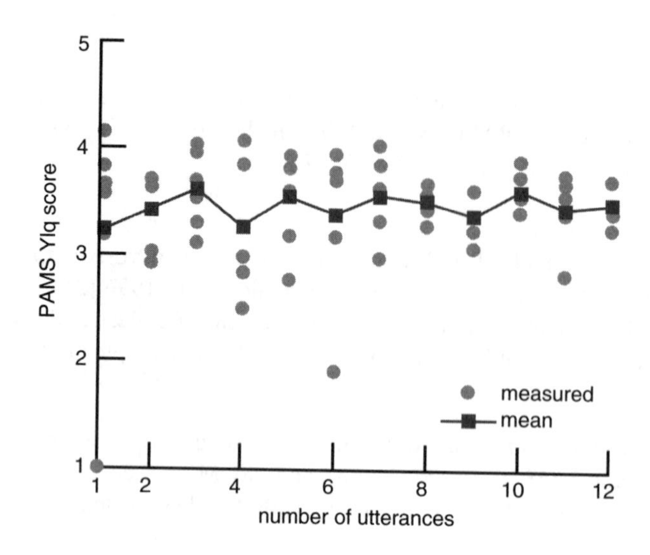

Fig 2.8 Measured quality variation versus signal length.

A non-filtered test signal cannot achieve optimum quality when used to test a network. Testing with such signals results in a general reduction in measured network quality. With the G.729 VoIP test bed, this reduction is around 0.5.

2.3.2.2 Test Signal Length

A common question is how does the length of the test signal affect the measurement consistency of objective speech quality tools. To illustrate this, a single scenario of PAMS speech quality testing for a G.729 VoIP test bed with 3% random packet loss is presented in Fig 2.8. Multiple measurements were performed with different length test signals. The test signal length is defined in terms of utterances[1]. In this example the 1-utterance test signal lasts 1.9 sec and the 12-utterance signal lasts 18.9 sec.

As shown in the graph in Fig 2.8 a problem with short-test signals is that the distortion can either destroy the speech or have no effect. Therefore the measured system, which would measure subjectively a speech quality of around 3.5, can perversely appear as perfect or broken. Similar behaviour is observed when measuring the effects of jitter. Conclusions from extensive studies on test signal length are that:

- an 8-10 sec test signal is recommended for testing networks;
- in laboratory-based testing, such as that used to determine operational thresholds, while it is best to use longer length signals, it is not recommended to perform

[1] An utterance is a continuous burst of speech containing no silent period longer than 200 ms.

assessment of signals greater than 16 sec — if longer signals are used, the assessment should be divided into approximately 16-sec sections.

2.3.2.3 Test Signal Talker Dependence

It is known that telephony systems work better in passing some voices than others. This is illustrated in Table 2.6 which shows the results of tests performed using recordings of different talkers. All other factors were kept constant.

Table 2.6 Test results using different talkers.

Test file	Quality
Talker A — male	4.30
Talker B — male	4.25
Talker C — female	4.11
Talker D — female	3.92

This shows the value of using a common set of test signals when comparing performance measures of a system. PAMS is shipped with standard artificial speech-like test signals, in part to help tackle this problem.

2.3.3 Measuring Delay

VoIP systems can exhibit variable delay — in fact the delay experienced by a user could vary considerably during a call.

A single measurement is inadequate and it is therefore essential to perform a number of measurements of delay during each call and consider both the average delay and its range [17].

A method of assessing delay for VoIP systems, defined in TIPHON [28], is the mean delay from at least 10 measurements or 90% of the largest delay measured, whichever is greatest. This delay measurement requires a test signal as shown in Fig 2.9.

The test signal contains periods of speech activity (Talk1 and Talk2) and periods of silence. Talk1 is an initialisation sequence, which allows any dynamic jitter buffers to converge.

Both Talk1 and Talk2 contain periods of talk spurts and silence intervals. This allows for jitter buffers to adjust their length during the silence intervals. The average talk spurt should be around 1 sec and the average pause 1.6 sec. It is further recommended that the silence intervals are at least 300 ms long.

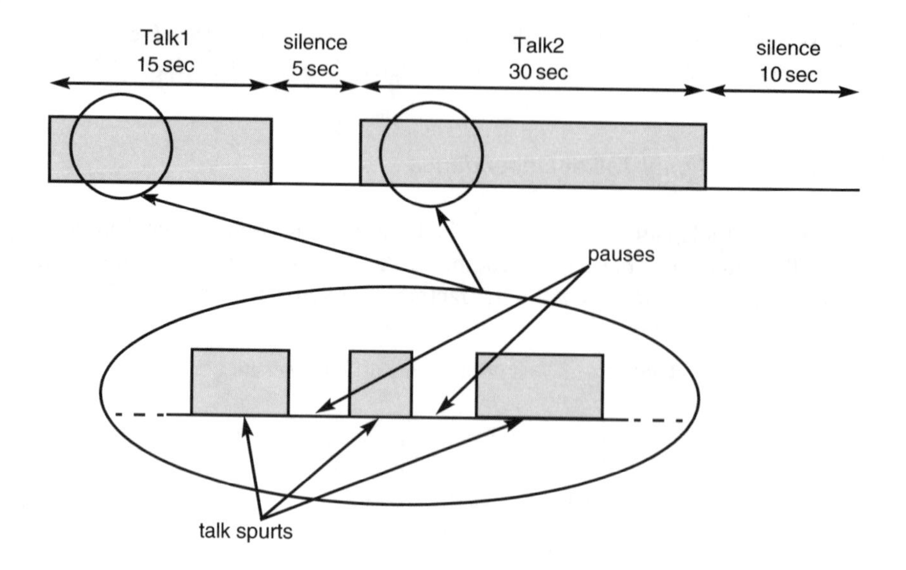

Fig 2.9 Delay measurement test signal composition.

2.4 Summary

The premise of this chapter was that a high-quality VoIP solution is not impossible, it is just an engineering challenge. In this context, the definition of 'good quality' is appropriately translated to 'just-good-enough quality' as over-engineering of solutions is not generally a cost-effective strategy.

Unfortunately, there is no simple recipe to follow to achieve this. In part this is because what may be 'just-good-enough' for one application may be significantly under- or over-specified for another. Furthermore, it is generally possible to achieve the equivalent level of quality in a number of different ways.

This chapter has presented information on how design choices ultimately affect, and potentially limit, the perceived quality of a VoIP solution. The emphasis has been to relate design choices to what users actually experience and hence what impact these choices have on the perceived quality of a solution. Specifically the chapter has discussed:

- choices of speech codec;
- packetisation efficiency;
- use of silence suppression;
- error concealment methods;
- jitter buffer implementation;

- codec tandem performance;
- delay;
- echo;
- speech level.

The intention of the chapter has been to provide a useful summary for anyone wanting to understand the potential impact of choosing various options for a VoIP system; it also comments on the design areas that offer developers the greatest opportunity to set their VoIP solution apart from a competitor's.

In addition to discussing how design and development decisions are likely to affect the perceived quality of the end solution, the chapter has also presented a summary of new measurement techniques to allow proper quantification of system performance.

References

1 ITU-T Recommendation G.114: '*One-way transmission time*', (February 1996).

2 ITU-T Recommendation G.102: '*Transmission performance objectives and Recommendations*', (November 1988).

3 ITU-T Recommendation G.131: '*Control of talker echo*', (August 1996).

4 ITU-T Recommendation G.168: '*Digital network echo cancellers*', (March 2000).

5 ETSI TIPHON: '*End-to-end Quality of Service in TIPHON Systems. Part 2: Definition of Quality of Service (Qos) Classes*', ETS 101 329-2 (July 2000).

6 ITU-T Recommendation G.711: '*Pulse code modulation (PCM) of voice frequencies*', (November 1988).

7 ITU-T Recommendation G.726: '*40, 32, 24, 16 kbit/s adaptive differential pulse code modulation (ADPCM)*', (December 1990).

8 ITU-T Recommendation G.728: '*Coding of speech at 16 kbit/s using low-delay code excited linear prediction*', (February 1992).

9 ITU-Recommendation G.729: '*C source code and test vectors for implementation verification of the G.729 8 kbit/s CS-ACELP speech coder*', (March 1996).

10 ITU-T Recommendation G.723.1: '*Dual rate speech coder for multimedia communications transmitting at 5.3 and 6.3 kbit/s*', (March 1996).

11 ETSI GSM: '*Digital cellular telecommunications system (Phase 2+); Full rate speech transcoding (GSM 06.10)*', ETS 300 961 (2000).

12 ETSI GSM: '*Digital cellular telecommunications system; enhanced full rate (EFR) speech transcoding (GSM 06.60)*', ETS 300 726 (1999).

13 Rix, A. W. and Hollier, M. P.: '*The perceptual analysis measurement system for robust end-to-end speech quality assessment*', IEEE ICASSP (June 2000).

14 Schulzrinne, H. et al.: '*RTP: a transport protocol for real-time applications*', IETF RFC 1889 (1996).

15 Degermark, M.: '*Requirements for robust IP/UDP/RTP header compression*', IETF Internet Draft (December 2000).

16 Garner, P. et al.: '*Robust noise detection for speech detection and enhancement*', IEE Electronics Letters, **33**(4), pp 270-271 (February 1997).

17 Rix, A, W. et al.: '*Perceptual measurement of end-to-end speech quality over audio and packet-based networks*', Audio Engineering Society 106th Convention, Pre-print No. 4873 (May 1999).

18 ITU-T Recommendation P.562: '*Analysis and interpretation of INMD voice-services measurements*', (May 2000).

19 ITU-T Recommendation P.56: '*Objective measurement of active speech level*', (March 1993).

20 ITU-T Recommendation P.800: '*Methods for subjective determination of transmission quality*', (August 1996).

21 ITU-T Recommendation P.861: '*Objective quality measurement of telephone band (300-3400 Hz) speech codecs*', (August 1996).

22 ITU-T Recommendation P.862: '*Perceptual evaluation of speech quality (PESQ), an objective method for end-to-end speech quality assessment of narrowband telephone networks and speech codecs*', (February 2001).

23 ITU-T Recommendation P.561: '*In-service, non-intrusive measurement device — voice service measurements*', (February 1996).

24 ETSI: '*Transmission and multiplexing (TM): Speech communication quality from mouth to ear for 3.1 kHz handset telephony across networks*', ETR 250 (July 1996).

25 Gray, P. et al.: '*Non-Intrusive speech-quality assessment using vocal-tract models*', IEE Proceedings — Vision, Image and Signal Processing, **147**(6), pp 493-501 (December 2000).

26 ITU-T Recommendation P.48: '*Specification for an intermediate reference system*', (November 1988).

27 ITU-T Recommendation P.830: '*Subjective performance assessment of telephone-band and wideband digital codecs*', (August 1996).

28 ETSI TIPHON: '*Technology compliance specification. Part 5: Quality of service (QoS) measurement methodologies*', ETS 101 329-5 (November 2000).

3

VoIP — THE ACCESS DIMENSION

D J Thorne

3.1 Introduction

As is apparent from the previous chapter, although there are many reasons for using VoIP (see also Chapters 1 and 13), transport efficiency, in the sense of payload to packet size ratio, is unlikely to be one of them. This can become a concern in any bandwidth-restricted part of the network due to the effects on perceived voice quality. Networks typically comprise a customer premises or local environment, an access network, and a core network. Since it is impractical to deploy large-scale public networks that are configured as a complete mesh, some degree of bandwidth constriction will always be necessary.

Traditionally access systems, in particular copper-based access systems, have been of limited bit rate when compared with the customer premises environment or the core network. This remains the case today even though we are now seeing the mass deployment of DSL systems that are very much faster than traditional voice-band modems. Any successful VoIP deployment must therefore address the problems of how to cope with relatively low-speed network access. This chapter quantifies the issues this raises and considers various ways in which they might be overcome.

In contrast to the situation in an access network, the local area network (LAN) environment is typified by the deployment of high performance data networks such as Ethernet. In this case, significant bandwidth over-provisioning is both common and cost-effective. VoIP transport efficiency is therefore not generally an issue for LAN distribution. However, extending LAN connections over a wide area network (WAN) will typically involve some form of access network connection that will present the issues identified above.

With one particular exception, VoIP therefore presents a less than attractive technology for voice transport outside the LAN environment. This exception is the case where very long distance voice networks are required — such as global voice networks for multinational corporations or transcontinental public telephone networks where there are long distances involved. In these types of network, the costs for transmission are typically related to the distance covered. So it is attractive to ensure that the available bandwidth is used as efficiently as possible. This has resulted in the use of voice compression and multiplexer technology that enables existing 64 kbit/s speech signals to be compressed to 32 kbit/s, 16 kbit/s or less. While this has worked very effectively for many years, this creates a problem where this traffic needs to be switched at a destination or intermediate point, since most switching systems are designed to operate only at 64 kbit/s. The solution usually adopted is to deploy similar multiplexer technology at the entry and exit ports of a 64 kbit/s switch. Unfortunately this not only increases the hardware cost of the solution but also introduces a tandem codec segment into the transmission path to decompress received speech signals before switching and then recompressing them for onwards transmission. In contrast, the alternative VoIP-based solution involves deploying gateways (discussed further in Chapters 1 and 7) at the extremities of the long-distance network and a call server to direct the VoIP call sessions. The intermediate switching node, additional pre-switching multiplexers and the associated tandem codec segment can therefore be eliminated — reducing cost and increasing voice quality.

VoIP, along with other 'derived voice' systems, offers the potential to deliver a significantly increased number of 'lines' to both the business and residential market. However, unless there is an acceptable way of distributing derived voice lines within the customer's premises, this additional capability is of little practical benefit — and therefore of little value to service providers for revenue generation. The distribution of derived channels is a particular concern in the residential environment where there is unlikely to be a suitable distribution infrastructure in existence — and very limited scope for retrofitting one. Therefore the second topic addressed in this chapter is an assessment of the various ways in which VoIP can be distributed in the customer's premises.

3.2 Qualifying the Issues

The current narrowband switched voice network has evolved over many years, and people now have very entrenched expectations about the quality, availability and performance of a voice service. In particular, the delivery of a successful, quality, voice service using digital transmission generally requires the following to be achieved:

- constraining delay;
- constraining jitter;
- providing an effective constant bit rate data stream;

- providing a sufficiently low error rate;
- preventing not only those calls for which insufficient system resources are available, but also those that would have an adverse impact upon calls in progress.

However, it is possible to move away from these expectations where there is sufficient perceived advantage. Mobile telephones are the case in point. In general, these expectations still have to be met for a commercially viable voice service (as discussed in Chapter 2). Contrary to popular belief, however, IP is not intrinsically well suited to the transport of voice. A simple consideration of the features of TCP/IP transport illustrates why the above requirements might be difficult to achieve, namely:

- long packets to make the large per-packet overhead small can give rise to significant delay and jitter;
- best-effort delivery is at odds with delivering a constant bit rate media stream;
- the use of congestion back-off as a way to match network demand and resources adversely affects constant packet rate applications.

While UDP rather than TCP would normally be used for voice transport, the UDP voice traffic would typically be sharing a network with a lot of TCP traffic, and so be affected by these features. Various mechanisms have been devised to overcome such problems, but as yet these are not widely implemented. In fact, the most common way of providing quality of service (QoS) in IP networks is to significantly over-provision bandwidth. This is simply not an option in a bandwidth-restricted copper access network. The other common approach to deliver a constant low bit rate stream is to buffer a lot of data so as to smooth out the variations in instantaneous bit rate. However, as discussed in Chapter 2, this typically involves buffer delays of several seconds which are completely unacceptable for voice.

3.3 Dimensioning the Problem

3.3.1 Voice Delay Constraints

There are three main reasons why delay matters in a voice network — the effects of echo, the impact upon conversations, and the requirements for achieving interconnection between networks.

- Echo

  Reflections occur at various points in a voice network, and these give rise to echo. Once the echo delay exceeds a few milliseconds, the speaker starts to hear a delayed version of their own speech. This can become very objectionable to the point that it can become very difficult, if not impossible, to carry on a conversation. The ITU therefore gives guidelines as to how much delay is acceptable in a voice network, and recommends that echo delays need to be kept below 5 ms. Above this figure, echo cancellation needs to be employed.

- Conversational interaction

  Even if echo cancellation is used, the delay cannot be made arbitrarily long. While this is true just to constrain the complexity of the echo cancellers, a more important reason is to avoid problems with conversational interaction. Pauses are used in a conversation to invite a response. If a response is not received within a certain time, the original speaker may continue to speak. For example, they may clarify something that has been said, assuming that the lack of response is due to a lack of understanding rather than network delay. Therefore once the delay exceeds a certain value, the natural rhythm of a conversation can become upset. Again there are ITU guidelines, and the suggestion is that end-to-end delays need to be kept to below 150 ms to avoid this problem. There is a nominal budget breakdown of this figure with 50 ms being allowed for the access component. This is not a hard cut-off in the sense that the conversation will break down at 151 ms, although it does rapidly become more difficult to hold a conversation with end-to-end delays above this figure. Conversely, improvements in perceived quality increase as the delay is reduced below 150 ms.

- Interconnect agreements

  The third reason why delay needs to be constrained is so that there can be an appropriate partitioning of delay as part of any network interconnect agreement. This is necessary, as the overall delay needs to be known in order that the required end-to-end service quality can be achieved.

3.3.2 Delay Budgets

It is commonly assumed that delay is only due to packetisation effects. However, this is untrue since there are many components of delay in a packetised voice system (as discussed in Chapter 2). Even though packetisation can in some situations be the largest single component, it may constitute less than half of the total delay — it is important to remember this when considering delay optimisation techniques. The main elements of delay include:

- encoding, the time taken to digitally encode the voice samples — this can be significant for aggressive compression schemes, particularly those that use block encoding techniques;
- packetisation, the time taken to accumulate a packet's worth of data;
- serialisation time, the time taken to clock the packet's contents on to the transmission line;

- packet queuing — as VoIP systems support multiple simultaneous conversations, any given packet may have to wait for access to the transmission system, and furthermore, as the service offering would normally be voice and data, voice packets may have to queue behind data packets;
- DSL transmission delay — the 'speed-of-light' delays across an access network are typically very small, and so the DSL transmission delay is dominated by the transmission system itself;
- core network delay — the time taken to transfer the packet to the far-end access system or interworking unit;
- decoding/transcoding delays at the far end or some intermediate point(s);
- dejittering buffer — although some of the above delay components are fixed, some are variable, e.g. queuing and network delays vary with load; it is therefore necessary to provide a jitter buffer that essentially allows for the worst-case variable delay, otherwise output buffer starvation can occur which results in voice samples not being available for play-out; this can have a very bad effect on the perceived quality.

In order to quantify the above factors, it is necessary to make some assumptions. Typical values for simple voice transport schemes might be:

- 64-kbit/s PCM encoding;
- 40-byte header per voice packet (20 bytes IP, 8 bytes UDP, 12 bytes RTP);
- 40-byte payload;
- transmission technology, consisting of either:

  — HDLC framing with a 5-byte overhead, or

  — Ethernet framing with a 26-byte overhead, or

  — ATM AAL5 framing with 8 bytes AAL5 overhead + 5 bytes overhead per cell and any cell padding required;
- ADSL transport with a 256-kbit/s upstream rate;
- SDSL transport with a 512-kbit/s upstream rate;
- two active voice channels (this constrains the voice queuing jitter);
- 1500-byte data packets;
- no sub-IP layer multiplexing nor fragmentation.

The associated various delay components are shown in Table 3.1. The packet lengths are in bytes, and the delays in milliseconds. The voice and data queuing delays depend on the number of voice channels that are simultaneously active and the voice/data traffic mix. The jitter on these has been assumed to be the serialisation time for the appropriate packet type. This will be an underestimate where there are more than two derived voice channels. Finally, it just so happens that the total packet lengths for this set of assumptions are the same for both Ethernet and ATM AAL5 encapsulation; therefore the results for these two have been combined into a single column.

Table 3.1 Typical simple voice transport schemes.

Delay element	HDLC		Ethernet/ATM	
	ADSL	SDSL	ADSL	SDSL
Payload (bytes)	40	40	40	40
Total packet length	85	85	106	106
Encoding delay	0.7	0.7	0.7	0.7
Packetisation	5	5	5	5
Voice queuing	0.3	0.2	0.4	0.2
Voice queue jitter	2.7	1.3	3.3	1.7
Data queuing	6.0	3.0	6.0	3.0
Data queue jitter	47.7	23.8	47.7	23.8
Serialisation	2.7	1.3	3.3	1.7
DSL transmission	4	1.2	4	1.2
Packet network	5	5	5	5
Packet network jitter	2	2	2	2
Decoding	0.7	0.7	0.7	0.7
Total delay	76.7	44.2	78.1	44.9
Jitter buffer	52.4	27.1	53.0	27.5
Bit rate/channel — kbit/s	136	136	170	170

Several things are obvious from this analysis:

- the dominant delay component is the jitter buffer needed to allow the mixing of voice and data traffic;
- this budget is not strictly access only, as it includes some core elements — however, the total is close to or exceeds the ITU 50 ms access recommendation;
- even if the data mixing is removed, the delays are still well in excess of the 5 ms limit that needs to be adhered to if cancellation is to be avoided.

In addition to the large delays, this simple scheme is very inefficient in terms of the bit rate required per 64-kbit/s speech channel which ranges from 136-170 kbit/s per channel. Even with the most efficient HDLC framing it would not quite be possible to fit two VoIP channels in a 256-kbit/s ADSL system. Such a system would therefore be at a disadvantage when competing with more efficient systems.

As a consequence, this simple way of encoding and encapsulating VoIP is inappropriate for a DSL access system. A more sophisticated system is therefore needed. It transpires that delay and efficiency are closely related, since the simplest way of increasing efficiency is to increase the packet payload. This of course increases packetisation, serialisation and queuing delays. It is therefore important to ensure that efficiency is increased without increasing the already high delay to a completely unacceptable level.

3.3.3 Increasing the Efficiency

There are five ways of increasing the efficiency in terms of reducing the bit rate required per voice channel:

- increasing packet length;
- using low bit rate coding;
- sub-packet multiplexing;
- header compression;
- silence suppression.

The impact of each of the above strategies on transmission efficiency is now considered in turn.

- Increasing packet length

  As noted above, simply increasing the packet length on its own is not likely to be acceptable as the delay is already on the point of being too great with 40-byte packets. Assume, however, that the data queuing delay can be reduced or eliminated, which may be achieved by not supporting a voice and data mix. Suppose that the packet length is quadrupled to 160 bytes, and the data queuing is eliminated. The delay budget then becomes as shown in Table 3.2. Note that in this case the ATM and Ethernet framings are different. This is because padding is required in the ATM case, since the payload and AAL5 overhead do not fit into an exact number of ATM cells.

  This approach has significantly improved the efficiency when compared with the simple case. However, it is still not very good in the sense that, depending on the encapsulation, it still only supports two or three voice channels in a 256-kbit/s link. While further increases in efficiency could be achieved by using even longer packets, this approach is unfortunately subject to the law of diminishing returns. In particular, the delay is already at or close to the ITU 50 ms limit, and it presumes that the data queuing delay can be made negligible. This is therefore about the limit of improvement that can be achieved by this method, and so does not really provide an acceptable solution.

- Low bit rate encoding

  There are in fact two different strategies that may be considered when selecting the basic voice encoding scheme. These either aim to:

Table 3.2 Effects of increasing the packet length.

Delay element	HDLC		Ethernet		ATM	
	ADSL	SDSL	ADSL	SDSL	ADSL	SDSL
Payload (bytes)	160	160	160	160	160	160
Total packet length	205	205	226	226	265	265
Encoding delay	0.7	0.7	0.7	0.7	0.7	0.7
Packetisation	20	20	20	20	20	20
Voice queuing	0.6	0.4	0.9	0.4	1.0	0.5
Voice queue jitter	6.4	3.2	7.1	3.5	8.3	4.1
Serialisation	6.4	3.2	7.1	3.5	8.3	4.1
DSL transmission	4	1.2	4	1.2	4	1.2
Packet network	5	5	5	5	5	5
Packet network jitter	2	2	2	2	2	2
Decoding	0.7	0.7	0.7	0.7	0.7	0.7
Total delay	46.0	36.4	47.4	37.1	50.0	38.4
Jitter buffer	8.4	5.2	9.1	5.5	10.3	6.1
Bit rate/channel — kbit/s	82	82	90.4	90.4	106	106

— reduce the encoding rate while maintaining land-line toll quality, or

— reduce the encoding rate so as to obtain the 'required' number of channels.

There is probably general agreement that 32-kbit/s ADPCM (G.726) encoding gives land-line quality. So taking this as the assumption for the first case, and still assuming no data queuing delays, gives the budget shown in Table 3.3. For the sake of conciseness, the packet network delay and jitter, and the DSL transmission delay have been removed as line items, but are still included in the total delay. Since the packetisation delay increases with decreasing bit rate, and the total delay in the previous example was still of concern, the packetisation delay has been kept constant, which requires the packet payload to be halved to 80 bytes.

Table 3.3 Effects of reducing the encoding bit rate to 32 kbit/s.

Delay element	HDLC		Ethernet		ATM	
	ADSL	SDSL	ADSL	SDSL	ADSL	SDSL
Payload (bytes)	80	80	80	80	80	80
Total packet length	125	125	146	146	159	159
Encoding delay	1	1	1	1	1	1
Packetisation	20	20	20	20	20	20
Voice queuing	0.5	0.2	0.6	0.3	0.6	0.3
Voice queue jitter	3.9	2.0	4.6	2.3	5.0	2.5
Serialisation	3.9	2.0	4.6	2.3	5.0	2.5
Decoding	1	1	1	1	1	1
Total delay	41.3	34.4	42.7	35.0	43.6	35.5
Bit rate/channel — kbit/s	50	50	58.4	58.4	63.6	63.6

This approach provides four channels of capacity in 256 kbit/s, with the delay being below 50 ms. So this is an acceptable solution provided a suitable way can be found to constrain the data queuing jitter. In contrast, the same encoding using native ATM AAL2 encapsulation (as is done in BLES architectures [1]) requires only 38 kbit/s per channel. This can provide six channels in 256 kbit/s — and with a significantly lower delay.

In order to increase the number of channels, significantly higher levels of compression may be used. In practice, the limit for compressing voice is probably 6.3 kbit/s (G.723.1) — resulting in 'mobile' rather than land-line quality. G.723.1 uses a block encoding technique that works on 30 ms blocks of data with an additional 7.5 ms look-ahead delay. The total encoding delay per block is therefore 37.5 ms. This replaces the packetisation delay previously considered for 'sampling' codecs. However, the packets need to be restricted to a single block, or the effective packetisation delay will become excessive. The delay budget for this scenario is shown in Table 3.4.

Table 3.4 Effect of high compression codecs.

Delay element	HDLC		Ethernet		ATM	
	ADSL	SDSL	ADSL	SDSL	ADSL	SDSL
Payload (bytes)	24	24	24	24	24	24
Total packet length	69	69	90	90	106	106
Encoding delay	37.5	37.5	37.5	37.5	37.5	37.5
Voice queuing	0.3	0.1	0.4	0.2	0.4	0.2
Voice queue jitter	2.2	1.0	2.8	1.4	3.3	1.7
Serialisation	2.2	1.0	2.8	1.4	3.3	1.7
Decoding	1	1	1	1	1	1
Total delay	54.1	49.0	55.5	49.7	56.6	50.2
Bit rate/channel — kbit/s	18.1	18.1	23.6	23.6	27.8	27.8

Depending on the encapsulation strategy adopted, this approach can provide a much larger number of channels. In this case in the region of 9-14 channels may be supported within a 256-kbit/s access link. The delay is, however, fairly high, and indeed the voice queue jitter would be much higher if a large number of channels were actually being used. Using a high compression codec, such as G.723.1 encoding, therefore provides a large number of channels, but is on the limits of acceptability from the point of view of quality and delay.

- Sub-packet multiplexing

  One of the reasons for the inefficiency in VoIP systems is the large per-packet overhead of IP and associated protocols. In a full-blown, end-to-end VoIP architecture where the source and destination addresses are being used in the normal fashion, this may be unavoidable, as every channel could have its own source and destination address. However, there is a 'trunking' variant where a number of channels are trunked to a common destination, using sub-packet multiplexing. This technique involves a number of VoIP packets being multiplexed together into a single packet, and transported to an interworking unit — typically to be broken out into the PSTN as shown in Fig 3.1.

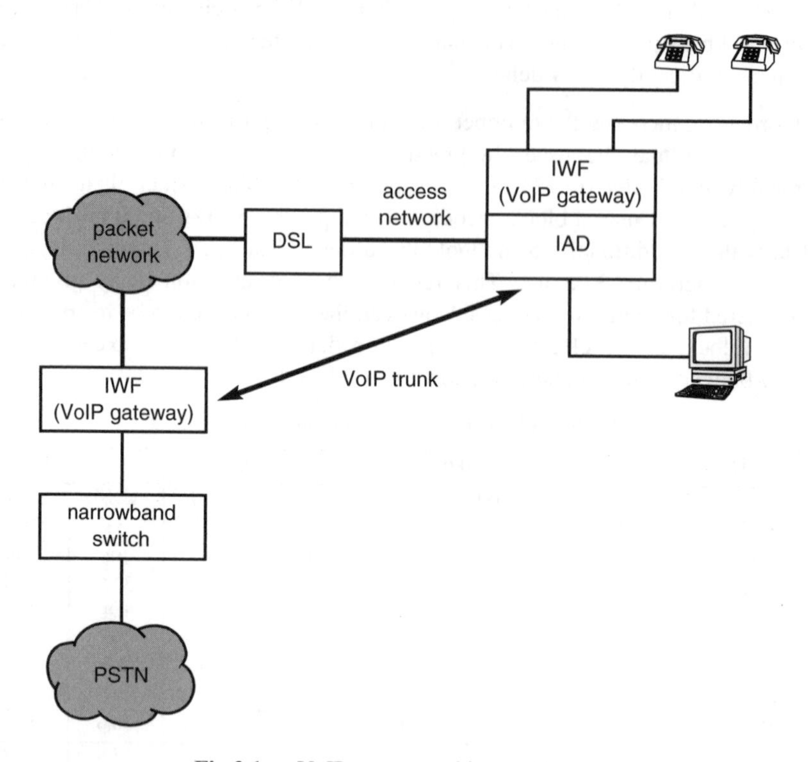

Fig 3.1 VoIP access trunking architecture.

Within this architecture, each packet has the IP, UDP and FRF-11 header structure shown in Fig 3.2. Each mini-packet has a compressed RTP header of 2 bytes. In practice, this scheme can become comparable in efficiency to ATM AAL2 if there are several active voice channels.

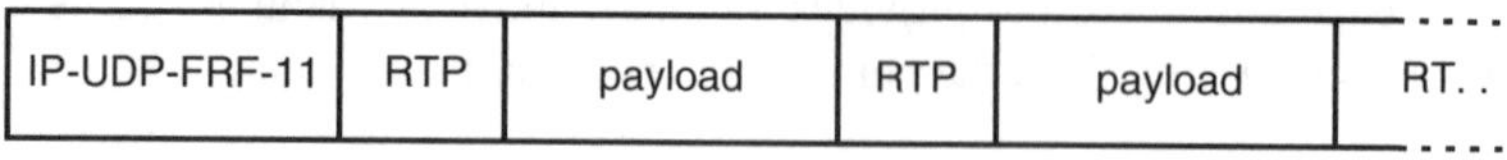

IP-UDP-FRF-11	RTP	payload	RTP	payload	RT. .

Fig 3.2 VoIP trunking packet structure.

In many ways this is not really VoIP in the sense that IP addressing and routing are not being used to route the voice data. Rather it is really the IP equivalent of the ATM-based BLES architecture [1]. However, it does exist as a commercial implementation.

- Header compression

  An alternative approach to reducing the packetisation inefficiency is to use header compression. There are a number of different methods of header compression that are defined in various request for comment (RFC) documents published by the Internet Engineering Task Force (IETF), but the general principles of operation are very similar and essentially comprise the following elements:

  — the full header is sent with the first datagram of the communication and stored by the receiver;

  — each field can be classified as UNCHANGING, RANDOM changes, DELTA changes or inferred as DEFAULT;

  — the only segments of header information that need to be sent in every header are fields that change often and randomly, such as checksums or authentication data;

  — for fields that are incremented from the previous value (DELTA), only the delta increment is sent.

  Compressed packets are sent until particular header fields change — as specified in the appropriate IETF RFC describing the specific implementation. In which case an uncompressed packet is sent. The new uncompressed header is placed at the top of a store in the receiver and the one at the bottom dropped out. The receiver can then reassemble the header using the latest stored header and the decoded change information from the received packet.

  Header compression is configured during the link set-up phase, such as during the network configuration phase of the point-to-point protocol. The stations at the ends of the link need to agree to operate header compression. The stations may also set the store size which in turn determines how many full headers can be held in memory.

  The particular compression scheme that is appropriate to VoIP is defined in RFC 2508. This compresses the combined IP/UDP/RTP header which results in a greater efficiency than compressing each header individually. For IP/UDP/RTP the combined header is 40 bytes — 20 for IP, 8 for UDP and 12 for RTP. Applying header compression, as in RFC 2508, reduces the header to 4 bytes if UDP checksums are sent and 2 bytes if they are not. RFC 2508 is designed to fit into the framework of header compression laid down in RFC 2507 by classifying IP/UDP/RTP as a third class of transport. RFC 2508 does not apply compression to the real-time control protocol (RTCP) because it accounts for only a small percentage of the total traffic. IP/UDP/RTP header compression is configurable using the PPP control protocol as per RFC 2509.

A common feature of header compression techniques is that they only work over a point-to-point link (e.g. a PPP link). So all VoIP calls need to be trunked across the same link. It would therefore be necessary to have a 'PPP trunk' to an intermediate gateway, and use IP routing to reach a destination once the headers have been decompressed.

The second concern with the use of header compression in an access network environment is the effect of burst errors (induced by impulse noise). If these affect a compressed header, the reconstructed header will be incorrect until the next full packet is received, which could result in a significant period of missing data.

Header compression is therefore of somewhat limited applicability, and may not be particularly appropriate for use by VoIP systems.

- Silence suppression

  In general, only one of the two speakers engaged in a conversation will be active at any one time. So further bandwidth efficiencies can be achieved by simply not sending voice samples — or suppressing the silence — when a speaker is not actually speaking. Silence suppression may be applied in addition to any of the above techniques and can give rise to a theoretical 50% saving in bandwidth when averaged over the send and receive path of the link. Further slight efficiencies can be achieved by not sending packets during the pauses in conversation, and even the gaps between words. However, this does require some additional complexity in the system, such as voice activity detection functions and the insertion of comfort noise. This last feature is required as the complete absence of background noise can be very disconcerting. To overcome this problem, random noise may be inserted at the far end in order that the listener does not believe that the line has 'gone dead'. In practice, however, it is not easy to quantify the gains due to silence suppression. Further discussion on silence suppression and comfort noise can be found in Chapter 2.

The simplest approach of using sampled 64-kbit/s PCM encoding provides a very inefficient solution, and will severely limit the number of derived voice channels that can be supported. Therefore one or more of the above techniques need to be used. The most effective and predictable technique is to use low bit rate encoding. This can be combined with silence suppression if further efficiency gains are needed, although this will not give land-line quality.

3.4 Quality of Service

Constraining latency is just one example of the need to provide at least some rudimentary QoS across the bandwidth-constrained access network. This is true even when only voice traffic is being supported, but becomes much more critical when there is a voice/data mix. As was shown in Table 3.1, the jitter caused by interleaving voice and data packets can lead to the need for a very large jitter buffer. This makes the delay budget unacceptably large. At the very minimum two things are needed to overcome this deficiency:

- data packet fragmentation, ensuring that large data packets do not create excessive queuing delays;
- voice packet prioritisation, ensuring that the voice packets have sufficient priority over data packets.

Data packet fragmentation can be done either pre-emptively or reactively, i.e. data packets can be fragmented regardless of whether there is any voice traffic or fragmented only if there is a voice packet waiting to be sent. The simplest method of pre-emptive fragmentation is of course to set a low MTU size comparable with the voice packet size. However, this means that data packetisation becomes very inefficient. An alternative is to use some kind of suspend/resume scheme, which interrupts the transmission of data packets to send a voice packet. If there is a cell-based layer 2 (such as ATM), the problem is automatically solved, since this provides pre-fragmentation without an unacceptable decrease in efficiency.

3.4.1 Priority Mechanisms

Even when the data fragmentation problem has been solved, it is still necessary to ensure that the voice packets have priority over data packets. There are a variety of layer-2 and layer-3 mechanisms for achieving this. Some of these are based on relative priority. Fortunately, as the problem is mainly across the access link, the normal concerns about such mechanisms — that they are not quantified and do not work well end-to-end — do not really apply here. Cruder techniques, such as over-provisioning of bandwidth, which can be used in other parts of the network, remain inappropriate. There are some observations that are worth making on these different approaches.

3.4.1.1 Layer-2 Mechanisms

If there is an ATM-based layer 2, the voice and data traffic can be carried in separate PVCs. In this case it is not necessary to implement a full set of traffic management mechanisms to achieve the required prioritisation for this simple traffic mix. Per-PVC queuing, with the voice queue served to exhaustion before data cells are sent, would suffice. The disadvantage of this approach is the need to provision an additional PVC.

Similarly if the access were bridged Ethernet, the simple priority tagging scheme in IEEE802.1p/q could be used. However, this option is not really appropriate if routed Ethernet is being used.

3.4.1.2 Layer-3 Mechanisms

Prioritisation can also be done at the IP layer, using diffserv for example. While this is a simple scheme that uses relative priority, it would be quite sufficient for a simple voice and data mix. However, as the pinch point is the DSL modem rather than the end device, the modem would need to have the appropriate layer-3 routing capability in order to implement this.

3.4.2 Ensuring Sufficient Voice Bandwidth

Voice traffic requires a constant bit rate (CBR) stream, even when averaged over a fairly short time period. It is therefore important to ensure that the total number of derived voice channels does not exceed the available bandwidth of the access link. The only alternative is to use whatever degree of compression is required to fit all the requested channels into the available bandwidth. The problem with this approach is that the quality will be both undefined and changing — even within a call. This kind of variable quality is generally a bad idea, as human nature is such that customers complain when the quality deteriorates, rather than being grateful when it improves. Such variability is therefore a highly undesirable feature from the point of view of both the service provided and any on-going support costs.

To address this problem, there ideally needs to be some kind of connection admission control (CAC) that prevents more than the supportable number of calls being initiated. This can be achieved relatively simply if the DSL terminating unit also inherently limits the number of calls in some way. For example, with an integrated access device that uses analogue telephone presentation (i.e. a gateway), there is a hard limit on the number of calls that can be supported. This is because the number of telephone sockets on the gateway is fixed, which provides automatic protection as long as there is more access bandwidth provisioned than the number of ports require. However, where soft-phones are used, or the DSL modem is connected to a LAN, there is currently no such easily imposed limit as any number of soft-phone clients or Ethernet phones can, in principle, be attached. However, solutions, such as MIDCOM (see Chapter 5), are emerging to address this need.

In most cases the access link will be required to support a mix of voice and data traffic. The best approach in this case is to limit the total bandwidth allowed for voice traffic to somewhat less than the total link bandwidth. The data bandwidth can then flex to use the available bandwidth using normal TCP mechanisms. There are various ways in which this limiting can be done. However, some system entities need to have a view of the aggregate state of calls in progress and prevent further calls once a certain limit has been reached. Mechanisms for achieving this range from the simple port limit with provisioning as described above, to a call server attempting to keep a tally of active calls for a particular end-user network. There are, however, many problems with this latter approach.

3.4.3 Mixing Voice, Data and Video

In the current business environment, a simple voice/data mix may be sufficient. In contrast, the residential environment is more complicated due to the desire to simultaneously support voice, data and video for interactive applications. This presents a significantly more difficult challenge than a simple voice/data mix because of the properties of compressed video. In particular, there are two areas of concern — interleaving delay and QoS/admission control.

3.4.3.1 Interleaving Delay

DSL systems are designed to run over access network cables that have poor performance at high frequencies, since the cables were designed for voice-band operation. As a consequence, there are a number of issues this creates and, in particular, a problem with impulse noise. Impulse noise is caused by both natural (e.g. lightning) and man-made (e.g. light dimmers, starter motors) electromagnetic transients. These can induce very high common-mode voltages on cables, some of which appear as differential mode signals due to poor cable balance at high frequencies. The magnitude and duration of this noise are such that many successive bit errors may be caused — resulting in a so-called burst error. Highly compressed video is very susceptible to burst errors, as this can cause the loss of a significant number of video frames — depending on the type of frame corrupted. The resulting glitch in the programme is very obvious and annoying to the customer. In video-on-demand trials, customer complaints have occurred when one or more such errors have been seen per day.

It is not possible to use retransmission techniques to overcome errors such as these in streaming video, due to the very large amount of data that would need to be buffered while the retransmission was taking place. Error protection therefore needs to be built-in. Simple Reed-Solomon (R-S) correction does not work due to the potentially large number of successive corrupted bits. To overcome this, a technique known as interleaving is used. Using this technique, blocks of data have R-S codes computed in the normal way. However, successive bits are not sent in their normal time sequence. Rather, they are delayed, so they are well spaced out in time. An impulse noise event will still cause multiple consecutive bits to be errored, but these no longer correspond to the same R-S block. The small number of errors per block that remain after de-interleaving can then be corrected by the R-S code.

The required delay is related to the characteristic length of the impulse noise events. These are generally of the order of a few milliseconds. Interleaving delays therefore typically result in additional DSL transmission delays of tens of milliseconds. This makes the delay budgets discussed above significantly worse, and it is highly unlikely that it will be possible to obtain good voice performance across a DSL link that is operating in interleaved mode. Conversely, without interleaving, the video performance is unlikely to be acceptable.

This would seem to preclude the simultaneous support of video and derived voice, which would be a considerable blow for the residential market. However, there are two potential solutions. ADSL transmission standards specify a 'dual latency' mode that allows for the simultaneous support of an interleaved and non-interleaved channel on the same physical link. This is the ideal solution, but as yet has not been implemented by any equipment manufacturer. There are also considerable difficulties with the way in which capacity is assigned and reassigned to the two channels. The only pragmatic solution would seem to be to find some compromise interleave depth that provides adequate protection for compressed video, while at the same time does not add an excessive amount to the delay budget for derived voice. Whether or not such a compromise value exists is still under active investigation.

3.4.3.2 QoS and Connection Admission Control

Introducing video into the service mix also makes it harder to provide the necessary QoS and prevent damaging interactions between applications. This is because video has a much higher time-averaged (constant) bit rate requirement than voice — although, with the exception of interactive video, the delay requirements may be less onerous.

It is therefore no longer possible to apply the very simple prioritisation scheme that was appropriate for voice and data, nor to assume that the CBR traffic is only consuming a small proportion of the link bandwidth. Significantly more sophisticated traffic control and admission techniques are therefore required, but these are beyond the scope of this chapter.

3.5 Derived Voice Distribution Options

VoIP may well be applicable to both business and residential environments. One of the main drivers for any kind of derived voice service is the ability to deliver more than one voice channel down a single access link. Once delivered to the customer's premises, however, there needs to be a suitable way of distributing these channels to multiple simultaneous users.

There are a variety of ways in which this can be done, with the main distinction being whether analogue or digital distribution is used. For the analogue distribution case, the derived voice channels are converted back into baseband, analogue PSTN-type signals at an interworking unit (i.e. gateway), and so any existing extension wiring may be able to be used. Digital distribution, on the other hand, relies on there being some kind of local area network (LAN). The situation in the business environment is somewhat different to the residential environment and so these are considered separately.

3.5.1 Business Distribution

Most businesses of any significant size will have a distribution infrastructure that can be reused for VoIP. These include:

- a dedicated voice wiring system with a separate point-to-point connection to each telephone;
- a star-wired LAN infrastructure, with a point-to-point connection to each PC.

Either, or indeed both, of these can be used for VoIP. In the first case there needs to be a gateway that converts the VoIP back to analogue baseband voice. This can essentially replace, or be connected to, an existing PABX (see Chapter 4). The advantage of this approach is that it makes possible the reuse of both the existing telephone instruments, and the wiring infrastructure. This allows the convenience of a simple telephone call to be retained without the current high cost of a hard IP telephone (e.g. an Ethernet phone).

However, there are several disadvantages. The expectation of a service is often based on the user interface. In this case it is the very familiar telephone; yet this has certain associations in the mind of the user, in particular in terms of quality and functionality. It is both possible and likely that a VoIP service will be significantly different in both respects to the current PSTN service. This could give rise to considerable confusion in the mind of the user, which could have an adverse effect on help-desk costs and acceptance of the service. Further, since this is no longer end-to-end VoIP, the advantages of voice/data integration may be lost.

An alternative approach is to share an existing LAN. This requires a new terminal 'device', e.g. a soft-phone client on a PC or a hard IP telephone, but again reuses an existing wiring infrastructure (see Chapter 4). The problem here is how to ensure that the voice traffic gets sufficient priority over the LAN. The overwhelming majority of business LANs are Ethernet, either 10 Mbit/s (10BaseT) or 100BaseT. Basic logical shared-medium Ethernet has no concept of priority or constrained delay. So exactly how well VoIP distribution would work is impossible to predict as the LAN load varies over time. However, the voice traffic is likely to be small compared to the data traffic, and simply ignoring the problem will probably work quite well in a lot of cases.

The situation will, however, get worse with time as LAN loads increase, and the number of derived lines goes up. There are though, a number of fairly simple ways in which such a LAN can be upgraded to improve VoIP support. These include:

- upgrading to 100-Mbit/s (100BaseT) Ethernet — this simply throws bandwidth rather than control at the problem, but should work well for all but the most heavily loaded LANs;
- migrating to switched Ethernet by replacing a simple repeating hub with a switch — this again increases the total capacity of the LAN, which will improve the QoS for a given loading;

- adding IEEE802.1p/q prioritisation to the switched Ethernet, and prioritising the voice traffic above the data.

In general, any 100BaseT or switched Ethernet LAN is likely to provide acceptable support for VoIP, and is unlikely to be the limiting quality (of service) factor. However, none of these solutions are without cost.

The only other issue with LAN-based distribution is the address scheme to be used. Many current IP services use DHCP and/or NAT to overcome the limited availability of public IP addresses. This means that any given item of CPE may either have a temporary IP address, or a private one. In the latter case, the address will be hidden by a router and not visible to the outside world. If the voice terminals use IP addresses, this gives rise to a problem with being unable to route incoming calls to the correct item of CPE — or worse.

3.5.2 Residential Distribution

In contrast, the residential situation is quite different to the business environment. While the number of lines to be distributed is likely to be smaller than in the business situation, there is not the same opportunity to reuse existing infrastructure.

Many homes do have a network of telephone cables connected to extension telephones in the bedroom, kitchen, study, etc. However, in general these systems are used to connect multiple instruments in parallel to a single line.

This topology is not compatible with the point-to-point (star) wiring required to distribute multiple derived lines to separate telephones. There are, though, a number of ways in which this could be used for the distribution of VoIP.

- Convert to point-to-point wiring

  In practice, the situation is not quite as bad as it may first appear since many homes do have at least two cables returning directly to the master socket (MS).

  Table 3.5 shows an extract from a wiring survey of UK homes from which it can be seen that 72% of homes had at least two cables returning to the MS. With a minor modification, these could be used to support two separate telephones, but going beyond that would require extensive rewiring, which, although it remains a possibility, is probably best avoided.

Table 3.5 Number of analogue ports that could be supported by existing extension wiring.

No. of cables returning to master socket (including any device directly attached to the MS)	2	3	4	5
Percentage of homes in survey	72	20	3	3

- Use the extension wiring as a shared media LAN

  Analogue distribution also has the same disadvantages as in the business environment — the loss of end-to-end VoIP functionality and the possibility of confusing the customer with regard to the service. An alternative approach is to use digital distribution, but the challenge in the residential case is the lack of LAN infrastructure. Various systems have been developed that allow LANs to be run over existing telephone extension wiring. The most successful of these are now known as HPNA systems, after the name of the Home Phoneline Networking Alliance [2], which is the industry consortium that has been working on their standardisation.

  HPNA systems provide an Ethernet-like service, but of course over a true shared-medium. The point-to-multipoint architecture results in a transmission channel that is poorly defined, which limits the data rates that can be achieved. Second generation HPNA2 systems can operate at up to 7 Mbit/s, but this is still somewhat bandwidth limited, and so there are concerns about the QoS that can be achieved. On the other hand, as this is not actually Ethernet, and therefore has no installed-base backward-compatibility issues, it was possible to add a MAC layer prioritisation scheme. This has been done, but its effectiveness in providing voice support has yet to be tested. There is at least one commercial company that has a derived voice distribution scheme that is based on HPNA2.

- Frequency division multiplexing

  A variant on the above approach is to use FDM techniques to provide multi-channel voice distribution by using the spectrum between that used by ADSL and HPNA (approximately 1-3 MHz). Again there is a commercial offering, but the problem with this approach is that it is currently proprietary and uses unregulated spectrum. This means that there is no guarantee that someone else will not use this spectrum for another, conflicting purpose at a later date.

- Powerline distribution

  The reuse of telephone extension wiring is somewhat limited, as this wiring is far from universal in UK homes. In contrast, mains wiring is far more ubiquitous, and a number of LAN-like systems are being developed to run on AC mains systems. While this is again a transmission challenge, it has the additional problems of being a high-noise environment with significant safety issues. Consequently these systems have so far lagged behind phoneline systems in both performance and state of development, and so do not seem to be an immediate prospect for derived voice distribution.

- Wireless distribution

  Wireless systems have long been seen as the ideal solution for home networking, but have so far been hampered by their relatively high price and poor performance. There is, however, no doubt that a wireless voice distribution system remains a very worthwhile aim, especially if it were based on an existing system such as DECT. Wireless IP phone systems do exist, but are currently prohibitively expensive for the residential market.

3.5.3 Distribution Options — Summary

Distribution of VoIP within the customers' premises is not likely to be a major problem in the business environment. If LAN-based distribution is used, it may be necessary to upgrade the LAN to either 100BaseT or switched Ethernet. These are relatively cheap and simple upgrades that can be done as and when necessary. However, the residential environment presents a much more difficult challenge. While there are a number of possible approaches, there is currently no clear favourite. The best approach is possibly to use HPNA2 plus some limited new wiring in the short term, with a longer-term migration to wireless distribution.

3.6 Core Design Issues

The major issues that need to be addressed when designing any VoIP system that runs over a bandwidth-limited access network include the following.

- Choice of codec

  Choose a low bit rate codec which will give sufficient voice quality for the proposed service, and support the desired number of derived lines, given the access bandwidth constraints.

- Transmission efficiency

  Determine the voice packet size, which also has an impact on the above efficiency, such that the delay budget comes within the range of the echo cancellers, and if possible, leads to an access latency of < 50 ms.

- Network design

  Provide a prioritisation mechanism that ensures that voice traffic gets priority over data traffic across the access network. Depending on the access speed, it may be necessary to fragment data packets to prevent excessive jitter.

- Network characterisation

  Do an end-to-end jitter analysis in order to dimension the jitter buffers.

- Interconnect design

  Ensure that the latency budget does not exceed that allowed in any interconnect SLA.

3.7 Summary

VoIP is just one of a number of potentially exciting applications for broadband access, which is now much more widely available with the roll-out of DSL. However, even when VoIP is considered in isolation, the nature of the access network is such that considerable care has to be taken over the network design if an efficient service of acceptable quality is to be delivered. When VoIP is just one of a mix of different services, then the challenges become much greater. Although the distribution of VoIP in customer premises in the business environment is not likely to be a major issue (apart from the firewall and NAT-related issues discussed in Chapter 5), in the residential environment it poses some significantly greater problems that have not yet been fully solved.

References

1 DSL Forum: '*Broadband Loop Emulation Service (BLES) — TR-036 Annex A*', (August 2000) — http://www.dslforum.org/

2 Home Phoneline Networking Alliance — http://www.homepna.org/

4

IP TELEPHONY SOLUTIONS FOR THE CUSTOMER PREMISES

A B Catchpole and C J Middleton

4.1 Introduction

The convergence of voice and data networks [1] has brought new opportunities and threats to traditional suppliers of telephone switching systems in the customer premises equipment (CPE) market. Defined as describing a total end-to-end solution including quality of service from the user's perspective, IP telephony gives the opportunity for advanced communications applications. As IP telephony is an end-to-end application, it potentially opens up the opportunity for not just voice but also economically providing multimedia communications such as data collaboration or even videoconferencing. Voice over IP (VoIP) technology, and IP-based telephony systems in particular, therefore offer the promise of a fully converged communications network supporting a rich set of multimedia applications beyond simple telephony and so potentially yielding real business benefits.

This chapter addresses IP telephony solutions focusing specifically on the business customer premises market including CPE products and associated business applications.

4.2 Market Approaches to VoIP CPE Solutions

The technology media have been heralding VoIP, IP-PBX, and IP telephony for several years now and the hype has produced a ground swell of interest and anticipation among both customers and the communications industry at large. For some time there has been a viable market for VoIP for what might be labelled the transport element of the technology. Data equipment manufacturers have largely generated this market by developing the necessary hardware to replace traditional voice TDM trunk circuits. Typically these products have between targeted for use between individual locations for allowing voice and data to share the same WAN bandwidth. However, there are many different ways of replacing the traditional mechanism for inter-site traffic including voice over frame relay, voice over IP over frame relay, voice over ATM, and voice over IP over ATM. Each method has its own individual demands — some of which are discussed in Chapter 3.

Once the basic transport of voice had been addressed, it was only natural that the next target would be call-control, the bastion of the traditional PBX. Small start-up companies that proliferate in the Internet world developed the initial IP-PBXs during the late 1990s. The most promising of these companies were snapped up by data manufacturers keen to develop the VoIP story and expand their markets into the voice business.

This has led to two clear markets emerging — the evolutionary market and the revolutionary market:

- the evolutionary market-place requires many of the desirable features and much of the functionality of a traditional PBX to be taken into the IP world — this approach is likely to be desirable for a customer who has a large installed base of traditional PBX equipment that has room for expansion and is still being paid for;
- in contrast, the revolutionary market opportunity is characterised by the completely green-field environment, where switched-circuit telephony systems have not previously existed (typically including the equipping of new offices or other locations where there are no legacy infrastructure commitments to consider) — this approach is also suited to customers who have networks made up of disparate PBX systems and want to rationalise their network infrastructure; this may be achieved by installing a common telephony system providing similar features and functionality across the complete network which it may currently be lacking.

The decision regarding which of these alternatives is the more appropriate option can only really be made once a complete analysis of the surrounding business and existing infrastructure requirements have been completed. In particular, it is recommended that the pros and cons of both approaches should be carefully considered before coming to any decision. This is because it has been proven that what may be more cost effective in one scenario may prove to be more costly in another.

In either case, the underlying requirement that must be met to ensure success is the delivery of the all-important IP infrastructure. If the IP infrastructure is not stable, reliable, designed effectively and managed efficiently, then new unified voice and data networks will stumble and falter. The market opportunities are therefore not just for the supply and management of the call-control environment, but for the complete communication infrastructure for voice, video, and data that all businesses increasingly need.

4.3 VoIP Solutions for the Enterprise

The customer premises equipment market consists of the residential mass-market and the more specialised business market (ranging from small and medium sized enterprises through to the very largest corporate companies). In the residential market the basic telephone function has remained largely unchanged over many decades. However, there have been increasing demands for additional functionality from the business community — for example, from the simple key system through to large networked PBXs. From both a VoIP and IP telephony perspective BT has largely focused on the business market for a number of reasons, some of which are given below:

- potential for significant savings on call charges;
- IP network ubiquity within the corporate business community;
- opportunity for converged infrastructure;
- video- and data-conferencing becoming accepted;
- promise of 'value-add' applications;
- developments in soft PBXs and IP telephones.

The residential market has been driven by a different set of stimuli and opportunities — these are described in Chapter 13.

As suppliers of business CPE began to develop their product ideas, it became clear that there were two distinct categories for IP telephony customer premises equipment — products addressing the evolutionary market and products addressing the revolutionary market, reflecting the two market segments identified in section 4.2.

4.3.1 Solutions for the Revolutionary Market

The new-wave revolutionary products first started to appear in 1997. Notable examples were produced from companies such as Vienna Systems and Selsius Systems. These were small innovative companies that started from scratch without any constraints of an existing TDM PBX product line to consider.

The existing traditional PBX vendors ignored these small start-ups and it was left to the larger Internet companies to see the opportunity and to acquire both the technology and, perhaps more importantly, the people.

Cisco Systems [2] set the trend in October 1998 by buying Selsius Systems, followed by 3Com buying NBX, Ericsson buying Touchwave and Nokia buying Vienna Systems.

4.3.2 Solutions for the Evolutionary Market

In contrast, the traditional PBX vendors, who had designed very functional and robust PBXs, were publicly unconcerned by this immature and unreliable VoIP technology. By 1999 however, almost every major PBX vendor had a VoIP migration strategy for their PBX. These strategies may be characterised as allowing existing users to seamlessly migrate to VoIP while, most importantly, protecting the investment in the PBX (and, of course, the PBX suppliers' equipment upgrade revenue).

The initial approach was to IP-enable existing products to take advantage of the trend towards voice and data network convergence. This happened in two distinct areas of the network.

Firstly, the development of trunk-side gateways allowed companies to combine their voice and data traffic and route voice calls across their wide-area network (WAN). This enabled the rationalisation of WAN links, thus displacing PSTN and IDD call charges and the need for a separate private voice network with potentially increased expenditure on data networking services. The second area addressed the terminal or line-side of the PBX, allowing IP telephones to be deployed on an existing data local area network (LAN). In doing so, it created the potential to reduce infrastructure costs, and allow telephones to be deployed across the corporate data network, irrespective of however dispersed they may be. More importantly though, these two developments gave most of the advantages of a true IP-PBX, while at the same time still providing all the existing PBX functionality and reliability.

4.4 IP Telephony Products

Customer premises IP telephony solutions can be categorised as being either:

- VoIP gateways (stand-alone);
- IP trunk-side (exchange lines);
- IP line-side (telephone side); or,
- IP-PBX.

Of these the first three areas have generally become dominated by the traditional PBX suppliers who have enhanced their existing PBXs to support IP telephony on the trunk-side and/or on the line-side connected to IP telephones or PC-based soft-phones. In contrast, it has largely been the new players in the telephony market that have seized the pure IP-PBX opportunity.

4.4.1 Enterprise VoIP Gateways

The first VoIP gateways were stand-alone gateways that provided point-to-point replacement for conventional PBX networking protocols such as the digital QSIG standard and analogue E&M circuits. With this approach a VoIP supplier can provide cost-effective solutions for a wide range of existing customers' PBX networks and importantly avoid the need to upgrade the PBX. The downside was that most PBX manufacturers have provided enhanced inter-PBX functions using proprietary protocols that are not generally supported by stand-alone gateways. To overcome this, BT has developed methods, such as the Meridian Customer-Defined Network (MCDN), to tunnel protocols between gateways.

It should be noted that, for the business market, VoIP calls are routed over private intranets rather than the public Internet that today does not offer sufficiently reliable quality of service to deliver the required business-quality voice.

4.4.2 IP-Enabled PBX Trunk-side

Today most major PBX suppliers have an option to provide IP telephony trunks directly within the PBX itself (see Fig 4.1). The advantages of this approach are that the existing investment has been protected and generally there is no loss of features and functionality. The downside for IP-enabling a PBX is that often the PBX requires a major software upgrade that can require considerable investment in equipment and which may have an uncertain future.

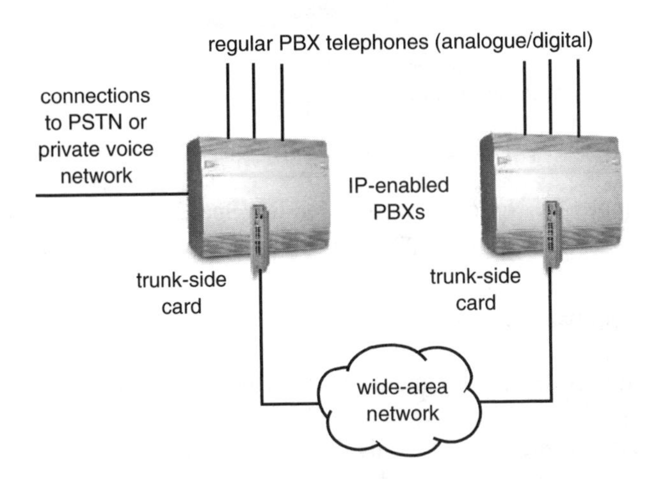

Fig 4.1 IP-enabled PBX (trunk-side).

A common disadvantage of both stand-alone and integrated PBX gateways is that there are still separate voice and data networks within the customer's premises and there is no opportunity to easily provide multimedia applications.

4.4.3 IP-Enabled PBX Line-side

The disadvantages of the trunk-side solution are removed when using IP line-side (extension-side) devices that allow a fully converged network to be realised. IP telephones and/or PC-based soft-phones can be deployed on local area networks or potentially anywhere on the corporate IP network — even remote sites and home locations are possible (see Fig 4.2).

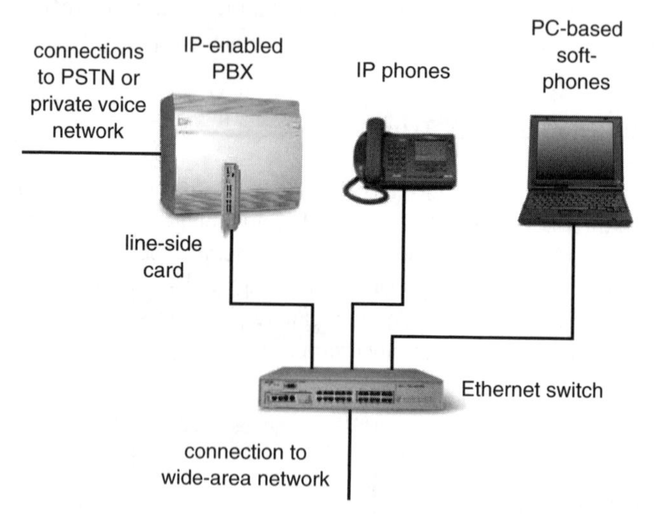

Fig 4.2 IP-enabled PBX (line-side).

As an example, the Nortel Networks Meridian Succession PBX [3] has recently been IP enabled to support IP telephony on both line-side and trunk-side.

4.4.4 The IP-PBX

An IP-PBX differs from a traditional PBX in two important ways. Firstly, call processing is performed by a call server application running on an industry-standard computing platform such as a PC. This is usually an industrial rack-mounted fault-tolerant server, although a standard PC can also be used for smaller systems. Secondly, the voice and signalling is both transported and switched in an IP network, rather than a separate TDM-based telephone network and switch infrastructure as with a PBX. This, in theory, allows the replacement of conventional business telephone equipment and wiring by integrating data and voice traffic over a single unified communications network.

An IP-PBX system consists of the following functional components, which are in addition to the data network itself:

- call server(s);
- gateway(s);
- user terminals;
- value-add applications.

In a true IP-PBX, each element only requires an IP data connection to communicate between other elements for both voice transport and signalling. Importantly, this means that devices such as gateways and IP telephones can be located anywhere within the corporate data network (given sufficient bandwidth — see Chapter 2).

Unlike a conventional PBX, there is no switching matrix required to connect the voice paths. Voice paths are effectively switched as IP packets within the existing data network element, typically comprising Ethernet switches and routers. Therefore a separate voice and data infrastructure is no longer required.

A unified communications infrastructure to the desktop allows common delivery of voice and data with the potential of reduced installation and maintenance costs. It also potentially lowers the cost of telecommunications when connecting remote offices.

Since voice and data share the same data network, voice can easily be extended to a company's WAN. In practice, the local and wide area networks would, in most cases, have to be upgraded to support the additional real-time bandwidth requirements and quality of service.

Other potential savings can be realised by using speech compression, which means that as less bandwidth is required to transmit voice, it is used more efficiently. Coupled with silence suppression techniques (see Chapter 2), further bandwidth savings can be made.

Most of the traditional PBX features and functions are provided by the IP-PBX call server, but with the addition of new multimedia applications, such as data and video collaboration which become much easier to incorporate in an IP-PBX environment and at a lower cost than ISDN-based solutions.

4.5 The IP-PBX Examined

The architecture of the IP-PBX solution is key to providing a fully scalable and robust solution that will be able to compete both technically and economically alongside the existing state-of-the-art circuit-switched PBXs (Fig 4.3).

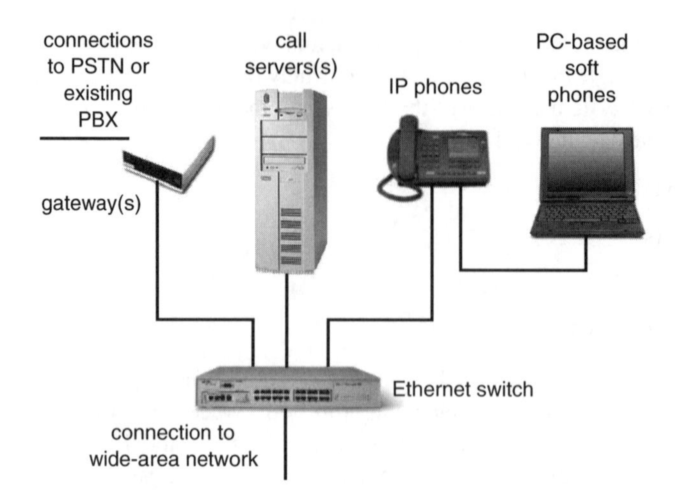

Fig 4.3 IP-PBX architecture.

4.5.1 Call Server

The call server function of an IP-PBX provides all the essential call control and signalling normally provided by the software of a traditional PBX. The call-processing software resides on a dedicated telephony server connected to the IP network and provides call handling (e.g. routing of internal and external calls) as well as supplementary features such as call transfer and call divert. It does not normally switch or process the actual speech information itself, although there are exceptions to this, including applications such as multiparty conferencing and voice mail.

4.5.1.1 Scalability

As an example of the scalability of an IP-PBX, when Selsius Systems first launched their IP-PBX in 1997 it could support up to 100 IP telephones and PC clients on a single NT server. This was fine for initial deployments that were mainly for exploratory use and small technical trials in the laboratory. By 1999, after Cisco's acquisition of Selsius, the same server could support up to 500 IP telephones. By early 2000 a single server, using Windows 2000 and an SQL database, could support up to 2500 IP telephones. Using a multiple call server architecture, where call processing is distributed across clustered servers, a maximum of up to 10 000 IP telephones is now possible. In theory, using inter-cluster signalling, an IP PBX system is scalable well beyond this figure to support the very largest corporate voice networks.

However, it is just as important to remember the vast numbers of smaller sites. Using a single call server, it is now both practical and economical for a small enterprise to have the same PBX functionality and features as larger users — an important advantage for a company as it grows. With traditional PBXs it is often necessary to replace the key-system or PBX as the company grows, which wastes investment in equipment, user training and any applications that may have been integrated with the telephone system. A good IP-PBX architecture and design should, in theory at least, protect this investment and allow the company to grow unhindered by the telephony technology.

4.5.1.2 Resilience

A major concern of IP PBXs is the anxiety over the resilience of the call server platform as it relies on general-purpose operating systems such as NT and Windows 2000. Server outage could occur due to server hardware failure, operating system crashes, power failure, data network failure, building damage (e.g. flooding or fire) or for planned maintenance reasons.

Concern over resilience can be addressed by providing redundant call servers for additional resilience, e.g. $N+1$. If a call server were to fail, the user terminals would need to be able to detect this and automatically re-host themselves to a pre-known secondary call server in the cluster. If the secondary call server were also to fail, then a tertiary server would be used, and so on. This is only possible because all terminals are connected to the call servers using an IP network and can easily reconnect from one call server (and IP address) to another. In contrast, a traditional PBX telephone is physically hardwired to the PBX line interface cards and the PBX call processor(s). Only the very largest PBXs have dual processors, which provide very high fault-tolerant systems, but are still susceptible to some major failures. While it is unlikely that a single IP-PBX server will ever be as robust as a traditional PBX it could potentially provide a far more resilient solution when multiple servers are deployed.

4.5.2 IP-PBX Features

One of the biggest criticisms of IP telephony from PBX vendors has been the lack of functionality. Indeed a simple IP telephony terminal, either PC-based or a physical IP telephone, can be limited to all but the very basics in telephony, such as making and clearing a call. This is probably adequate for low-cost mass-market telephony where the user has no interest in any supplementary features such as calling line identity (CLI) or call transfer. However, for a business user these PBX features become imperative to the business operations and must be supported by an IP-PBX.

In most IP-PBXs today the majority of PBX functionality is now widely supported by equipment vendors. Features typically supported include:

- calling line identity (CLI);
- calling name identity;
- attended call transfer;
- unattended (blind) call transfer;
- call divert (all calls, no reply, or on busy);
- conference calls (*ad hoc* and meet-me);
- call detail records;
- computer/telephony integration (CTI).

However, there are features that the new IP-PBX vendors have generally declined to support such as 'ring-when-free' and 'hunt groups'. Also one of the biggest drawbacks today is the limited support provided for voice mail systems, either as part of the IP-PBX architecture or through integration with existing voice mail systems. Until these issues are resolved, many businesses may be unwilling to migrate to an IP-PBX solution.

4.5.3 IP-PBX Gateways

An IP-PBX gateway allows calls from IP telephones to be connected with the traditional public and private telephone networks. Gateways can be either digital — connecting to the ISDN network (basic or primary rate ISDN) — or analogue — connecting to regular PSTN lines.

In a true IP-PBX architecture, gateways are devices connected anywhere on the corporate data network and can therefore be deployed in any location where there is adequate bandwidth to support the required number of voice calls. For example, a headquarters site may have one or more primary rate ISDN30 gateways, a remote office, a basic rate ISDN2 gateway or small analogue gateway.

Connections to public telephone networks are adequately supported by most IP-PBX vendors with protocols, such as ISDN30 (I.421) and ISDN2, supported by the gateways. However, in the UK many large corporate businesses have existing private voice networks using the digital private network signalling system (DPNSS) for inter-PBX communications. Unfortunately while it is highly functional, DPNSS is not well supported by the majority of IP-PBX suppliers at this time. This can be a limiting factor in the deployment for large corporate voice networks in the UK. Although the internationally accepted ITU QSIG standard has been supported by some suppliers in some of their gateways, this is normally limited to simple call control only and does not support any supplementary PBX features. A commonly requested feature when migrating from a digital PBX to an IP-PBX is to support 'calling name' display between the existing PBXs and the new IP-PBX. This feature is dependent on compliance to one of the inter-PBX signalling systems such as QSIG or DPNSS. In most cases a PBX has to be upgraded (at a cost to the customer) to support QSIG and, in fact, there is currently no advantage in using QSIG over that of an I.421 protocol for interconnecting IP-PBX solutions unless additional features are supported by the vendors.

On the IP telephony side of the gateway are a number of different IP protocols including H.323, H.450, MGCP, SGCP, MEGACOP and H.248. These are described in greater detail in Chapters 7 and 10.

Note that in a totally IP telephony world (without circuit-switched PBXs), gateways using private signalling systems, such as DPNSS and QSIG, are not required. Rather these protocols may appear in an IP-transported form for communicating between distributed call servers, such as H.323 and H.450 that define the PBX supplementary features.

4.5.4 IP Phones

As far as a user is concerned, the expectation is likely to be that an IP phone should not look or behave any differently from a conventional PBX feature-phone, even though technically it may be very different — see Figs 4.4 and 4.5.

Fig 4.4 Nortel Networks i2004 IP phone.

Fig 4.5 Cisco Systems 7960 IP phone.

The main difference is that an IP phone is connected into the company's data LAN, whereas normal PBX telephones require a separate telephone cabling system (although it is now common practice to use a structured cabling system for both data and telephony). Also, unlike a standard telephone, an IP-based telephone can be installed almost anywhere on the corporate LAN/WAN network and easily moved to another part of the network, while retaining its unique identity. Setting up an IP phone is more akin to setting up a PC's network configuration requiring host and server IP address details. It does not require a dedicated connection to the PBX.

4.5.4.1 Features

Now that the telephone is connected to an IP network, providing advanced services on it is much easier using programmable soft-keys and enhanced displays to present both textual and picture information. This allows easier access to PBX features such as call diversion as well as other applications, e.g. personal and corporate directories, voice mail messages and other information services.

4.5.4.2 Line Powering

Some IP phones require an external DC power supply unit; however, this is not ideal. An alternative approach by some suppliers is to provide line powering over the structured cabling system. This can be achieved retrospectively by installing an in-line powered patch panel in the communications room or alternatively by using new line-powering Ethernet switches available from data network suppliers such as Nortel, 3Com and Cisco. Unfortunately, as of yet there are no agreed standards on how this is done or what voltage to use. This means that items of line-powering equipment from different suppliers may not be compatible with each other, making it more attractive to use IP phones and LAN products from the same supplier. Standardisation should resolve this problem.

4.5.4.3 Ethernet Repeater Ports

Secondary Ethernet repeater ports allow an IP phone and a PC to share the same data port and switch port normally supporting 10 and 100 Mbit/s. The IP phone connects to the LAN and the PC is re-connected to the secondary port on the back of the IP phone.

A secondary Ethernet port does not sound very important; however, it potentially reduces the number of structured cabling outlets and LAN switch ports to one per desk. In a new building this could potentially halve the data network infrastructure investment compared with separate connections being required for IP telephony and computing devices.

4.5.4.4 IP Telephone Protocols

The signalling protocol employed between the call server and IP phone is either based on industry standards such as the ITU H.323 (see Chapter 10) or on proprietary protocols. Proprietary protocols enhance the functionality and features not possible using H.323 version 1 which was aimed at multimedia communications rather than PBX feature phones. IP phones, using proprietary protocols, are normally used exclusively in association with the same vendors' call servers and gateways. For example Nortel Networks have an IP telephone (Fig 4.4) that works exclusively with all its 'Succession' portfolio of IP-enabled PBXs and server-based IP-PBXs. Conversely Cisco has attempted to open up the IP phone market by publishing their 'skinny client' protocol. A small number of other suppliers have now produced IP devices using this protocol and other manufacturers are sure to follow.

Competing with H.323 and the proprietary protocols is the newer session initiation protocol (SIP). The relative merits of both SIP and H.323 are discussed in Chapters 10, 11 and 12. However, since an IP-phone profile for Megaco has now been developed in the IETF MeGaCo Working Group (see Chapter 7), it is anticipated that this protocol may yet emerge to challenge the use of both SIP and H.323 in this area. This is because the Megaco IP-phone solution is more consistent with the stimulus signalling approach desirable for PBX feature set control.

4.5.4.5 Voice Codecs

Most IP telephones will support the ubiquitous 64 kbit/s G.711 codec, which is fine for calls that stay on the same local area network. However, for calls across a wide-area network or for remote access over a modem where bandwidth is at a premium, one of a number of compression codecs are used such as G.729 or G.723 which compress the voice down to 6.3 kbit/s. Using silence suppression can make further savings of 40-60%. Voice compression and the associated silence suppression is an important economic advance for IP telephony in that a given bandwidth can be used more efficiently than a standard G.711 voice call. Voice codecs and silence suppression, including the effect on the voice quality of these techniques, are discussed in more detail in Chapters 2 and 3.

4.5.4.6 Costs

Currently the number of IP telephone vendors is limited which has the effect of driving up the price. At the top end is Cisco's latest 7960 model IP phone (Fig 4.5) which has full hands-free capability and a large LCD display. This currently retails at around $600 including the associated software licence (not normally required for a traditional PBX telephone). If the user is willing to sacrifice the large LCD display and Ethernet switch port then the price can fall by half. Siemens has produced an

H.323-compliant IP phone that works both with their RC3000 IP-PBX and with any other H.323-enabled IP-PBX. Unfortunately the price is similar to Cisco's 7960 and does not offer many of the features or an Ethernet repeater port. Other suppliers, including Nortel and Mitel, have lower cost IP phones, but these are still proprietary and will only work with their own IP-PBXs. Since early 2001, there has been a marked rise in the number of new IP phones being announced and appearing on the market using either the Cisco 'skinny client' protocol or the H.323 standard. This is likely to result in a trend towards lower cost telephones over the next year.

The current cost of IP phones is seen as one of the major issues for any company considering IP telephony. However, when comparing IP telephony solutions it is essential to compare both the capital and whole-life costs rather than individual components such as the telephone.

4.5.5 IP Terminal Adapters

Many of the IP-PBX suppliers currently have IP phones for use with their product; however, some suppliers have decided only to provide terminal adapters to support simple analogue telephones. In theory this should provide some of the economic advantages of a converged network. However, this is not an elegant solution and would require an associated PC software application to provide any advanced features that are inherent in an IP phone. In an office environment, it is unlikely that the corporate user will replace existing digital feature-phones with low-functionality analogue telephones.

4.5.6 Soft-Phones

Soft-phones are PC-based IP telephony applications that exploit the PC's multimedia capabilities such as sound card, loudspeakers and microphone (Fig 4.6). Soft-phones can provide all the benefits of desktop CTI applications, the major difference being that the soft-phone application also handles the audio streams (voice) as well as call control. Therefore it is no longer necessary to have an expensive feature-phone.

Soft-phones also allow users to log on and remotely access a company's telephone system from any remote location having IP connectivity back to the call-processing server. Soft-phones should be able to support all the features that a conventional PBX telephone supports and in addition can have even more user/system programmable features.

A limitation of PC-based soft-phones that could affect user acceptance is the lower voice quality that is achieved in practice, specifically due to the higher audio delays produced within the PC itself.

Fig 4.6 PC-based soft-phones.

A number of developments have been made to overcome this performance problem:

- Sound cards

  Special telephony sound cards are available that provide enhanced audio capabilities with on-board hardware-based speech processing, full-duplex speech, and echo cancellation. These cards also provide additional connections for a regular analogue telephone handset and headset if required. Unfortunately replacing sound cards incurs additional cost in buying the hardware and installation.

- USB headsets

  Some of the traditional headset suppliers have developed USB versions of their headsets. In theory, these USB headsets are a good idea and potentially replace the need for the PC speakers and microphone. However, in practice, a user may require both devices and, importantly, be able to switch quickly between them, e.g. speakers for alerting a new call and headset to hold a conversation.

- USB phones

  Alternatively USB phone devices have started to appear as a replacement for the more call-centre-oriented headsets. The cost of a USB phone is expected to be significantly lower than the equivalent IP phone, possibly as low as around $20. USB phones (like IP phones) have the potential advantage of simplicity, i.e. the telephone rings and the user picks up the handset. Most of the sophisticated features can easily be incorporated in the soft-phone application software running inside the host PC.

All of the above IP telephony PC peripherals do require a PC and the associated software-based soft-phone whereas an IP phone does not.

4.5.7 IP Telephony Trials

As with any relatively new communications technology, the real problems generally do not emerge until there has been a serious attempt to deploy it. Significant among these have been a number of successful internal experimental and public customer trials that have been carried out by BTexact Technologies for the BT Group [4, 5]. Building upon earlier pilot schemes (see Chapter 1), BT's most extensive internal trial of IP-PBX technology started in 1998. This provided invaluable practical experience and demonstrated that IP telephony in the business environment has a real future. The first real customer was a forward-thinking public sector organisation deploying a two-site trial. These trials included all the main elements of a complete IP telephony service — namely VoIP transport over a router-based WAN link, gateways to the PSTN, IP phones and soft-phones, and an IP-PBX providing the PBX call control.

As expected, many problems were encountered owing to the lack of maturity in the equipment available. These resulted in software being frequently updated and enhanced both to overcome known problems and also to introduce richer functionality missing from earlier versions. Extensive testing was carried out using sophisticated voice-quality test equipment. The quality of calls were monitored and scored using a mean opinion score (MOS) under a range of network configurations and conditions to demonstrate the effects on the IP voice experience.

The results proved that, when the network is designed effectively and configured with an efficient QoS schema, voice quality is acceptable and appropriate for normal business use. In particular, the trials have clearly proved the importance of IP switch and router configuration to the successful performance of any installed voice applications.

A blend of voice and data networking skills has also been found to be essential for an IP telephony installation to be successful. The days are fast disappearing when the voice and data networks can be treated as distinct and separate entities.

4.6 Value-Add Applications

IP telephony provides the possibility for a number of interesting value-add business applications. For the IP-PBX concept to be successful, suppliers must demonstrate the advantages offered over and above that of a conventional PBX. In the future, it may be that applications generate new revenue opportunities beyond the supply of customer premises equipment and simple transport of IP packets. Applications can be provided:

- as part of the IP-PBX solution;
- an additional option;
- from a third-party supplier who has integrated additional functionality with the IP-PBX platform.

Described below are some possible value-add business applications.

4.6.1 Data-Conferencing

With an IP-PBX and a PC-based soft-phone, the communications medium is no longer restricted to voice. Both the IP telephone and soft-phone are able to add data-conferencing to any IP-to-IP call at any time during a call. This is achieved by the IP telephone working in association with a PC-based soft-phone. Typical data-conferencing tools include:

- application sharing;
- white-boarding;
- file transfer;
- text chat;
- videoconferencing.

These collaborative working tools are useful for reviewing documents or showing slide presentations remotely. By adding a relatively cheap video camera (around $50) to a desktop PC, videoconferencing also becomes an inexpensive and simple-to-use reality and the IP-PBX now becomes a multimedia-PBX (indeed, parts of this chapter were reviewed and edited by the authors using the voice, video and data collaboration features of an IP telephony system across the BT Intranet between Adastral Park and London).

4.6.2 Voice Mail

A voice mail service allows users to record their own personalised greetings on the (IP-PBX) call server or a separate adjunct voice mail server. Any unanswered calls are forwarded to the voice mail server that plays a pre-recorded greeting and records the caller's message. The recorded message is then picked up either through the telephone or sent to the user's e-mail account as a sound file. Unified messaging builds on the voice mail concept and brings together a number of today's separate messaging systems such as voice mail, fax, e-mail, etc, into a single integrated set of functionalities. The user then has a choice of how to retrieve the messages — either by telephone, Web browser, or Internet mail client. It is now possible for users to listen to their e-mails via the telephone using text-to-speech technology.

4.6.3 IP-ACD

An IP-based automatic call distributor (ACD) allows a customer browsing the Internet to initiate a multimedia call to a call centre associated with a company's Web page by clicking on a 'Call' button, as with the example above. However, with an IP-ACD, the customer and the call centre advisor can, in theory, talk over the data network using VoIP. This is only achievable because an IP-ACD architecture does not use a traditional 'circuit-switched' PBX or ACD system. Instead, VoIP is used to provide end-to-end connectivity between the customer and the call centre advisor that allows full multimedia (voice, video and data) communications.

4.6.4 IP-IVR

An interactive voice response (IVR) system can be used to automate the routing of calls to the most appropriate person by prompting the caller to key in DTMF tones to select the required option. An IP-IVR provides the same functionality but in the IP domain and can be used in conjunction with the IP-ACD.

4.6.5 Instant Messaging and Presence Awareness

Instant messaging and presence awareness provide the means for users to send messages to other connected users in real time as well as notify other connected users of their availability and status.

This offers the user the ability to control incoming calls, chat sessions and conferencing by informing users about their current status (e.g. busy, willing to accept, not in the office). Users can also send instant messages to other connected users in real time, while enjoying the ubiquitous nature of e-mail. To date, users of instant-messaging and presence-aware applications have been confined to the mobile networks and the Internet. There is now the opportunity to expand this into the business IP telephony world.

4.6.6 Advanced IP Phones and Applications

The launch of IP phones with larger displays, such as Cisco's 7960 and 7940 phones, has for the first time enabled much more information to be presented to the user. For example, it has now become much easier for a PBX feature phone user to access corporate directories and search for other users.

However, the real step forward is the availability of configurable 'services'. Although it is expected that some of these services will be offered by the IP-PBX vendors themselves, such services may be enabled by either the system administrator or the end-user. However, it is also predicted that many of the more innovative services will come from independent companies — known as communications application service providers — that offer customised application services to complement the IP-PBX itself.

Some of the possible 'service' applications aimed at the business market include real-time share price information and graphs, business news bulletins and briefings, messaging services such as SMS and e-mail, telephone directories (e.g. white pages and yellow pages), simple data entry and collection.

One of the first companies to offer services for Cisco's IP phone users was Berbee.com, who provide US-based information, such as stock prices, news information, and weather forecasts.

Other services may be more appropriate for the residential market. In the future, advertising through the phone display could be used to supplement some or all of the telephone call. However, this requires the IP phone market to mature beyond its current state, which is largely entrenched in the PBX replacement market. For example, although the current industry-leading IP phones, such as the Cisco 7940/7960, do incorporate an LCD display capability, they tend to be limited in their ability to display text and graphics at a resolution sufficient to complement Web browsers on personal computers. Such terminals also tend to lack the real-time handling performance required for more advanced applications.

It will be interesting to see how the IP phone market matures and whether having applications and services downloadable via the telephone will deliver real business benefits over using a standard desktop PC browser.

4.7 Core Design Issues

4.7.1 Infrastructure Issues

There are some interesting challenges that need to be faced when voice is put into the IP environment. Putting voice, or indeed any real-time data, into an IP network puts a specific set of demands on that network. In particular, voice data cannot cope with any delay at all well. The agreed delay budget to which an IP call can be subject, before quality is affected to any noticeable degree, is approximately 150 ms telephone to telephone (see Chapter 2). Another important factor in the perceived quality is the consistency of delay, i.e. it is better to have a consistent delay, throughout a call, of 150 ms than to have a random delay that fluctuates wildly between the limits of the delay budget, as explained further in Chapter 2. A company may have a network that is designed to ensure that the 150-ms delay budget is never exceeded across its network from any one point to any other. However, if a call is made to another company who also have an IP telephony system and network designed to the same guidelines, the end-to-end delay could be up to 300 ms. It may be possible to set up special routing and compression rules for such cases where it is known that certain numbers are destined for other IP telephony systems to limit the delay to a more acceptable level.

The most important, and sometimes totally ignored, part of any IP telephony network is the IP infrastructure over which the IP calls will pass. The network must be able to deliver the IP packets without delay. There are two ways in which this can be achieved:

- deploy large amounts of bandwidth so that no IP packet, either normal or real-time, is ever delayed because of congested bandwidth;
- employ a mechanism that allows real-time voice data to receive preferential treatment over all other data — these mechanisms are generally referred to as quality of service (QoS) mechanisms.

Understanding the IP network itself, designing, and implementing suitable network components, such as switches and routers, are all crucial to the successful deployment and support of an IP telephony solution. The IP telephony provider needs to be in control of the IP infrastructure to be able to provide realistic service level agreements and guarantees (SLAs/SLGs) against voice quality, as the voice is only as good as the underlying network. Where this is not possible, the ability to show where any voice transport difficulties lie becomes even more important to the IP telephony supplier since they will receive the bulk of the fault reports, even though the quality problems may be caused by something outside their control.

Further, most IP telephony systems require configuration of other network services such as the domain name service (DNS), BOOTP servers, dynamic host control protocol (DHCP) servers, firewalls, network address translators (NATs) and NT domains. The IP telephony supplier must have competencies in all the areas.

4.7.2 Voice Quality of Service

As described in Chapters 2 and 3, QoS is a fundamental design challenge for enabling next generation services based upon VoIP. As has been discussed, any QoS assurances are only as good as the weakest link in the chain between the sender and receiver. To guarantee a successful IP telephony deployment, QoS techniques must be used from end-to-end so that voice traffic can be treated properly under all network conditions. This is an essential requirement for a fully converged network of voice, video and data. If it is not achieved, voice packets will be treated as 'best effort' and, as a result, not necessarily deliver business-quality telephony.

In the LAN environment typical of many IP-PBX installations, the avoidance of congestion through the over-provision of bandwidth is going to be more viable than in the WAN because of the cost of WAN bandwidth — although this cost is being driven down by market forces. There are other factors that affect the delay suffered by real-time data if sent across most existing data networks. For example, in today's data networks, it is not unusual for data packets to be 1500 bytes long. If a voice data packet, which may be only 20 or 30 bytes long, gets stuck in a transmit queue behind one of these large packets, significant transmission delays will ensue. To overcome this problem, the transmitted packet size can be reduced. However, doing so will have an adverse effect on the other data traffic — total packetisation overhead goes up and network devices have to sustain the higher processing requirements of increased volumes of frames for the same information content. A balance therefore needs to be struck between the differing requirements of voice, data and the network as a whole, as discussed in Chapter 3.

4.7.2.1 Quality of Service Techniques in the Office Environment

Although QoS offers the ability to actually promote voice packets ahead of data packets it is apparent that not only does QoS need to be employed in the network, it should also be extended and suitably managed on the access side all the way to the customer premises and into the IP telephony devices and associated applications themselves. This is a considerable challenge that largely remains to be resolved. However, bringing QoS functionality closer to the desktop gives more flexibility in configuring networks to support IP telephony products. There are various methods that can be deployed within the CPE devices to help accomplish this.

In the case of the LAN infrastructure, the IEEE 802.1q standard is a potentially useful tool that can be used to help mitigate some of these issues. IEEE 802.1q is significant because it embodies a design specification for the construction of virtual LANs (VLANs). The VLAN concept is a means of dividing a single network into logical groups as if they physically resided on a dedicated network of their own. This allows IP telephony devices to exist in a different logical LAN to other devices such as PCs.

If the IP phones and other CPE devices support 802.1q, then it is beneficial to logically separate the voice and data traffic into separate logical VLANs. For example, IP phones would default to the voice VLAN irrespective of the physical Ethernet port to which it happened to be connected. However, a problem occurs with soft-phones that, because they are hosted in the PC, will default into a data VLAN and not a voice VLAN as with the IP phones.

In layer-2 switching, Ethernet frames arriving at LAN switch ports are classified according to class-of-service (CoS) bits set within the 802.1p Ethernet header. This classification determines into which of the Ethernet switch input queues the packets are placed. Most IP phones now support 802.1p to set the CoS for the voice traffic — thus the voice packets are promoted ahead of best-effort packets. As some IP phones also have Ethernet repeater ports, it is also important that the PC's data packets have a lower CoS and are not prioritised with the voice. In contrast, a soft-phone has the downside that it will have its voice and data packets all treated as lower priority data.

Within the IP packet header (layer 3) the type-of-service (ToS) field also allows packet priority. The ToS field contains an IP precedence field (bits 0-2) which marks the packet priority from 0 (low) to 7 (high). An additional 3 bits also mark the packet for delay sensitivity, throughput and reliability. Most IP telephones set the IP precedence bits in the IP ToS to 5 with the delay sensitivity being set to high, with throughput normal and reliability normal for UDP traffic. Other devices, such as call servers and gateways, also need to support these QoS methods.

The Nortel and Cisco IP phones, both of which BT has in its portfolio, all support QoS prioritisation using 802.1p/q and IP precedence/ToS. However, there is now a move to migrate these devices to support the newer diffserv QoS method that enhances the IP precedence/ToS field. Quality of service is described in detail in Chapter 2 and in Willis [6].

4.7.2.2 Voice Performance Analysis

Currently lacking from the IP telephony industry are the tools to provide an expert analysis as to the suitability of a customer's network for deploying IP telephony. A number of tools do exist to provide basic health check services, but these cannot predict the impact of deploying real-time voice traffic. BTexact Technologies at Adastral Park has been developing IP telephony simulation tools to replicate and predict weaknesses in a data network before IP telephony is actually deployed — see Fig 4.7. Once an IP telephony network has been deployed, there are limited tools that allow an IP telephony system to be non-intrusively monitored and reports generated on the end-to-end voice quality. For IP telephony to be acceptable as a business communications system these support tools must be available to support these deployments.

4.7.3 Management Systems for IP Telephony

The management systems of any new product have historically lacked functionality and the emerging IP telephony solutions are no exception. Telecommunications managers and service providers are used to having access to an extensive range of management capabilities and statistics that traditional PBXs have built up over many years. IP PBX management functionality is lagging behind the core technology development. However, the open nature of IP and the availability of call data records that are captured by the new systems will make the task of catching up a much quicker experience as third-party developers produce software packages that provide full call statistics and billing facilities.

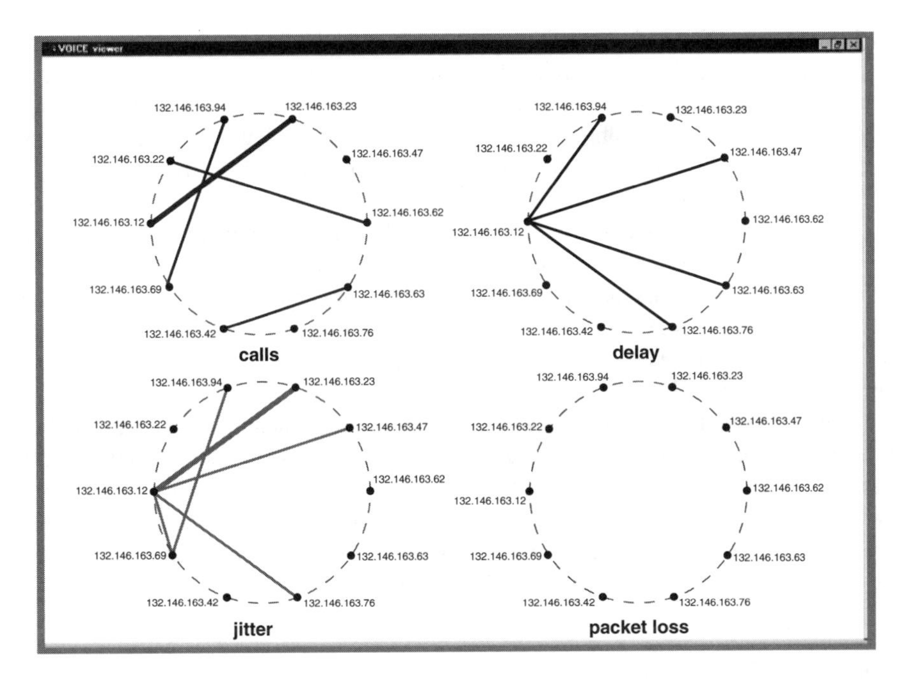

Fig 4.7 VoIP network performance tool.

The remote management of IP devices is usually done using SNMP. Solutions providers have spent a huge amount of resource, in both time and money, in recent years developing and deploying large network management systems to enable them to manage the large IP networks under their control. With the advent of the converged voice and data networks, there will be an opportunity to use the generic SNMP management systems to manage the new voice applications that are being deployed.

In the managed network environment, essential elements of any contract are the SLA and/or the SLG — as converged networks develop further, they will be important differentiators between suppliers. Voice quality will be one of the key parameters in these agreements with management systems needing to be able to prove that the service has been delivering voice to an agreed level. While doing this, the system must not be burdening the network with excessive amounts of extra data.

In many organisations that implement IP telephony there is the requirement to manage both new and old world telephony devices. There is also the need to interwork them. This is probably the most difficult area for design and operational teams. The integration of the dial-plans in the two separate environments requires careful co-ordination, as does the change request regime, to ensure continuity in number structure. This is complicated because of the different management systems necessary. To provide a complete managed solution for these converged networks three separate management systems will be required — one to manage the traditional circuit-switched PBX, another to manage the new IP telephony environment, and a third to manage the IP infrastructure. The extent to which these systems can be integrated will initially be low. The focus will be on providing training on the new IP telephony management systems and developing the processes necessary to overlay these three disparate systems and run an effective network.

The converged or unified network has huge potential for consolidation of management systems, which will simplify the whole task of managing the complete communication solution and therefore reap significant benefits for both the suppliers of the network and the customer. The converged network becomes the corporate intranet, and, with the use of directory systems such as LDAP, the integration of the whole communication infrastructure for the business or corporation is an attainable goal.

Utopia for the IP infrastructure would be the complete management of the quality of service from end to end within a network. This would include a management system that can function with a configurable level of autonomy, adjusting network parameters, queuing mechanisms and bandwidth allocations to ensure that predefined SLA/SLG contracts are attained. Suppliers are working on policy-based networks that will go some way towards this ideal.

4.8 Summary

This chapter has introduced a number of approaches to adopting and using VoIP technology in the business environment. It has been shown how equipment manufacturers have embraced these opportunities and that products are now available to satisfy a wide range of these needs. These products increasingly enable new communications-based applications to be created that offer the potential for generating business efficiencies and cost savings. However, irrespective of the attractive offer of new applications enabled by the IP infrastructure, it may prove difficult to fully realise a completely IP-based environment since switched-circuit PBX systems are widely deployed today and serve their users needs' adequately. A hybrid VoIP solution may therefore emerge where some users may be connected via a conventional PBX and use non-IP telephones. Therefore in reality, not all calls will be IP end to end and most scenarios will need to break out to the switched-circuit world via a gateway device at some point. Once this happens only voice telephony is realistically possible since data and video exchange are not generally available on switched-circuit voice networks. Even so, there are now many more options to address the communications needs of the business community. As a consequence, VoIP has created additional degrees of complexity in the market-place that require the business community to appreciate even more the core requirements and drivers behind any purchasing decisions they make.

References

1 Catchpole, A. B.: '*Voice/data convergence and the corporate voice-over-IP trial*', British Telecommunications Eng J, **17**(4) (January 1999).

2 Cisco Systems — http://www.cisco.com/

3 Nortel Networks — http://www.nortelnetworks.com/

4 BT Retail, Business Information Systems — http://www.business-systems.bt.com/avvid

5 Ignite Syncordia Solutions — http://www.syncordia.bt.com/

6 Willis, P. J. (Ed): '*Carrier-scale IP networks: designing and operating Internet networks*', The Institution of Electrical Engineers, London (2001).

5

INTERNATIONAL STANDARDS FOR VoIP

G Travers and R P Swale

5.1 Why Do We Need Standards for VoIP?

This chapter aims to explore what standards are, why they are required for realising voice over IP (VoIP), and how they are created. The trite answer to why we need standards for VoIP is that it is based upon IP and IP is a standard and so you cannot have VoIP without it. However, there are numerous, and potentially less obvious, reasons why standards are essential to VoIP. Some of these are a consequence of the natural separation of transport and application that occurs in VoIP systems; others derive from the establishment of any voice service, especially one that crosses the boundaries between the domains of different service or network providers and so encounters differences in administrative policy — in other words, the real world is much more complex than the laboratory bench.

While a service is operating in your domain, you can control it, accepting the normal expectations of occasional failures of the technology. However, as soon as it leaves your domain, control passes to the service provider that owns the domain to which the service has passed. So for example, a voice call from the UK to Australia will cross from the UK domain to the Australian domain at some point, and the Australian domain will have control over parts of that call.

So that the Australian domain can handle the call correctly, it has to be able to accept it and understand what it requires, e.g. what telephone address (number) it is trying to reach. It therefore helps if the call and its requirements are specified in a format that the Australian domain can 'understand'. Such a format, if agreed privately between the two domains, could be called a standard. However, its usefulness would be limited to calls between the UK and Australia. Any gateway device between the two domains would therefore be specific to that purpose and the full cost of developing, manufacturing and maintaining the gateway would most likely fall to the two domains in question. Consequently it makes more sense to have a globally agreed format, applicable to calls between any domain. These globally agreed formats are called international standards, and are the basis of the existing world-wide telephony services, such as international direct dialling (IDD).

5.2 Interface Requirements

The way in which global telephony applications are structured presents a need for international standards to be specified for two kinds of interface.

The first kind is a more general case of the inter-domain example given above, called the 'horizontal interface'. The second is the set of interfaces between technologies that are 'stacked' on top of one another in a series of layers, to provide an overall 'service stack'. These can be referred to as 'vertical interfaces', as shown in Fig 5.1.

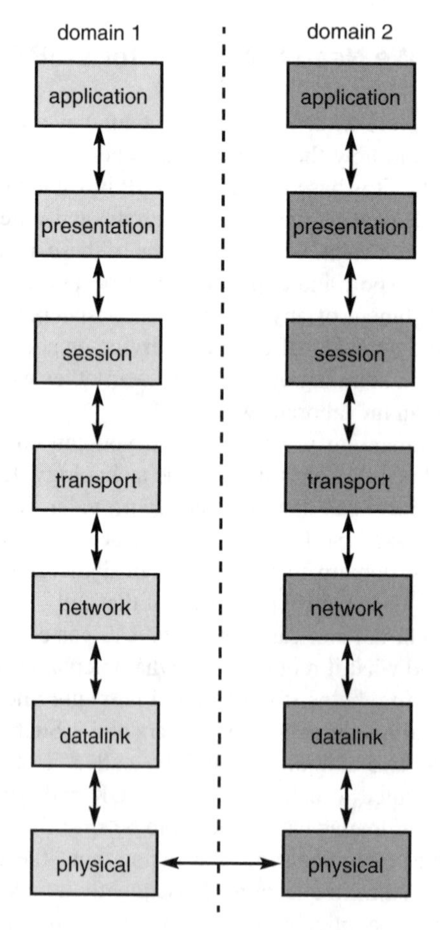

Fig 5.1 Horizontal and vertical interfaces.

5.2.1 Horizontal Interfaces

5.2.1.1 Inter-Domain Interworking

The simple example given above referred to the interface between the domains of different service providers. To request, connect, maintain and terminate a call across two or more domains, the data necessary for these operations has to be passed between the domains concerned. Both the format of the data and the way in which it is passed between domains have to be standardised. To date, VoIP solutions deployed in the market have relied upon the existence and capabilities of the public switched telephone network (PSTN) to provide these functions. It is apparent that these issues will have to be resolved for VoIP-based protocols in order for them to supplant the existing PSTN circuit-switching technology.

5.2.1.2 Inter-Technology Interworking

As well as the need for standardised interfaces between domains under the control of different service providers, the requirements for interworking between the different technologies, which each of these domains could be using, have to be considered. Different technologies are likely to handle data about a call in a different manner in addition to any technology-specific handling of the associated media streams. Additionally, the domain of a service provider may comprise several sub-domains, each operating a different set of technologies that need to communicate in a standard way (see Fig 5.2). While an internal technique could be agreed to resolve this within the domain, use of the standard methods is preferable, since it:

- eliminates the need to develop and maintain two separate techniques;
- allows the service provider to procure equipment from a pool of suppliers who conform to the international standards;
- enables suppliers to recoup development investments more quickly by building products for a much larger, global, market-place;
- enables flexibility in commercial relationships between and within service providers.

It is important to recognise that interworking between technologies increases the complexity of the specifications needed for the passage of data between domains. Standardised methods are therefore desirable for dealing with the meta-data associated with each call, such as the type of terminal to which it needs to connect, which cannot automatically be assumed to exist in a technologically different domain.

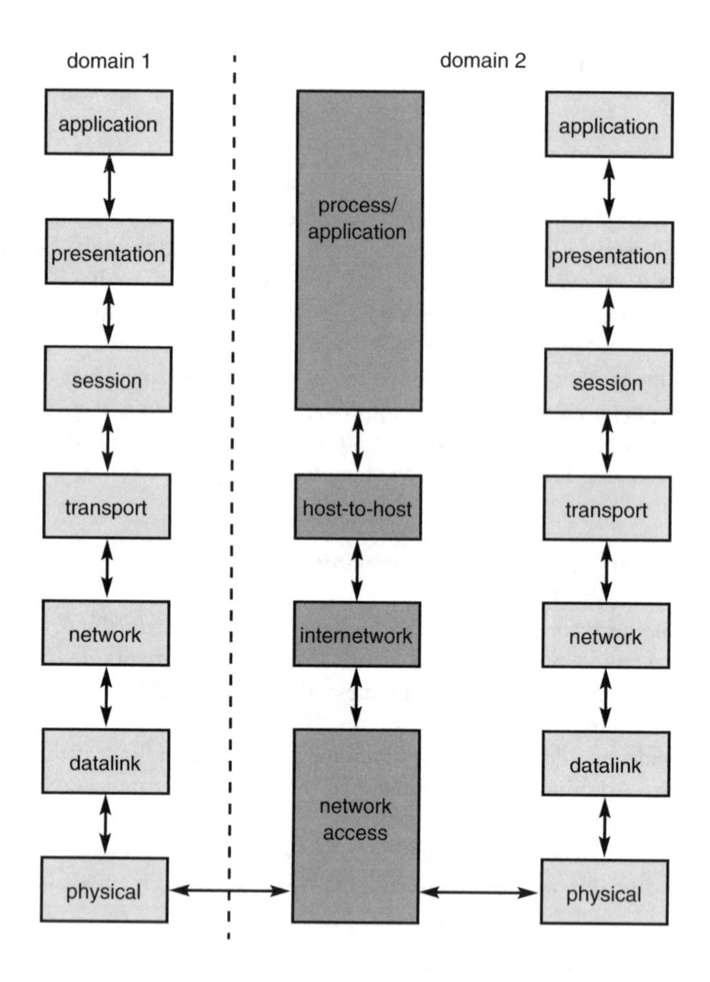

Fig 5.2 Inter-domain and horizontal inter-technology interfaces.

5.2.2 Vertical Interfaces

5.2.2.1 Inter-Technology Interworking

In addition to interworking between different technologies in different domains or sub-domains, there are requirements for interworking between technologies within the same domain. For example, IP is said to work over any lower level data link technology, but it is necessary for each technology that supports IP to 'understand' it, and to handle it in the correct way. The way that this is achieved is for IP to exploit a standard interface to any supporting technology. Each technology that is developed to support IP is then developed to handle the standard interface.

This is a simple example of the layered approach to the 'service stack' that considers only two layers — an application plane and a transport plane. Most modern models of the service stack have more than two layers, and there is a similar requirement for interworking between each layer, as shown in Fig 5.3. For example, when considering VoIP applications, the higher layer application protocols invariably differ in some way. So even if two VoIP applications are running, there is no implicit guarantee that they will interwork at all.

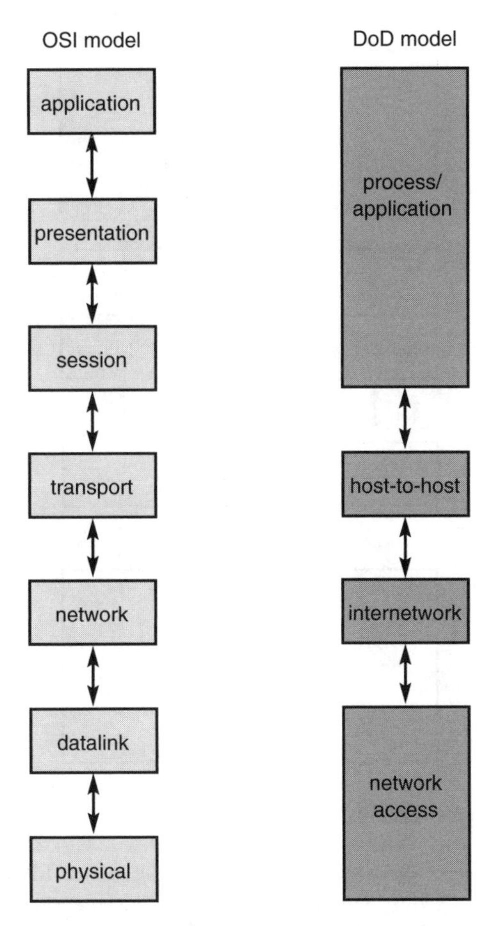

Fig 5.3 Vertical inter-technology interfaces.

5.2.2.2 Interworking Between Service and Network Providers

Traditional telephony services are provided by a vertically integrated protocol stack. This arrangement effectively ties the telephony application to the underlying transport network. Although there have been a number of technologies [1, 2] that have sought to separate this type of close integration, such a separation is an implicit feature of a VoIP network by virtue of its use of IP (see Fig 5.4).

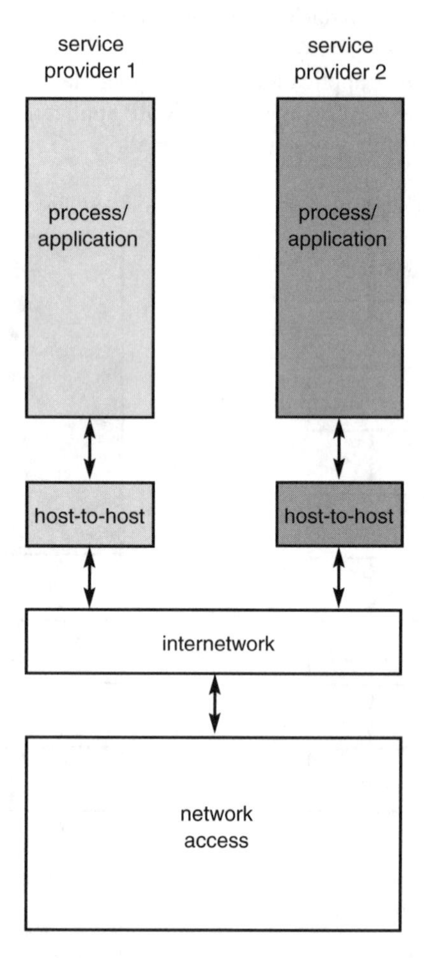

Fig 5.4 Interworking between service and network providers.

The separation of service from transport offers the possibility that there may be a number of service providers supplying services over the same physical network infrastructure. Each of these will purchase the transport service that it needs to transport its calls from a network provider. Similarly a particular network provider may sell transport services to many voice service providers. The most efficient way of interconnecting between these is to use an interface that is standard for all.

The provision of standardised interfaces to third party service providers [3, 4] can also be a regulatory requirement for network providers within the UK and elsewhere.

5.3 Standards Bodies

As it is of great importance to the industry to establish appropriate common interfaces, a number of the main standards groups are active in areas relevant to VoIP. While it has often been said that the great thing about standards is that there are so many to choose from, the greater the number of competing protocols, the less likely it is that end-to-end services can be provided at scale. The cost of developing both equipment and services will also be higher where a multiplicity of protocols offering much the same functionality have to be supported. It is therefore important to appreciate the work of each standards initiative and its role relative to others. In practice, there are two types of standards-related organisation — formal standards bodies and industry groups. The main ones relevant to the VoIP industry are described in the remainder of this chapter.

5.4 IETF

The Internet Engineering Task Force (IETF) [5] may be considered as the technical home of IP. It is a large, open, international community of network designers, operators, vendors and researchers concerned with the evolution of the architecture and smooth operation of the Internet. It has developed, and continues to develop, where necessary, the Internet protocol, and its related family of protocols. These include the transmission control protocol (TCP) and the user datagram protocol (UDP) which provide the basic communications infrastructure upon which most IP applications are based today. The IETF operates under the auspices of the Internet Society (ISOC) [6], and is open to any interested individual rather than company or other entity. This aspect sets it apart from other groups within the industry where representation is based upon company or national membership. A second aspect that distinguishes the IETF is its specification development process which is predicated upon the existence of 'rough consensus and running code'. This generally results in the development of specifications being implementation led.

The actual technical work of the IETF is done in a number of Working Groups (WGs) organised under a number of 'Areas', with each 'Area' dealing with the development of a particular aspect of Internet technology. The current IETF Areas comprise the applications, general, Internet, operations and management, routing, security, sub-IP, transport, and user services. Of these, the majority of work pertinent to VoIP is conducted in the transport area. Area directors, of which there are usually two per Area, collectively form the Internet Engineering Steering Group (IESG), and have the responsibility of guiding the WGs, including recommending their establishment or termination, and reviewing Internet-Drafts and Requests for Comment. Although the IETF Areas are largely static, the individual Working Groups have a varying life span. Some are quite long-lived, while others start up, achieve their goals and close down again in a relatively short time. The total number of groups at one time is typically over one hundred and much of the work in developing specifications is handled via electronic mailing lists.

There is a mailing list for each Working Group, for technical discussion relevant to it. There is also an IETF general discussion list and an IETF announcement list, the latter receiving the following types of messages:

- IETF meeting logistics;
- agendas for Working Group and BOF sessions at IETF meetings;
- Working Group actions;
- Internet-Draft announcements;
- IESG Last-Calls;
- IESG protocol and document actions;
- RFC announcements.

Anyone who subscribes to this list will automatically receive an e-mail message announcing these, as they are notified.

The WGs operate under the overall guidance of the Internet Architecture Board (IAB) [7], which has the responsibility of directing and co-ordinating the work of the IETF, and its sister body, the Internet Research Task Force (IRTF) [8]. The full list of the currently active WGs can be found on the IETF Web site [5], but the activities of some of the more important for VoIP are summarised later in this chapter.

5.4.1 IETF Documents

IETF documents fall into two major categories, Internet-Drafts (often abbreviated to I-Ds) and Requests For Comment (RFCs).

5.4.1.1 The Internet-Draft

Anyone can write and submit an I-D. Provided that it is in the correct format, the IETF Secretariat will publish it and notify the IETF Announcement List.

An Internet-Draft is a 'working document', and has little status in the larger IETF scheme of documentation. For informal review and comment, it is placed in the IETF's Internet-Drafts directory, which is replicated on a number of Internet hosts. It is not archived, and is automatically deleted after six months if it has made no further progress. It is possible for an I-D to have a longer existence, as it may be re-submitted, or updated to a newer version, which will be given another six months to progress. A successful I-D will be published as an RFC (see section 5.4.1.2).

An I-D submitted by an individual contributor is usually allocated a name in the form 'draft-submitter's name-technical area-subject-00.txt'. The '00' denotes the version number. It is also possible for an IETF Working Group to issue an Internet-Draft, in which case it will bear the identifier of the Working Group, thus 'draft-working group abbreviation-technical area-subject-00.txt'.

Working Groups will sometimes adopt I-Ds that have been submitted by individuals, but it is more common to amend them, or to amalgamate several I-Ds into a Working Group I-D. Normally, an I-D adopted and issued by a WG has more credibility than an I-D issued by an individual. I-Ds may also be submitted by the IESG, IAB or IRTF.

5.4.1.2 The RFC

Each version of an IETF specification is published as a Request for Comments (RFC). The Request For Comments is no longer what the name suggests, but a document that is part of the official archived output of the IETF. RFCs can be obtained from a number of Internet hosts using anonymous FTP, gopher, World Wide Web, and other Internet document-retrieval systems.

Each RFC is reviewed by the IESG, and, if approved, is announced and disseminated by the RFC Editor, via the IETF Announcement List. An RFC will be given a name in the form 'RFC number on subject'.

It is a common misconception that all RFCs are IETF Standards, but this is not so. RFCs cover a wide range of topics in addition to Internet standards, from early discussion of new research concepts to status memos about the Internet, and they can be classified as one of several categories, the names of which are self-explanatory:

- Standards Track;
- Best Current Practice;
- Informational;
- Experimental;
- Historic.

It is possible for an RFC to hold a different status at different times. For example, an RFC might change to 'historic' status, if superseded by a more current technology or method. The status of Internet protocol and service specifications is summarised periodically in an RFC entitled 'Internet Official Protocol Standards' (STD1).

Standards Track RFCs define Internet Standards. When a specification is adopted as an Internet Standard, it is given the additional label 'STDxxx', but it keeps its

RFC number and its place in the RFC series. Just as all RFCs are not standards track documents, not all standards track documents reach the level of Internet Standard.

Best Current Practice RFCs standardise the IETF community view about a principle or the best way to perform some function. When a specification is adopted as a BCP, it is given the additional label 'BCPxxx', but it keeps its RFC number and its place in the RFC series. Not all RFCs that describe current practices are reviewed and approved as BCPs.

Specifications of protocols or services for the Internet, which are not intended to become Internet Standards or BCPs, are not subject to the rules for Internet standardisation. A non-standards track specification may be published as an 'Experimental', 'Informational' or 'Historic' RFC, at the discretion of the RFC Editor in consultation with the IESG. Documents with these labels are not Internet Standards.

The 'Experimental' specification typically denotes some research or development effort. It is published for the general information of the Internet technical community and as an archival record of the work. An Experimental specification may be the output of an organised Internet research effort (e.g. a Research Group of the IRTF), an IETF Working Group, or it may be an individual contribution.

An 'Informational' specification is published for the general information of the Internet community, and does not represent an IETF consensus or recommendation.

A specification that has been superseded by a more recent specification, or is considered to be obsolete, is assigned to the 'Historic' category.

5.4.1.3 The Internet Standard

A specification subject to the Internet standards process is either a Technical Specification (TS) or an Applicability Statement (AS).

A Technical Specification describes a protocol, service, procedure, convention, or format. It may describe all the relevant aspects, or it may leave some options unspecified. A TS may be completely self-contained, or it may incorporate references to other documents, which might or might not be Internet Standards.

A TS does not specify requirements for its use within the Internet. Since these depend on the particular context in which the TS is incorporated by different system configurations, they are defined by an Applicability Statement.

An Applicability Statement specifies how, and under what circumstances, one or more TSs may be applied to support an Internet capability. An AS may specify uses for Technical Specifications that are not Internet Standards. An AS also specifies the circumstances in which the use of a particular TS is required, recommended or optional.

Although TSs and ASs are conceptually separate, in practice a Standards Track document may combine an AS and one or more related TSs.

5.4.1.4 IETF Standards Track

A specification on the Standards Track has to proceed through several steps before final approval. Formally known as 'maturity levels', these steps are 'proposed standard', 'draft standard', and 'standard'.

Proposed Standard

To be given the status of Proposed Standard by the IESG, a specification has to be complete and credible, and its use must have been demonstrated. It will maintain this status for at least six months, but no longer than two years, after which it must be elevated to the next status, deprecated as unfit to be elevated, or recycled for further work.

A Proposed Standard is generally stable and well understood. It has resolved known design choices, received significant community review, and enjoys enough community interest to be considered valuable. Usually, neither implementation nor operational experience is required for a specification to be designated as a Proposed Standard, but such experience is highly desirable, and is usually a strong factor in its favour.

A Proposed Standard should have no known technical omissions in respect of the requirements it aims to satisfy. However, the IESG may allow a specification with known technical omissions to advance to the Proposed Standard state, if it is considered to be useful, necessary and timely.

Draft Standard

Elevation to Draft Standard is a major advance in status, indicating a strong belief that the specification is mature and will be useful. To be elevated to this status, a Proposed Standard must have been proved by sufficient successful operational experience in two or more independent and interoperable implementations. The Working Group chairperson is responsible for submitting it to the Area Director with documentation about the implementations, which qualify the specification for Draft Standard status, and the testing of their interoperation.

A Draft Standard may still require additional field experience, since it is possible for implementations based on Draft Standard specifications to demonstrate unforeseen behaviour when subjected to large-scale use in production environments. A Draft Standard is normally considered to be a final specification, and changes are likely to be made only to solve specific problems encountered.

A Draft Standard will maintain its status for at least four months, but no longer than two years, after which it must be elevated to (Internet) Standard, deprecated as unfit to be elevated, recycled for further work, or reduced to the status of Proposed Standard.

Internet Standard

To achieve the final status of (Internet) Standard, it must be demonstrated that a document is the basis of significant and successful operational experience. It can then remain as an IETF Standard forever, or can at some time be relegated to 'Historic'. In theory, any new version of the same Standard must negotiate the whole standards track from the beginning. In practice, Standards have been known to have minor changes incorporated, if they have proved to be practicable in real operational environments.

5.4.1.5 Best Current Practice (BCP) RFCs

Since the Internet is composed of networks operated by a great variety of organisations, the operators and administrators of the Internet must follow some common guidelines for policies and operations.

The BCP sub-series of the RFCs is used to define and ratify the IETF community's best current thinking on a statement of principle or on the best way to perform some operation.

A BCP document follows the same basic set of procedures as a Standards Track document, but undergoes a much shorter process. The BCP is submitted for review by the IESG, including a Last-Call on the IETF Announcement List. Because a BCP is regarded as expressing community consensus but is processed more quickly than standards, it requires particular care. Once the IESG has approved the document, the process ends and the document is published.

5.4.1.6 Experimental and Informational RFCs

Unless they are the result of IETF Working Group action, documents to be published as Experimental or Informational should be submitted directly to the RFC Editor, who will publish them as Internet-Drafts, if they have not already been published. After this publication, the RFC Editor will wait two weeks for comments before proceeding further.

To prevent these types of documents circumventing the Internet Standards Process, the RFC Editor will refer to the IESG any document submitted for Experimental or Informational publication which may be related to work expected to be done within the IETF community. The IESG will review it, and recommend either that it be published as originally submitted or referred to the IETF as a contribution to the Internet Standards process.

5.4.2 IETF Standards Process

Entering a specification into, advancing it within, or removing it from, the Standards Track must be approved by the IESG. The first step is to post it as an Internet-Draft, unless it has previously been published as an RFC, and not changed since. It then remains as an Internet-Draft for not less than two weeks, so that it can be reviewed by the IETF community. After which a recommendation for action may be initiated.

Such a recommendation is made to its Area Director by the IETF Working Group responsible for the specification, and copied to the IETF Secretariat. If the specification is not associated with a Working Group, an individual can recommend it to the IESG. The IESG will review the specification to assess whether it satisfies the applicable criteria, and whether or not its technical quality and clarity is consistent with the proposed maturity level.

The IESG will send via electronic mail to the IETF Announce List notice that it will consider the document, to permit a final review by the general community. Anyone may comment on this 'Last-Call' notification. The Last-Call period must be no shorter than two weeks, except in cases where the proposed standards action was not initiated by an IETF Working Group, in which case the Last-Call period must be no shorter than four weeks. If the IESG believes that the community interest would be served by allowing more time for comment, it may decide on a longer Last-Call period.

The IESG is not bound by the action recommended when the specification was submitted. For example, the IESG may consider its publication in a category different from that requested. The IESG could also decide to change the publication category based on the response to a Last-Call. If this decision would result in a specification being published at a 'higher' level than the original Last-Call indicated, a new Last-Call must be issued giving the IESG recommendation. If there is a significant controversy in response to a Last-Call for a specification which does not originate from an IETF Working Group, the IESG may recommend the formation of a new Working Group.

At the end of the Last-Call period, the IESG will decide whether to approve the standards action, and will notify its decision via electronic mail to the IETF Announce List.

5.4.2.1 Publication

If a standards action is approved, notification is sent to the RFC Editor and copied to the IETF with instructions to publish the specification as an RFC. At that time, the specification will be removed from the Internet-Drafts directory.

An official summary of standards actions completed and pending appears in the Internet Society's newsletter. Periodically, the RFC Editor publishes an 'Internet Official Protocol Standards' RFC, summarising the status of all Internet protocol and service specifications.

5.4.2.2 Advancing in the Standards Track

A Proposed Standard remains at that level for at least six months, and a Draft Standard stays at that level for at least four months, or until at least one IETF meeting has occurred, whichever comes later. These periods are measured from the date of publication of the corresponding RFC, or the date when the IESG approval is announced.

A specification is likely to be revised as it advances through the Standards Track. At each stage, the IESG can modify the recommended action. Minor revisions are expected, but a significant revision may require the specification to accumulate more experience at its current maturity level before progressing. Finally, if the specification has been changed very significantly, the IESG may recommend that the revision should be treated as a new document, re-entering the Standards Track at the beginning.

A change of status will result in re-publication of the specification as an RFC, unless there have been no changes at all in the specification since the last publication. Generally, desired changes will be stored up for incorporation at the next level in the Standards Track. However, an important typographical error, or a technical error that does not represent a change in the overall function of the specification, may be corrected immediately. In such cases, the IESG or RFC Editor may be asked to republish the RFC (with a new number) with corrections, without resetting the start date for the status period.

When a Standards-Track specification has remained at the same maturity level for twenty-four months without reaching the Internet Standard level, and every twelve months thereafter until the status is changed, the IESG will review its viability and the usefulness of the technology. The IESG will either approve its termination or continuation, and whether to maintain the specification at the same maturity level or to move it to Historic status. This decision will be communicated by electronic mail to the IETF Announce List.

5.4.2.3 Revising a Standard

A new version of an established Internet Standard must progress through the full IETF Standards process like a new specification. Once the new version has reached the Standard level, it will usually replace the previous version, which will be moved to Historic status. However, in some cases both versions may remain as Internet Standards to honour the requirements of an installed base of implementations. In this situation, the relationship between the previous and the new versions must be explicitly stated in the text of the new version or in another document.

5.4.2.4 Retiring a Standard

As the technology changes and matures, it is possible for a new Standard specification to be so clearly superior technically that one or more existing Standards Track specifications for the same function should be retired. When it is felt that an existing Standards Track specification should be retired, the IESG will approve a change of status of the old specification to Historic, with the same Last-Call notification procedures used for any other standards action. A request to retire an existing Standard can originate from a Working Group, an Area Director or some other interested party.

5.4.3 Working Groups Relevant to VoIP

The activities of some of those WGs more important for VoIP are summarised here. The full list of the currently active WGs can be found on the IETF Web site [5].

5.4.3.1 Audio/Video Transport (AVT)

The AVT Working Group concentrates on the definition of the real-time transport protocol (RTP) (see Chapter 1) and the mapping of codecs into this protocol. The work of this group is therefore important for the interworking of media streams between differing technologies.

5.4.3.2 Multiparty Multimedia Session Control (MMUSIC)

The MMUSIC Working Group is developing a framework of protocols (see Fig 5.5) for the provision of multimedia communications over the Internet. The main protocols developed by MMUSIC so far have included the session initiation protocol (SIP) (see Chapters 1 and 11) and the session description protocol (SDP).

5.4.3.3 Session Initiation Protocol (SIP)

The MMUSIC Working Group developed the session initiation protocol as part of an overall conferencing architecture for the Internet. SIP has an increasing following in the industry as a candidate technology for wide-scale use in network-deployed systems.

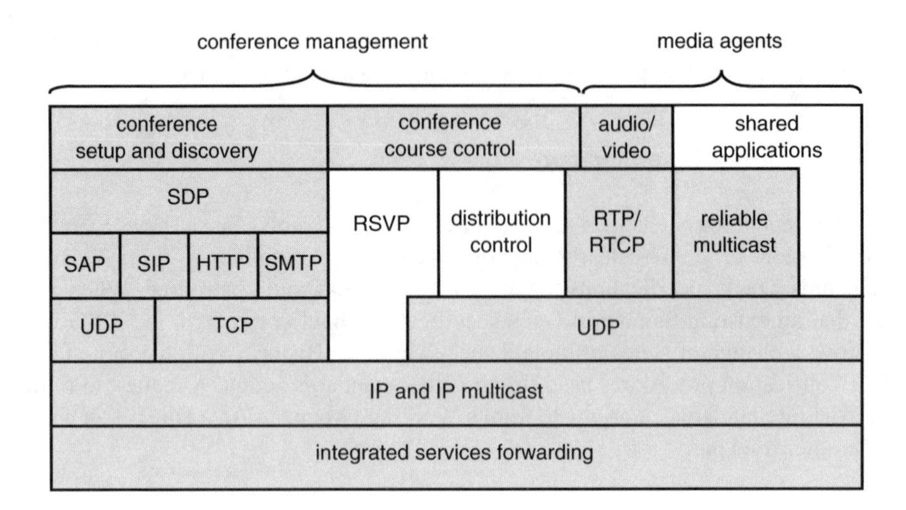

Fig 5.5 IETF multimedia conferencing architecture

SIP is important for a number of reasons, not least because 3GPP (see section 5.6.10) has elected to use it as the user-to-network protocol for 3rd generation networks (see Chapter 11). At present, implementations are very immature and it is likely that significant extensions and modifications will be required to make it a more practical proposition. Even so, there is a clear indication that future call control signalling will be based upon SIP.

5.4.3.4 IP Telephony (IPTel)

The IPTel Working Group is working closely with the SIP Working Group on defining the use of SIP for IP telephony applications. It has developed a protocol (telephony routing over IP — TRIP) for locating gateways and routing VoIP calls, and a scripting language, known as the call processing language (CPL), for providing call-handling applications [3].

5.4.3.5 Media Gateway Control (MeGaCo)

The MeGaCo Working Group has been working with the ITU-T (see section 5.5.3) on the definition of a control protocol for use between media gateways (MGs) and media gateway controllers (MGCs). The Megaco protocol can also be used to control IP devices such as Ethernet telephones. Users should note that there are two encoding schemes used for the gateway control protocol — text encoding (ABNF) and binary encoding (ASN.1). Although the two encoding schemes are almost identical (see Chapter 7), expect subtle differences between the two, and hence the need to buy gateways and controllers that match. Further details concerning the results of the MeGaCo Working Group are given in Chapter 7.

5.4.3.6 Signalling Transport (SigTran)

The SigTran Working Group has initially been driven by the need for separating C.7 (SS7) signalling gateways from the underlying physical modem terminations and the intermediate controller (see also the MeGaCo Working Group) in dial modem network access servers (NASs). The main deliverable from the Working Group at present is the stream control transmission protocol (SCTP) — previously known as the simple common transport protocol (in drafts prior to draft 7 of the specification). A number of adaptation layers are being defined for SCTP to enable existing telecommunications signalling protocols to be transported over IP networks. These are further described in Chapter 6. In addition, it is possible that SCTP may emerge as a replacement for the existing TCP.

5.4.3.7 Middlebox Communications (MIDCOM)

The MIDCOM Working Group is initially concerned with establishing requirements for protocols to be used in enabling end-to-end media and signalling flows through IP networks where firewalls, network address translators and other forms of 'middle box' appear in the packet path. The outcome of the work in this group is likely to have a major influence on the long-term viability of VoIP as a true multimedia application and has been strongly influenced by the results of the TIPHON project (see Chapter 16).

5.4.3.8 PSTN and Internet Internetworking (PINT)

The PINT Working Group addresses IP applications that need to invoke services in the PSTN. The work has largely been driven by applications from USA long-distance telephony companies seeking to use Web-initiated dialling to eliminate the buy/sell/buy model of normal long-distance telephony peculiar to that country. This application uses SIP to pass IN service requests from IP-based clients through an intermediate server to the PSTN. In doing so, it is driven by a regulatory-defined price opportunity.

5.4.3.9 Services in the PSTN/IN Requesting Internet Service (SPIRITS)

The SPIRITS Working Group complements the work of the PINT Working Group by addressing the PINT scenario in the reverse direction — events in the PSTN initiate actions in the IP domain. Internet call waiting has been the main application driving this work.

5.4.3.10 Telephone Number Mapping (ENUM)

The ENUM Working Group is concerned with the use of a DNS-based architecture and protocols for mapping a telephone number to a set of attributes (e.g. URLs) which can be used to contact a resource associated with that number. ENUM is important for a number of reasons, not least because the results from this Working Group could be used in advanced VoIP services. However, there are concerns about the practicality and utility of this approach.

5.5 ITU-T

The International Telecommunication Union [9] is an international organisation established under the auspices of the United Nations, within which governments and companies co-ordinate global telecommunications networks and services. Within the ITU, the ITU Telecommunication Standardisation Sector (ITU-T) [10] studies technical, operating and tariff questions and publishes Recommendations about them with a view to standardising telecommunications globally. A distinguishing feature of the work of the ITU-T is therefore to establish and agree requirements and approaches to communications technology prior to implementation work commencing.

The ITU-T has a very formal structure, which reflects the formal participation of the countries of the world. The World Telecommunication Standardisation Assembly (WTSA), which takes place every four years, defines general policy for the Sector, establishes the study groups and approves their work programme for each study period of four years, and appoints the study group Chairmen and Vice-Chairmen.

The Telecommunication Standardisation Advisory Group (TSAG), reviews priorities, programmes, operations, financial matters and strategies for the Sector, follows up on the accomplishment of the work programme, restructures and establishes ITU-T study groups, provides guidelines to the study groups, advises the Telecommunication Standardisation Bureau (TSB) Director, and elaborates Recommendations on organisation and working procedures.

Fourteen ITU Telecommunication Standardization Study Groups (SG) and their Working Parties develop Recommendations by the study of 'Questions', i.e. areas for study.

5.5.1 ITU-T Documents

Like the IETF, the ITU-T has its own ephemeral documents, which are known as Temporary Documents (TDs). These are the typical input to ITU-T meetings that are working on developing ITU-T standards, known as Recommendations.

Like RFCs, ITU-T Recommendations can take several different forms:

- Recommendations in force — available for purchase and published in different forms — paper, electronic files, entire collection on CD-ROM, etc;
- pre-published Recommendations — these are approved Recommendations, available in electronic form in the original language only, which have not undergone the final phase of editorial processing and translation, and can therefore be subject to editorial changes before the final version;
- superseded Recommendations — available as paper booklets and electronic files.

5.5.2 ITU-T Standards Process

The ITU-T has two approval processes, the traditional approval process (TAP), and the alternative approval process (AAP). Generally, Recommendations in ITU-T Standardisation Domain 04 (numbering and addressing) and Domain 11 (tariff, charging and accounting) are assumed to fall under TAP, while Recommendations not in Domains 04 or 11 are assumed to fall under AAP.

5.5.2.1 Traditional Approval Process

This is the process for the approval of Recommendations which require the formal consultation of Member States. Formal consultation asks Member States to delegate authority to the competent study group (SG) to proceed with the approval process at a formal meeting of the SG. The SG may also seek approval at a WTSA. The status of Recommendations is the same, whether approved at an SG meeting or at a WTSA.

Determination

A draft Recommendation is 'determined' when a study group or working party meeting, or exceptionally a WTSA, determines that work on a draft Recommendation is sufficiently mature. Each SG should establish an editing group to review the text of each Recommendation in each of the working languages — English, French and Spanish. The text of the draft Recommendation must be available to the TSB in a final edited form, with a summary, in at least one of the working languages. The summary is a brief outline of the purpose and content of the draft Recommendation and the intent of the revisions. No Recommendation is considered complete and ready for approval without this summary.

At the request of the SG chairman, the Director of the TSB will announce to all Member States and Sector Members, when convening the meeting of the SG and at least three months before the meeting, the intention to apply the approval procedure.

The text of the draft Recommendation must be distributed in the working languages at least one month prior to the announced meeting.

Approval may only be sought for a draft Recommendation within the SG's mandate as defined by the questions allocated to it. Where a draft Recommendation falls within the mandate of more than one SG, the chairman of the SG proposing the approval should consult any other SG chairmen concerned before proceeding with the application of the approval procedure.

In certain cases, the approval of new or revised Recommendations should be deferred for consideration at a WTSA. These cases include:

- an administrative Recommendation concerning ITU-T as a whole;
- a Recommendation for which the SG concerned wishes the WTSA to debate and resolve particularly difficult or delicate issues;
- a Recommendation on which the SG has failed to agree, due to non-technical issues such as differing views on policy.

Consultation

The consultation period for Member States, during which they must decide whether they will assign authority to the SG to approve the Recommendation, starts with the Director's announcement of the intention to apply the approval procedure, and finishes seven working days before the beginning of the SG meeting.

The Director of the TSB will notify the Directors of the Radiocommunication Bureau (BR) and the Telecommunication Development Bureau (BDT), as well as recognised operating agencies, scientific and industrial organisations and international organisations participating in the work of the SG, that Member States are being consulted on a proposed Recommendation. Only Member States are entitled to respond.

If a Member State decides that consideration for approval should not proceed, they should give their reasons and indicate the possible changes that would facilitate further consideration and approval of the draft Recommendation.

If 70% or more of the replies from Member States support consideration for approval, or if there are no replies, the SG may make the necessary technical and editorial changes and consider approval. If less than 70% of the replies received by the due date support consideration for approval, it may not proceed at that meeting. Any comments received with the responses to the consultation will be collected by the TSB and submitted as a temporary document to the next meeting of the SG.

Procedure at the Study Group Meeting

After reviewing the draft text, the SG meeting may accept editorial corrections or other amendments not affecting the substance of the Recommendation. The SG should assess the summary statement for completeness, and its ability to convey the intent of the draft Recommendation to a telecommunications expert who has not participated in the SG work.

The meeting may only make technical and editorial changes arising from written contributions, results of the consultation process, or liaison statements. Where proposed revisions are justified, but have a major impact on the intent of the Recommendation or depart from points of principle agreed at the previous SG or working party meeting, consideration of the approval procedure should be deferred to another meeting — unless the chairman of the SG, in consultation with the TSB, considers that the proposed changes are reasonable for those Member States not represented adequately under the changed circumstances, and that the proposed text is stable.

Under this procedure, the decision to approve the Recommendation must be unopposed. However, the report of the meeting may record a delegation's reservations about one or more aspects, in a case where it does not elect to oppose the approval of a text. Such reservations must be mentioned in a note appended to the text of the Recommendation concerned. Exceptionally, but only during the meeting, a delegation may request more time to consider its position. A Member State which requests more time, and which then indicates disapproval within a period of four weeks from the end of the meeting, must give its reasons and indicate the possible changes that would facilitate further consideration and future approval of the draft Recommendation. Unless the Director of the TSB is advised of formal opposition from the delegation's Member within the four weeks, he will commence Notification.

If the Director is advised of formal opposition, the SG chairman, after consultation with the parties concerned, may schedule consideration of approval, without further determination at a subsequent working party or study group meeting.

If a delegation decides to abstain from the decision to apply the approval procedure, its presence shall be ignored for that purpose. Such an abstention may subsequently be revoked, but only during the course of the meeting.

Decision

When a draft Recommendation has been approved, it is said to be 'decided'.

TAP Notification

Within four weeks of the closing date of the SG meeting, the Director of the TSB will notify the status of the text, by circular, and he will arrange for this information to be included in the next ITU Notification. He will also ensure that any Recommendation agreed during the SG meeting is available on-line in at least one working language, with an indication that the Recommendation may not be in its final publication form.

The TSB may correct clear oversights or inconsistencies in the text, or make minor editorial amendments, with the approval of the SG chairman.

The Secretary-General will publish the approved Recommendation in the working languages as soon as practicable, indicating a date of entry into effect. Minor amendments may be covered by corrigenda, rather than a complete reissue.

Additional Changes

Once a Recommendation has been approved, further amendment of the new text or the revised portion should not normally be sought within a reasonable period of time (at least two years in most cases), unless the proposed amendment complements rather than changes the agreement reached in the previous approval process, or a significant error or omission is discovered.

Any Member State considering itself to be adversely affected by a Recommendation may refer their case to the Director of the TSB, who will submit it to the relevant SG for prompt attention.

Correction of Defects

When a study group identifies the need for implementors to be made aware of defects (e.g. errors, ambiguities, omissions or inconsistencies) in a Recommendation, an Implementors' Guide may be produced, as a historical document recording all defects and their status of correction, from their identification to final resolution. An Implementors' Guide must be approved by the SG and made available to the public as part of its COM series of documents.

Deletion of Recommendations

There are two alternatives for the deletion of Recommendations. The SG chairman can request that the WTSA should delete a Recommendation, or an SG meeting may agree to delete a Recommendation, because it has been superseded by another Recommendation or become obsolete. This agreement must be unopposed. Information about this agreement, including an explanatory summary of the reasons for the deletion, must be provided by a circular. If no objection to the deletion is received within three months, the deletion will take effect. In the case of objection, the matter will be referred back to the SG.

Notification of the result will be given in another circular, and the Director will inform the TSAG, and also publish a list of deleted Recommendations whenever appropriate, but at least once by the middle of a study period.

5.5.2.2 Alternative Approval Process

ITU-T Recommendations will be approved using the new AAP, except Recommendations that have policy or regulatory implications, which will be approved using the TAP. The AAP procedure is very similar to the TAP, but with the consultation period for Member States being replaced by a last call procedure. The main differences are indicated below.

Consent

When an SG meeting, working party meeting or a WTSA agrees that a draft Recommendation is sufficiently mature, it is considered to have 'consent'. This is the equivalent of the TAP's 'determined' status. The SG chairman will then ask the Director of the TSB to announce the intention to apply the AAP and to initiate the last call. The draft Recommendation will be electronically posted.

Last Call and Additional Review

The last call normally takes four weeks. If no comments from Member States or Sector Members are received by the end of the last call, other than comments indicating typographical errors, the draft Recommendation is considered approved, and the typographical errors are corrected.

If other comments are received, the SG chairman, in consultation with the TSB, will decide either to consider the draft Recommendation for approval at a planned SG meeting, or to ask appropriate SG experts to resolve comments, under the chairman's direction, using electronic correspondence or meetings. When all comments have been resolved and the edited draft is available, the SG chairman, in consultation with the TSB, will decide either to consider the draft Recommendation for approval at a planned SG meeting, or to begin an additional review.

The additional review takes three weeks. The revised text of the draft Recommendation comments from the last call must be available to the Director of the TSB, before he announces the additional review. If the TSB receives no comments other than those indicating typographical errors, the Recommendation is considered as approved, and the typographical errors are corrected by the TSB. Otherwise, the draft has to go for approval at an SG meeting. The Director of the TSB will announce the intention to approve the draft Recommendation at least three weeks before the SG meeting.

Procedure at the Study Group Meeting

The study group will review the draft Recommendation and its associated comments, and may then accept amendments to it. Changes may only be made as a consequence of written comments from the last call, additional review, contributions, or liaison statements. Where proposals for revisions are justified but will have a major impact on the Recommendation, the AAP should not be applied at this meeting.

If approval of the Recommendation is unopposed at the meeting, it is considered to be 'decided'.

If unopposed agreement has not been reached, the Recommendation is considered to be approved if, after consultation with the Sector Members present, there is unopposed agreement of the Member States present in the meeting. Otherwise, the SG may authorise extra work on the remaining issues.

AAP Notification

The Director of the TSB will notify the membership by circular of the outcome of the last call and additional review, and that any Recommendation agreed to is available on-line, normally within two weeks of the closing date of the SG meeting. He will also arrange for the information to be included in the next available ITU Operational Bulletin.

5.5.3 ITU-T and VoIP

Historically, the ITU-T has been the world authority for standardising telephony based on circuit-switched technologies. With the advent of VoIP, it has turned its attention to the issues raised by the new technology set, and has begun work on several areas which are complementary to the work of the IETF on VoIP technologies.

Notable among these are the issues of providing the underlying networks that are assumed to be there to transport IP around the world. In fact much of today's Internet traffic is transported over networks that have been constructed using technologies standardised by the ITU-T.

The ITU-T study groups most active in the VoIP arena are:

- Study Group 9, working on IP CableCom solutions for VoIP over cable TV networks;
- Study Group 11, developing bearer-independent call control (BICC) (see Chapter 8);
- Study Group 13, working on PSTN/IN interworking;
- Study Group 16, the lead study group on multimedia communications, which produced the H.3xx series of Recommendations for multimedia conferencing, including H.323 (see Chapter 10), and encompassing supporting specifications relating to gateway devices including H.246 and H.248 (the Megaco protocol) (see Chapter 7).

In addition to the above, the ITU is initiating a new project — known as MediaCom — that may consider the future direction of multimedia applications from the ITU perspective. This activity is in its formative stages at present.

5.6 ETSI

The European Telecommunications Standards Institute (ETSI) [11] is a non-profit making organisation whose mission is to produce telecommunications standards that will be used throughout Europe and beyond. Its 730 members from 51 countries inside and outside Europe include administrations, network operators, manufacturers, service providers, research bodies and users.

In Europe, as telecommunications standardisation is regarded as an important step towards building a harmonised economic market, ETSI enjoys the support of the European Commission (EC). ETSI produces voluntary standards, some of which may go on to be adopted by the EC as the technical base for Directives or Regulations. ETSI consists of a General Assembly, a Board, a Technical Organisation and a Secretariat. The Technical Organisation produces and approves technical standards. It encompasses ETSI Projects (EPs), Technical Committees (TCs) and Special Committees, which are collectively known as Technical Bodies. As well as the central Secretariat of ETSI, additional experts work full time in Specialist Task Forces (STFs) to accelerate standardisation. Such an STF is normally dedicated to the work of one Technical Body, but it will consult widely among other Technical Bodies, and outside ETSI, if required.

Work in ETSI is done on the basis of ETSI work items, approved by the Technical Body and adopted by the ETSI membership.

5.6.1 ETSI Documents

An ETSI work item results in one or more ETSI deliverables, each of which can be a standard, a specification, a guide or a report. They can be divided into two broad categories:

- specifications and standards

  TS — ETSI Technical Specification
  ES — ETSI Standard
  EN — European Standard (telecommunications series)

- guides and reports

  TR — ETSI Technical Report
  EG — ETSI Guide
  SR — ETSI Special Report

An ETSI Technical Specification (ETSI TS) or an ETSI Technical Report (ETSI TR) is adopted by the Technical Body responsible. A TS is normative, with a short time to publication. It may be converted later to an ES or EN, but is aimed at achieving the early delivery of a high-quality specification for an emerging technology, for which it may be too soon to define a completely stable text. A TR is mainly informative, and is the default deliverable, when a specification cannot be defined.

An ETSI Standard (ES) or an ETSI Guide (EG) is adopted after ETSI membership weighted voting. An ES is normative, and it is considered preferable, or sometimes necessary, that the whole ETSI membership approves it. An EG gives guidance on handling technical standardisation activity in the whole, or major parts of the, Technical Organisation.

An ETSI European Standard — telecommunications series (ETSI EN) is adopted after ETSI membership national-weighted voting. It is a formal Standard at European level, which is specific to Europe and needs transposition into national standards.

Alternatively, it may be the subject of an EC/EFTA mandate from the European Commission (EC) or the European Free Trade Association (EFTA).

An ETSI Special Report (ETSI SR) is an informational document used for various purposes, including publicising information not produced within a Technical Body, but of general interest to ETSI members or the public. It can also be a deliverable with dynamic content, generated by a software application on the ETSI Web site.

5.6.2 ETSI Standards Process

A Technical Body is free to organise its work in any way it wishes, within the rules of the Technical Working Procedures, including the creation of Working Groups which are given the tasks of drafting parts of the Technical Body's work programme.

Drafting usually takes place in a small team, or Rapporteur Group, led by a Rapporteur or editor. The work is largely done by exchange of documents via the ETSI DocBox server and LISTSERV e-mail exploder facilities. When the Rapporteur Group's draft is considered ready, it is handed over to the Working Group (when it exists) for approval. The formal approval for further processing or, in the case of ETSI Technical Specifications or ETSI Technical Reports, approval and adoption, can only be done by the Technical Body, either at a meeting or by correspondence. Some drafting activities for a Technical Body may be performed by an STF.

The adaptation of specifications from external bodies (publicly available specifications (PASs)) to the ETSI deliverable structure follows the same rules, but will normally be performed by the PAS provider, as defined in the Guidelines for adoption of publicly available specifications. While the drafting process is, in principle, the same for all ETSI deliverables, the process elements required for adoption depend on the type of deliverable being processed.

5.6.3 ETSI Technical Specifications (TS) and ETSI Technical Reports (TR)

The technical body approval and adoption take place at the same time, as one combined decision. Publication is then the only element in the adoption process.

The publication process element consists of final editing of the Microsoft Word for Windows version of the adopted TS or TR, archiving, and publication as Adobe Portable Document Format (PDF) electronic documents. The published deliverable will then be made available for distribution via the ETSI Web server, and constitutes a part of the ETSI Documentation Service (EDS).

5.6.4 ETSI Standards and ETSI Guides

After technical body approval, the draft ES or EG is made available to the ETSI Membership (full members and associate members) for voting in accordance with the Membership Approval Procedure (MAP) defined in the ETSI Rules of Procedure and Technical Working Procedures.

Voting with MAP is done via a member voting application with a Web browser interface. The voting period is 60 days, to comply with the World Trade Organisation requirement. The deliverable is adopted if at least 71 % of the national-weighted votes cast are for the draft. After adoption, the deliverable is finally edited, archived and published in PDF format.

5.6.5 European Standard — Telecommunications Series (EN)

After the Technical Body approval, a European Standard (telecommunications series) produced by ETSI (ETSI EN) is entered into one of the two approval procedures stipulated by the ETSI Rules of Procedure:

- one-step approval procedure (OAP);
- two-step approval procedure (TAP).

5.6.5.1 EN — One-step Approval Procedure

This procedure is used when the technical body considers that the draft is mature, or a new version of an ETSI EN. After editing, the draft is made available to each of the ETSI National Standards Organisations (NSOs) to establish its national position for the vote, i.e. to consult nationally in its territory. The implementation of this may vary from one NSO to another.

The vote is formally cast by the Head of National Delegation to ETSI. In most cases this person is not from the NSO, but may be when the NSO is a member of ETSI.

Often referred to as 'NSO voting', the procedure is basically the same as that for the Membership Approval Procedure, except that the voting period is 120 days. The deliverables are made available to the NSOs using file transfer via the Internet and CD-ROM, although the latter is being phased out. The NSO sends the national position for the vote to ETSI via a Web-based electronic voting application. The deliverable will be adopted if at least 71 % of the national-weighted votes cast are for the draft.

5.6.5.2 EN — Two-step Approval Procedure

This procedure, which is normally obligatory for so-called Harmonised Standards, involves the NSOs at two stages, with resolution actions taken by the Technical Body responsible for the draft, as necessary.

The first NSO involvement, 'NSO Public Enquiry', has a duration of 120 days. The second, 'NSO voting', is 60 days. Any comments received from the Public Enquiry are used by the Technical Body to decide whether changes should be made to the draft before it is sent to the NSOs for their consultation and establishment of their national position for the vote.

5.6.6 Combined Processes

To make the results of the work of the technical body available at an early stage, some of the above processes may be combined in such a way that two deliverables with identical content are processed/published in parallel.

For example, if the intention is to publish the draft as an ETSI EN (telecommunications series), but only after application of the two-step approval procedure, the editing of the ETSI EN (sub-process Editing prior to Public Enquiry) also covers the Publication of an ETSI TS with identical contents.

Parallel ETSI ES and ETSI TS processing is also possible, but there is less time gained in this case.

Detailed rules for the approval procedures described above may be found in the Technical Working Procedures (TWPs). TWPs also define the rules which apply to the previous regime deliverables (e.g. European Telecommunication Standards (ETSs)) maintained by ETSI.

5.6.7 TIPHON

The lead body working on VoIP-related technology within ETSI is Project TIPHON (EP TIPHON) [12]. TIPHON stands for 'Telecommunications and Internet Protocol Harmonisation Over Networks', with more than 40 ETSI members and other companies committing their support.

TIPHON's objective (see Fig 5.6) is to support the market for real-time communication between users, including voice and related voice-band communication (such as facsimile) over multiple network technologies. Specifically, it ensures that users connected to IP-based networks can communicate between themselves and also with users in switched-circuit networks (SCNs), which include PSTN/ISDN and GSM, and vice versa. In this regard TIPHON addresses the difficult, yet extremely important, area of multi-network interworking across multiple administrative and technology domains.

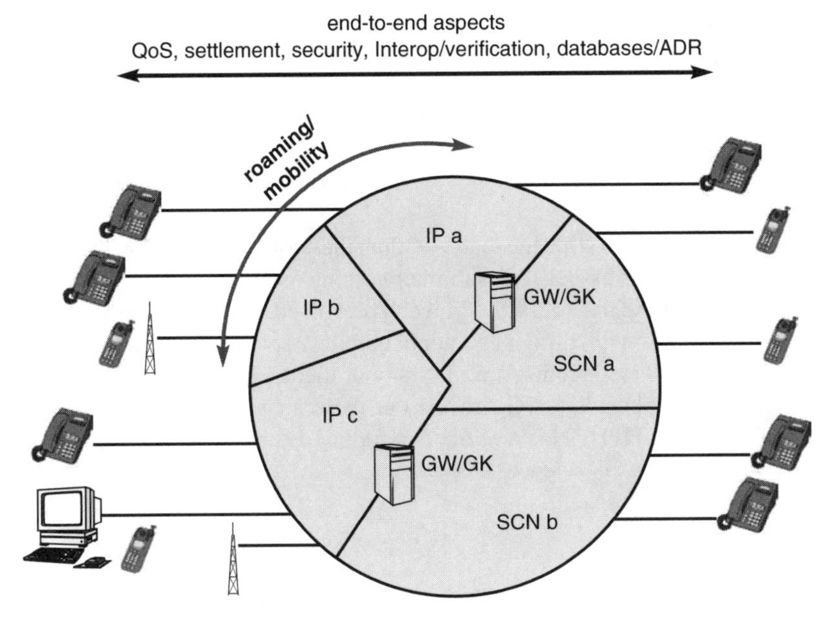

Fig 5.6 Scope of the TIPHON project.

To achieve this objective, TIPHON has identified the need for an overarching technology and domain-independent protocol framework — known as the TIPHON meta-protocol. By mapping this 'meta-protocol' to individual network technologies, TIPHON ensures a higher degree of end-to-end capability than would otherwise be possible.

TIPHON deliverables take the form of ETSI technical specifications and reports, the prime goal being to produce global standards. It recognises that co-operation with relevant groupings in ITU-T and IETF is essential, and it also co-operates closely with relevant forums, like the IMTC VoIP Activity Group and the Multiservice Switching Forum (see section 5.8.3). This co-operation supports the validation and demonstrations necessary to confirm the appropriateness of the solutions proposed.

TIPHON itself has eight working groups, tackling different themes for which VoIP requires solutions:

- WG1 — requirements for service interoperability, technical aspects of charging/ billing and security;
- WG2 — architecture and reference configurations;
- WG3 — call control procedures, information flows and protocols;
- WG4 — naming, numbering and addressing;
- WG5 — quality of service;
- WG6 — verification, demonstration and implementation;
- WG7 — mobility;
- WG8 — security.

TIPHON has acted as an incubator within the VoIP industry for many of the implementation issues associated with interworking VoIP systems, including inter-domain settlement. As the TIPHON Project has evolved, it has become apparent that there are synergies with third generation mobile networks such as those being developed by 3GPP (see section 5.6.8). Many of the issues considered by TIPHON are increasingly applicable to 3G networks as these networks increasingly embrace VoIP architectures. TIPHON is described in further detail in Chapter 16.

5.6.8 3GPP

The 'Third Generation Partnership Project' is a joint effort by a number of standards groups, including ETSI and the ITU, aiming to rapidly develop globally applicable specifications for third generation mobile networks. It is commonly known by the acronym 3GPP [13]. The partners involved are the various regional standards bodies from around the world. The full list can be found on the 3GPP Web site [13].

The partners have agreed to co-operate in the production of globally applicable Technical Specifications and Technical Reports for a 3rd generation mobile system based on evolved GSM core networks and the radio access technologies that they support. It is relevant to the VoIP industry since it has selected VoIP and IP-based technologies to form the basis of subsequent releases [14].

5.7 Registries

These are not standards bodies, as such, but they are more 'official' than the industry forums described in section 5.8, because they help to administer operational requirements imposed by the use of the standards. The need for standard formats for elements such as Internet protocol parameters and addresses led to the recognition that registries were needed to ensure that every new element is in the correct format, and unique from those that have already been used. Registries maintain these lists of unique elements, and control the assignment of new ones.

5.7.1 IANA

The Internet Assigned Numbers Authority (IANA) [15] is the central co-ordinator for the assignment of unique parameter values for Internet protocols. IANA is chartered by ISOC to act as the clearing house to assign and co-ordinate the use of numerous Internet protocol parameters.

5.7.2 ICANN

Within the last two years, some of the functions of IANA have been taken over by the Internet Corporation for Assigned Names and Numbers (ICANN) [16]. It is the non-profit corporation that was formed to assume responsibility for the IP address space allocation, protocol parameter assignment, domain name system management, and root server system management functions previously performed under US Government contract by IANA and other entities.

The intention is to make ICANN a truly international body, to reflect the status of the Internet as an international resource, no longer to be controlled by the USA. For this reason, there are strict guidelines that dictate that the directors must be recruited from several continents.

5.8 Industry Forums

There are a large number of industry groups active in the VoIP industry. At one time they seemed to appear on a daily basis. Any list of such groups is therefore invariably going to be incomplete in some way. Within the space constraints of this chapter, the following provides a sample of such groups.

5.8.1 Forums Related to Mobile VoIP Applications

As will be appreciated from its registration process (see Chapter 1), VoIP offers an inherent mechanism supporting a certain degree of mobility for end users. Not surprisingly, in addition to the formal 3GPP group, there are a number of industry groups that have adopted VoIP as the basis of their work towards third generation mobile networks [17], including the following.

- 3G.IP

  3G.IP [18] is a group of operators and vendors who share a common strategy to define a third generation network architecture based on packet technologies and encompassing:

  — IP telephony for simultaneous real-time and non-real-time services;

  — a common core network;

  — personal mobility;

  — interoperability between mobile and fixed networks.

- MWIF

  The Mobile Wireless Internet Forum [19] is another industry pressure group, whose mission is to drive acceptance and adoption of a single mobile wireless and Internet architecture that is independent of the access technology. Like 3G.IP, it has its own architecture, and is actively promoting it within 3GPP and other relevant forums, like TIPHON.

- WAP Forum

  The WAP Forum is the body that developed the wireless application protocol (WAP), which it describes as '... the *de facto* worldwide standard for providing Internet communications and advanced telephony services on digital mobile phones, pagers, personal digital assistants and other wireless terminals ...' [20]. It is not an international standards body, but has substantial support from the mobile industry. WAP is regarded as an interim solution, which will eventually be replaced by the 3rd generation systems that are under development.

Further details of these groups and how they relate to 3G networks are given elsewhere [14].

5.8.2 Forums Related to Cable Access VoIP Applications

As part of the development of multimedia interactive services for cable TV users, CableLabs in the USA have developed a suite of specifications for the provision of IP-based services over cable access networks. This initiative is known as PacketCable [21]. Although the development of the specifications takes place in a closed and proprietary environment, once completed the specifications are made available to the public and some have recently been submitted to the main public standards groups. While providing a focus for the work, this has had the unfortunate consequence of aspects being incorporated into the specifications which increasingly diverge from wider industry development. Most notable of these has been the premature adoption of a gateway control protocol resulting in the selection of the MGCP protocol (see Chapter 7).

5.8.3 Forums Related to Achieving Product Interoperability

In addition to any implementation support provided by formal standards groups, several industry forums address issues of implementation interoperability, the main ones relevant to VoIP being the IMTC and MSF.

- IMTC

  The International Multimedia Teleconferencing Consortium (IMTC) [22] is a non-profit corporation comprising more than 150 companies from around the globe. The IMTC is not a standards body, but its mission is to promote, encourage, and facilitate the development and implementation of interoperable multimedia teleconferencing solutions based on open international standards.

  Its activities include sponsoring and conducting interoperability test sessions between suppliers of conferencing products and services, as well as providing a forum for technical exchange that will lead IMTC members to make submissions to the standards bodies to further enhance the interoperability of multimedia teleconferencing products and services.

  The IMTC's work on achieving interoperability between implementations of specific technologies has developed an excellent track record of interoperability events, known as 'Superops', for the T.120 and H.323 series of ITU-T Recommendations. Its activity groups include 'iNow!', and its output deliverables are Interoperability Agreements (IAs) that profile standards.

- MSF

  The Multiservice Switching Forum (MSF) [23] seeks to accelerate the deployment of open communications systems, supporting a full range of network services using multiple infrastructure technologies. Its focus is on the development of architectures and industry agreements that enable interoperability and innovation in a rapidly evolving environment. It also defines requirements for multiservice switching platforms, the uses of which will include VoIP.

  The MSF was initially predicated on ATM but is now moving to be technology neutral [24]. It develops Implementation Agreements (IAs) which can be used by developers and network operators to ensure interoperability between components from different vendors.

- International SoftSwitch Consortium

  The International SoftSwitch Consortium (ISC) [24] seeks to promote and develop interoperable products and services in the 'SoftSwitch' arena. It comprises a mix of service providers and equipment manufacturers mainly from the USA. It has eight working groups at present, covering applications, architecture, carriers (covering interconnect issues), device control, legal interception, marketing, session management and SIP.

5.9 Industry Promotion

Although neither a formal standards group nor industry group *per se*, no consideration of the VoIP industry scene would be complete without recognising the efforts of Jeff Pulver's pulver.com [25] and Voice on the Net (VoN) [26] activities. Surrendering a position as a Wall Street analyst to concentrate on his Pulver and VoN activities, Pulver's organisation provides regular conference gatherings that have constituted the focal point for promotion of the VoIP industry over recent years.

In addition, pulver.com has been a leading experimenter through a series of 'freeworld dial-up' experimental VoIP networks.

5.10 Core Design Issues

International standards are needed to deal with many interworking issues. These are just a few of the more important issues currently being addressed.

5.10.1 Boundary Traversal

Traditional telephony calls are transmitted over a dedicated network. However, IP-based systems carry all sorts of data traffic, with which the voice traffic has to compete for carriage, and they are guarded by various devices to protect customers' valuable data, such as firewalls and network address translators (NATs).

Any VoIP call could have to negotiate an infrastructure that is constructed with zones of different administrative policy and control, separated by firewalls or similar devices. At the boundary points between these zones, it may be necessary to map IP addressing ranges and/or IP header parameters (e.g. type of service (ToS) field values) for both media and signalling packets. Failure to correctly map and police IP telephony traffic results in asymmetric transmission, inability to complete call signalling, and poor end-to-end performance due to the absence of appropriate packet admission control criteria. Crossing boundaries therefore has to address:

- firewalls;
- network address translation;
- quality of service;
- border gateways;
- security.

A common theme to these problems is the need to be able to route and map packets from one IP network to another within a profile determined by some call control messages. Without this, the real-time characteristics of the application cannot be ensured, except by resorting to massive over-provision of bandwidth which may not be sustainable.

The two key pieces are the entity that can handle the call control messages and the entity that can map and police packets flowing from one side of the device to the other based upon the direction of the call control entity. This suggests the need for a control relationship between these two entities and the requirements for a protocol to enable them to function correctly.

The initial definition of these issues began in the ETSI TIPHON Project [12, 27], and has progressed to the IETF, where the MIDCOM Working Group has been established to develop the requirements for a protocol which can be used to request border element passage for traffic, including VoIP. BTexact Technologies took a leading role in the TIPHON drafting, and is co-ordinating production of the IETF requirements draft.

5.10.2 Quality of Service (QoS)

The dedicated network used for traditional telephony calls is optimised for their passage around the world. However, IP-based systems are not optimised for voice. They have usually operated on a 'best-effort' basis, which is adequate for unidirectional data transfer applications, where the delivery time is not critical.

However, the requirements of 'real-time' interactive voice traffic are more stringent. Significant work is required to ensure that the end-to-end quality of a VoIP call, as perceived by the end participants in the call, can meet the standards that are generally associated with traditional voice calls.

Once again, as described above, the problems of inter-domain carriage and control of this voice traffic are central to ensuring end-to-end QoS. The ETSI TIPHON Project has developed the requirements for QoS, a framework for handling it and techniques for measuring it [12]. It has also defined QoS classes for speech, which are being harmonised with those being developed by 3GPP.

While the IETF is beginning to identify requirements for QoS over IP networks, TIPHON's solutions are ready to be adopted in overcoming the QoS problems associated with diverse transport networks, whether IP-based or using some other technology.

5.10.3 Terminal Specifications

Although IP devices continue to emerge on the market these are invariably highly proprietary in some respect due to the absence of mature specifications in this area. It is expected that further consideration of standards for IP-based terminals will be required — even at an informal level — to ensure a sustainable open industry.

5.10.4 Management

The previously mentioned problems of interworking between domains not only pose difficulties for the traffic and its control, but also for service management across the various domains involved. Typical service management issues include fault and performance management, billing and settlement issues, service level agreements.

Recognising this, a plethora of initiatives to address these issues began in various standards bodies — the ITU-T, the IETF, the Telecommunications Management Forum (TMF), ETSI Telecommunications Management Network (TMN), ETSI TIPHON, etc. Recognising that many different solutions would be counter-productive, these bodies agreed to work together in the 'JointNM' initiative.

As a contribution to this initiative, ETSI TIPHON is producing a service management framework, which covers fault, configuration, accounting, performance and security management [12].

5.11 Summary

VoIP offers the potential to radically change the structure and nature of the global communications scene. To do so will require a common set of well-defined and widely adopted standards. Specifications continue to emerge and be refined towards a common set although this is unlikely to reduce to a single specification, if for no other reason than that the approaches to the development of standards documents are not aligned across the various groups developing them. The need to interwork services across multiple technologies and multiple domains of administrative control is therefore fundamental to the development of the industry. Activities, such as ETSI's TIPHON project, have developed to address these requirements and the acceptance of such approaches will form an essential part of the VoIP environment over the next few years.

References

1 ETSI: '*ETS 300 347-1: V5.2 interface specification*', http://www.etsi.org/

2 Dufour, I. G. (Ed): '*Network intelligence*', Chapman and Hall, London (1997).

3 Cookson, M. and Smith, D.: '*3G service control*', in Clapton, A. J. (Ed): '*Future mobile networks: 3G and beyond*', Institution of Electrical Engineers, London, pp 99-119 (2001).

4 Stretch, R.: '*The OSA API and other related issues*', in Clapton, A. J. (Ed): *Future mobile networks: 3G and beyond*', Institution of Electrical Engineers, London, pp 121-135 (2001).

5 IETF Web site — http://www.ietf.org/

6 ISOC Web site — http://www.isoc.org/

7 IAB Web site — http://www.iab.org/

8 IRTF Web site — http://www.irtf.org/

9 ITU Web site — http://www.itu.int/

10 ITU-T Web site — http://www.itu.int/ITU-T

11 ETSI Web site — http://www.etsi.org/

12 TIPHON Web site — http://www.etsi.org/tiphon/

13 3GPP Web sites — http://www.etsi.org/3gpp/home.htm and http://www.3gpp.org/

14 Harrison, F. and Holley, K.: '*The development of mobile is critically dependent on standards*', in Clapton, A. J. (Ed): '*Future mobile networks: 3G and beyond*', Institution of Electrical Engineers, London, pp 35-45 (2001).

15 IANA Web site — http://www.iana.org/

16 ICANN Web site —http://www.icann.org/

17 Clapton, A. J. (Ed): '*Future mobile networks: 3G and beyond*', Institution of Electrical Engineers, London (2001).

18 3G.IP Web site — http://www.3gip.org/

19 MWIF Web site — http://www.mwif.org/

20 WAP Forum Web site — http://www.wapforum.org/

21 CableLabs Web site — http://www.packetcable.com/

22 IMTC Web site — http://www.imtc.org/

23 MSF Web site — http://msforum.org/

24 Ward, R.: '*The Multiservice Switching Forum — an Architectural Framework for the 21st Century*', Journal of the Institution of British Telecommunications Engineers, **1**(4) (Oct-Dec 2000).

25 Jeff Pulver Web site — http://www.pulver.com/

26 '*Voice on the Net*' Web site — http://www.von.com/

27 Swale, R. P.: '*Temporary Document 18TD028r1*', ETSI TIPHON meeting 18, Ottawa, (May 2000).

6

SS7 OVER IP — SIGNALLING TRANSPORT PROTOCOLS

K D King and G Rufa

6.1 Introduction

Until relatively recently the Internet and PSTN networks had been developing independently of each other. While each has evolved to provide globally accessible services, the design principles of each have been very different. The initial PSTN services have been focused on conversational, real-time, applications — principally telephony. The integrated services digital network (ISDN) has extended the PSTN model to support additional applications such as telefax and data transfer. Similarly, intelligent network (IN) extensions [1] have been added to support call-handling services. In contrast, IP networks, and the Internet in particular, have principally been developed for exchanging data between computers. In the case of the public Internet, the information accessible is both location-independent and enormous. With new, more complex, features continuing to be developed, mass-market services such as video-on-demand and on-line shopping become increasingly possible. VoIP represents one path for the possible convergence of these networks, which is starting for a number of additional reasons, including the following.

- Access

  Most people access the Internet via dial-up modem connections provided by the PSTN. It would therefore intuitively appear desirable for one network to support both IP and PSTN services. This network transition needs to be carefully managed as the Internet traffic increases due to the inherent effects of call holding times taking the PSTN outside its design envelope.

- Interworking

  The services offered by both networks are increasingly interacting and interworking with each other. Examples of this include:

  — the ability to use multi-functional mobile telephones to access Internet Web pages;

— enhanced connections between a telephone and a PC that enable advanced call handling and multimedia services;

— unified messaging applications involving e-mail message arrival notifications, text-to-speech, or short message service (SMS) messages that can be received on a mobile telephone;

— the use of VXML allowing the creation of voice network applications that retrieve and manipulate information accessed via the Internet from normal telephones.

- Cost

  Cost reduction options for existing PSTN switches and IN SCP platforms increasingly favour the use of common open computing platforms. These invariably provide, use or enable IP-based protocols for inter-machine communication.

- High-speed links

  To avoid flow control and congestion problems caused by increasing SCP INAP traffic and HLR MAP traffic, high-speed links between the SCPs/HLRs and the intermediate transfer nodes may be required. These could be, and often are, data links using IP-based protocols.

The services provided by conventional switched circuit networks are, however, effectively defined by the signalling messages that flow through the network. In contrast, IP-based data networks largely deliver services derived from the content of the user traffic that flows across the network. VoIP therefore potentially represents a convergence of voice and data networks for the provision of voice services. As the services provided by both existing switched-circuit and IP networks grow closer together, the two technologies increasingly converge and the reliability required by new IP services increases. This raises several questions. For example, can an IP network provide sufficient network signalling protection and reliability for a wide-scale converged public network? Will the IP network provide the reliability and transmission speed expected by SCN signalling protocols and associated services? This chapter outlines and explores the key issues in supporting signalling transport over IP, and, in particular, the problem of enabling Signalling System No 7 (SS7) transport over IP. It summarises issues documented in a more detailed SS7 over IP analysis [2]. The chapter aims to indicate how SS7 and IP network convergence could be realised. It therefore includes a description of the distributed IP telephony model, the framework architecture for signalling transport and the protocols that could be used. The chapter then describes specific SIGTRAN signalling transport protocol architectures and how IP-based high-speed links can be realised.

With any large-scale network engineering task, there is always the concern that it may be possible to transfer problems experienced in one network into another connected network. When considering the inter-connection of SCN and IP networks providing telephony services, the problems that may occur can rapidly compound due to the mix of technologies and subtleties in the implementation of the equipment concerned. In order to address this, and to provide required overall signalling performance, key network management functions need to be supported. This chapter therefore outlines how a seamless network management interworking function may be achieved. Even with these areas covered, problems may still arise, associated with the specific protocol and network architectures used. Some of the key known design issues are therefore highlighted in section 6.5.

6.2 Signalling in Public Telephony Networks

In modern digital telephone networks, inter-exchange signalling (network-to-network signalling) is implemented using the signalling system number 7 — known as SS7 or C.7 depending upon where you live. SS7 is defined by the ITU-T (see Chapter 5 for details of the work of the standards bodies). It is available in a number of variants depending upon the country, manufacturer and network operator in question. All of these have subtle differences that, on the whole, make operating and maintaining a public network of any reasonable size a perpetual engineering challenge.

The SS7 protocol stack comprises a signalling transport layer, which support application-specific layers providing telephone and related services.

Within the SS7 protocol stack, shown in Fig 6.1, the reliable signalling transport mechanism is provided by the SS7 message transfer part (MTP) functionality. This layer provides a generic signalling transport layer [3] and consists of three layers — or levels — known as MTP1, MTP2, and MTP3. The functionality provided by the MTP levels support the various SS7 connection-oriented and connection-less protocols providing the actual communications services offered by the network. These higher layer protocols comprise the various user parts, providing a full stack telephony application, and the signalling connection control part (SCCP) based protocols which provide non-circuit-related applications such as mobility and intelligent network functions [3].

6.2.1 Signalling Transport Protocols

As defined in the ITU Recommendations [4], the user part protocols '... define the functions and procedures of the signalling system that are particular to a certain type of user of the system'. In order to maintain the functionality expected by the SS7 user part protocols, the MTP3 functionality therefore needs to be either maintained or replicated. Alternatively an appropriate interface needs to be specified between the user part protocols and the underlying IP network transport protocols.

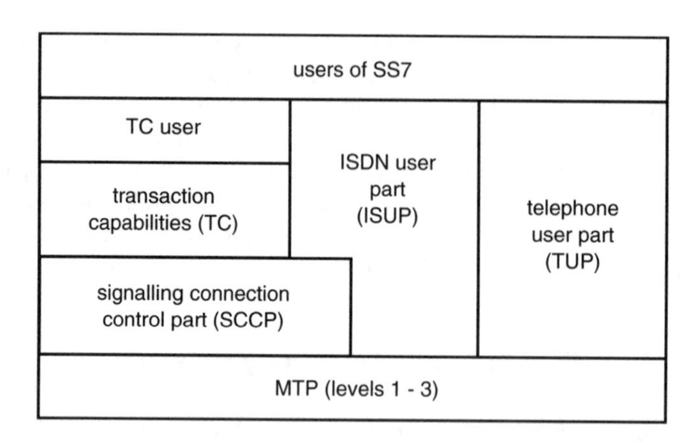

Fig 6.1 Architecture of SS No 7.

From the point of view of supporting SS7 over IP, the MTP2 and MTP3 layers are therefore of key interest. The MTP2 '... defines the functions and procedures for, and relating to, the transfer of signalling messages over one individual signalling data link. These functions include error detection and correction and signalling link failure detection'. Conversely, the MTP3 '... defines those transport functions and procedures that are common to, and independent of, the operation of individual signalling links ... including signalling message handling functions (directing message to proper signalling link or user part)' and 'signalling network management functions (control message routing, configuration of signalling network facilities and control reconfigurations and other actions to preserve or restore the normal message transfer capability) ...'.

As highlighted in the description of the MTP3 functionality above, it is apparent that the MTP3 provides much more than a primitive interface between the upper protocols and the lower layer transport. For any IP interface development, it is therefore necessary to ensure that the required elements of the MTP3 functionality are also provided.

The 'simplest' method to support the transport of SS7 user parts would be to provide the MTP functionality they expect. This is attractive since many hours of engineering effort have been invested in designing the functionality provided by MTP and the user part protocols that use it. However, a complete replication of MTP3, for IP, would give no benefit — why replicate when it may be possible to use MTP3 itself? Also, revisiting and revising the MTP3 protocol functionality for an IP environment would give the opportunity to review the protocol's behaviours and to 'fine-tune' them for a new network type. Unlike MTP3, much of the MTP2 functionality should be provided by the chosen IP transport protocol. This therefore does not need to be replicated.

The key focus of development for this work is being undertaken in the Internet Engineering Task Force (IETF) Signalling Transport Working Group (SigTran). Specifications coming from this work offer methods for supporting the required signalling transport functionality and are discussed later in this chapter.

6.2.2 Public Network IP Telephony Model

Signalling System No. 7 has been designed to meet the signalling and control infrastructure requirements of public telephone networks based on modern digital switching systems. This signalling system provides both bearer and call control signalling capable of scaling to control thousands of connections and enables access to many telephony, ISDN, mobility and IN services [3]. The core transport functionality required by the SS7 protocols includes:

- in sequence delivery of messages;
- no loss, corruption or duplication of messages;
- no excessive time delay in transmission.

If problems are detected in the network, it is essential that network management actions be taken promptly to ensure the required quality of network signalling is maintained. As the PSTN network and IP networks converge, a signalling system will be required to provide both call and bearer control on the 'new' network. The 'mature' IP protocols, such as the transmission control protocol (TCP), cannot fulfil these stringent requirements. Therefore, until recently, there has been no adequate IP signalling transport system to enable this, due to the high quality required of signalling transport by the SS7 protocols (e.g. ISUP, MAP). However, the signalling requirements for such a system have now been identified in the IETF [5] and these form part of a distributed public network IP telephony architecture. In particular the IETF SIGTRAN Working Group is currently defining protocols and methods that enable SS7 protocols and services to be supported across an IP network, and between IP network elements and PSTN signalling points.

The distributed IP telephony model aims to provide a framework enabling the convergence of PSTN and IP networks in an architecture that is capable of supporting both voice and data. This framework comprises the following network components (see Fig 6.2):

- the media gateway (MG) — providing a media interface between a PSTN and IP network. It performs the packetisation of speech signals received from the PSTN, enabling it to flow over an IP network (and vice-versa);
- the media gateway controller (MGC) — controlling the media gateways and providing call control (see Chapter 1);

- the signalling gateway (SG) — transferring signalling information between a PSTN node and a media gateway controller, or an IP signalling point (IPSP); the signalling gateway may contain MTP3 and SCCP functionality (although this may not always be the case).

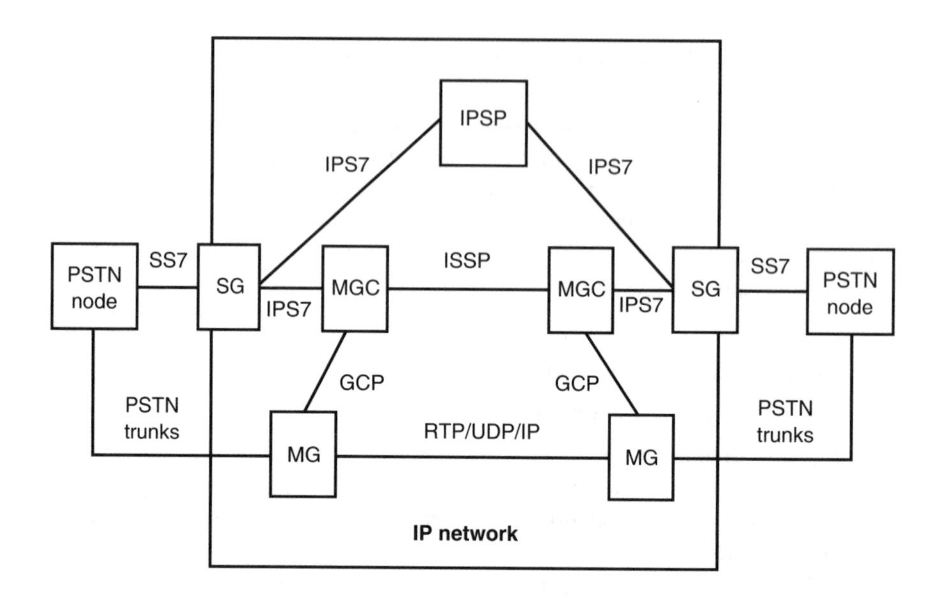

Fig 6.2 The public network IP telephony model.

In the version of this architecture described here, an 'Internet Protocol Signalling System No. 7' (IPS7) is used to transparently transfer SCN signalling messages between the signalling gateway and the media gateway controller. The SCN signalling messages are 'packaged' in datagrams and carried across an IP network. The inter-softswitch signalling protocol (ISSP) carries signalling information between media gateway controllers. Call control procedures are performed by the media gateway controller and the connection control required, e.g. to create, delete or modify a connection, is communicated to the media gateway using an appropriate gateway control protocol (GCP) (see Chapter 7 for more detail). In this model, SS7 over IP enables switched-circuit network (SCN) signalling messages to be transported between signalling gateways (SG), media gateway controllers (MGC) and IP signalling points (IPSP). Figure 6.2 also shows the separate functions of the SG, MG and MGC. These functions may be implemented in separate physical units or may be combined within a physical unit, e.g. SG/MG functions or MG/MGC functions implemented in one unit.

6.2.3 Drivers for the Public IP Telephony Network

One key driver for network operators to consider when adopting a distributed IP telephony network is that the existing SCN MTP level 2 protocol is largely based on assumptions of operating on a now antiquated processor technology. Furthermore MTP2 is optimised for the 64 kbit/s transfer rate of standard TDM circuits, thus limiting it to comparatively low bandwidth links when compared with modern data networks. An updated and cost-effective transport protocol for switched-circuit network messages is therefore required which reflects the ability to leverage general-purpose data networking technology as far as possible. Due to a relatively open development process and the comparable ease with which applications can be constructed, the Internet protocol (IP) has emerged as the *de facto* data network technology. It is therefore the candidate for use in the evolution of SS7 networks. As with many data networking technologies, in networks based upon IP, all communications are viewed as 'packets'. It therefore offers the potential of a fully converged network infrastructure. Due to the perceived cost benefits of such packet networks, some network operators may wish to migrate completely to this type of network.

Other reasons also exist for requiring an IP network to 'seamlessly' carry existing SS7 user part protocols. These include enabling the interconnection of new IP networks to existing PSTN networks, thus increasing network customer reach. This would also allow maintenance of existing PSTN services between PSTN customers, while transporting the information across an IP network. In such circumstances, it would also be possible to maintain existing PSTN services while migrating a PSTN network to an IP network infrastructure.

6.3 IP-Based Signalling Transport

The functionality of signalling transport protocols is key to providing both signalling transport reliability, and protecting the network from potential failure caused by occurrences such as network overload and congestion. Most IP signalling transport protocols provide a basic 'best-effort' service, with little in the way of network protection, and with little consideration of the requirements of any SCN protocols that may need to be carried. The IETF has been developing a framework to enable the convergence of the existing public switched telephone networks (PSTNs) and the Internet. It is also developing a signalling transport protocol that provides, and enables the preservation of, many of the facilities required both for signalling network protection and reliable signalling transport. Work is in progress on adaptation layers that enable SS7 protocols to be carried over the transport protocol with no change (at least in theory!).

The applications that could be supported through the use of SS7 over IP include:

- IP-based high-speed links between switching and control platforms (e.g. MGCs);
- migration of PSTN networks towards an IP-based infrastructure;
- convergence of signalling and data networks.

6.3.1 A Signalling Perspective of IP Protocols

Originally, the main focus for IP transport protocols was transmission of packets in either a connection-less or connection-oriented manner. These requirements gave rise to the user datagram protocol (UDP) and transmission control protocol (TCP) respectively. Both of which were developed with data applications in mind.

UDP provides an unreliable IP transport mechanism. This has the advantages of speed, as little processing delay is incurred through use of retransmission functions, etc. This makes it ideal for non-critical transport, such as certain aspects of voice transport. However, a lack of retransmission mechanisms does give obvious limitations with regard to the transport of critical packets, such as those containing signalling information that require some form of assured delivery. This is because it is essential that such information reliably reaches its target within a reasonable time. Failure to achieve this could result in the loss of signalling integrity, possible lost calls or ultimately network element failure.

In contrast to UDP, TCP provides a reliable connection oriented transport mechanism. It therefore contains many of the features required for the successful transport of signalling information. Mechanisms such as retransmission, and error checking with retransmission are included. Although it goes some of the way to providing the features required for carrying SS7 over IP, limitations still remain since it has not been specifically developed to carry the SS7 user parts.

The stream control transmission protocol (SCTP) [6] has grown from investigations into the limitations of TCP and research into distributed computing applications. Among other additional facilities, it can preserve message boundaries of message-based signalling protocols and provides selective retransmission capabilities. Adaptation layers, enabling SCN protocols to run over the top of SCTP, have been and continue to be developed for the SS7 user part, MTP2 and MTP3 layers. SCTP has been adopted by both the IETF SIGTRAN Working Group for use in SS7 over IP transport and the IETF Transport area for further study and development as a generic protocol on a par with TCP and UDP. Work continues in these areas and further specifications include adaptation for other higher layer protocols.

6.3.2 Stream Control Transmission Protocol (SCTP)

Among other things, the SCTP has been developed to provide the basic generic functionality required to reliably and efficiently transport SS7 user parts across an IP network. SCTP provides three main areas of functionality:

- sequencing — providing in-sequence delivery of messages on a stream-by-stream basis;
- reliability of associations, through message arrival and integrity — this is independent of sequential delivery;
- restructuring — fragmentation of large messages into a maximum datagram size, and bundling of small messages.

Key functionality within these areas include:

- error checking, through checksums;
- selective acknowledgement — highlighting only those datagrams that require retransmission;
- retransmission of missing or corrupted datagrams;
- cookie mechanisms (thus ensuring identity of both users prior to session set-up);
- preservation of message boundaries;
- support of multi-homing (more than one transport address that can be used as a destination address to reach that end-point);
- a method to escape head-of-line blocking (avoiding the destination delaying messages while awaiting receipt of a previous one missing in the sequence);
- heartbeat/keep alive mechanism — to check the 'reachability' of a particular destination transport address in the association;
- optional unordered delivery — no reordering of received datagrams prior to onward routing to end-point.

At initialisation of an SCTP session, an association is created. This is a logical channel between two SCTP supporting end-points and can be thought of as being similar in functionality to an SS7 link. The association can contain a number of 'streams' that, for example, could be chosen to represent an individual call. In itself, however, SCTP cannot directly transport SS7 signalling messages. In order to provide interfaces to, and the functionality expected by, the SS7 user parts, several adaptation layers have been specified. The functionality of these is described in section 6.4.

6.3.3 IP-Based Signalling Protocol Models

There are two important models that offer the potential of transporting signalling protocols over IP networks. These are the ITU-T SSCOPMCE model, which has largely been derived from an ATM background, and the IETF SIGTRAN model that is predicated on SCTP.

6.3.3.1 The ITU-T SG 11 SSCOPMCE Model

The service-specific connection-oriented protocol (SSCOP) [7] had been defined for use with the service-specific convergence sub-layer of the signalling ATM adaptation layer (SAAL), providing reliable delivery of protocol data units (PDUs). Extensions, such as the 'service-specific connection-oriented protocol in a multi-link and connectionless environment' (SSCOPMCE) extension, have been defined to enable its use in other network types. The SSCOPMCE target functionality is similar to SCTP and could provide reliable signalling transport between users supporting it.

The SSCOPMCE 'connectionless environment' mode of operation enables one or more links to the connectionless environment, e.g. IP or UDP, to be deployed. Links may then be added or removed during the life of the SCCOPMCE object.

The SSCOPMCE protocol data units (PDUs) containing the user information are passed to other users via the IP transport layers. These PDUs are mapped into the datagrams required by IP, with the additional IP header information added. On reaching the target user, the datagrams are mapped back into SCCOPMCE PDUs. The MTP3 adaptation required may be performed by the service-specific co-ordination function (SSCF), as with SAAL (see Fig 6.3).

6.3.3.2 The SigTran Model

The IETF SigTran Working Group has defined a framework architecture enabling SCN signalling message transport over IP using SCTP.

This is achieved through the use of the following components (see Fig 6.4):

- IP network layer;
- common signalling transport protocol;
- switched circuit network (SCN) adaptation modules.

The stream control transmission protocol, defined within the IETF SigTran and Transport area Working Groups, provides the common signalling transport protocol functionality. As previously described, SCTP provides a datagram transfer protocol that provides error checking, efficient retransmission, sequence and flow control suitable for enabling the reliable transport of signalling messages over an IP network. The SCTP is primarily supported directly over IP (although it can also be supported over UDP).

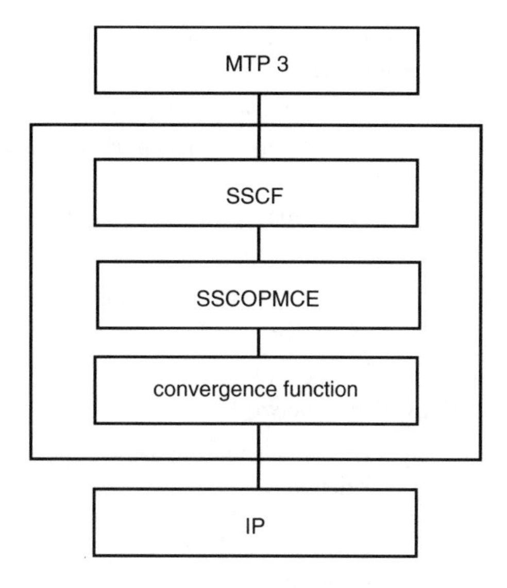

Fig 6.3 SSCOPMCE — SS7 over IP protocol stack.

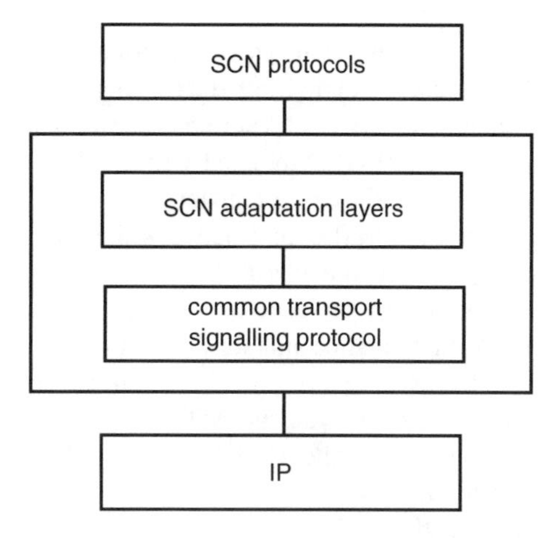

Fig 6.4 IPS7 signalling transport components.

All SCN protocols have similar basic transport requirements, with any protocol-specific requirements being met by the adaptation layers. The SCN adaptation layer specifications enable different upper layers to use the SCTP functionality without the need to alter any of their functionality, behaviour or services. This is achieved by each adaptation layer providing an 'interface' to the SCTP, such that the SS7 protocol used is unaware of any change to its transport layer. The specific user primitives are supported in such a way as to enable this. The following adaptation layers are currently being defined within SigTran:

- MTP2 user peer-to-peer adaptation layer (M2PA) [8] — enabling full support of MTP3 (message handling and network management capabilities) between two SS7 nodes over an IP network;
- MTP2 user adaptation layer (M2UA) [9] — enabling transport of MTP3 messages over an IP network;
- MTP3 user adaptation layer (M3UA) [10] — enabling the transport of MTP3 user protocols (e.g. ISUP, SCCP) over an IP network;
- SCCP user adaptation layer (SUA) [11] — enabling the transport of SCCP user protocols (e.g. TCAP) over an IP network;
- Q.921 user adaptation layer (IUA) — enabling the transport of ISDN Q.921 user protocols (Q.931) over the IP network.

6.4 Signalling Adaptation Layers

6.4.1 Signalling Transport using M2PA

It is generally understood that the chances of experiencing congestion and flow control problems will increase as signalling traffic increases. Common points in modern digital public telephone networks where these effects are increasingly found include the signalling gateways (SG) associated with modem banks servicing dial-up Internet traffic, IN services and mobile register look-ups. The use of high-speed links to the points receiving this traffic, and between the intermediate SS7 transfer nodes, could significantly reduce the possibility of these problems. These high-speed links are also important for the transfer of large amounts of data used, for example, for SS7 network maintenance and administration.

IP-based links could be used to provide these. However, in such cases it is necessary that these are able to provide the MTP2 equivalent functionality. This includes the interface between the MTP2 and MTP3 functionality and communication between the layer management modules. The IETF are developing an IP protocol specification that will allow full support of MTP3 messaging and network management between two IP signalling points. This is known as the SS7 MTP2 user peer-to-peer adaptation layer (M2PA), which runs over, and uses the facilities of SCTP (see Fig. 6.5).

In the example shown in Fig. 6.5, the signalling gateway (SG) supports both SS7 and IP links, forming an IPSP. The SCTP association provides an equivalent of an SS7 link between the IPSPs. In this case, the MTP3 may be carrying SCCP or any other SS7 user part. This is because the M2PA provides the interface interworking required between SCTP and MTP3.

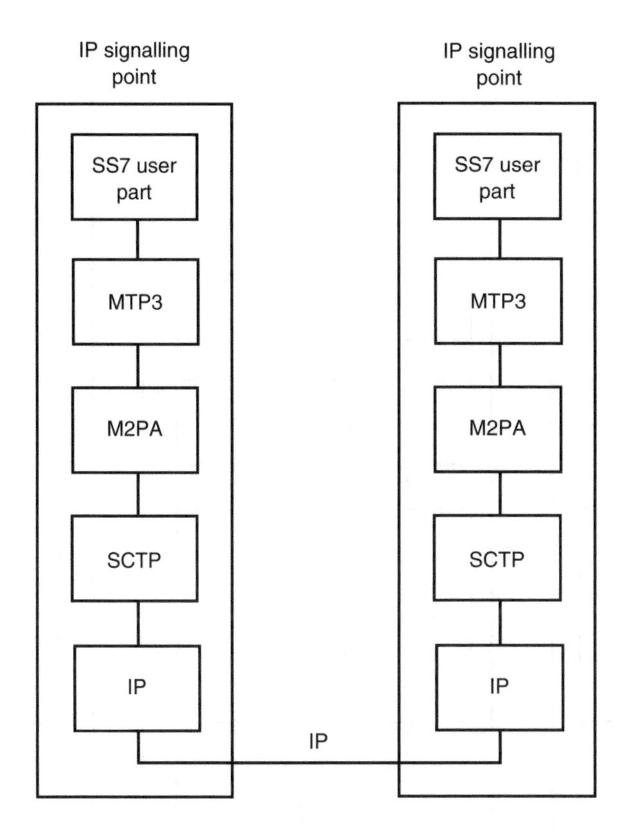

Fig 6.5 IP-based SS7 links.

When M2PA is used, the MTP3 and the SG can act as an STP (see Fig. 6.6) and perform the interworking between the IP domain and the SS7 network. SCCP can be in the SG, to provide required SCCP functionality, such as global title translation (GTT).

In either case, as MTP3 is supported, the IPSP must have an SS7 signalling point code, as outlined in the MTP specifications.

6.4.2 Signalling Transport using M2UA

To enable the transport of SS7 signalling messages between an IP SG and MGC, the IETF SigTran Working Group has developed another specification. This is called the SS7 MTP2 user adaptation layer (M2UA). When M2UA is used, the SG does not contain the MTP3 and SCCP functionality as shown in Fig 6.7.

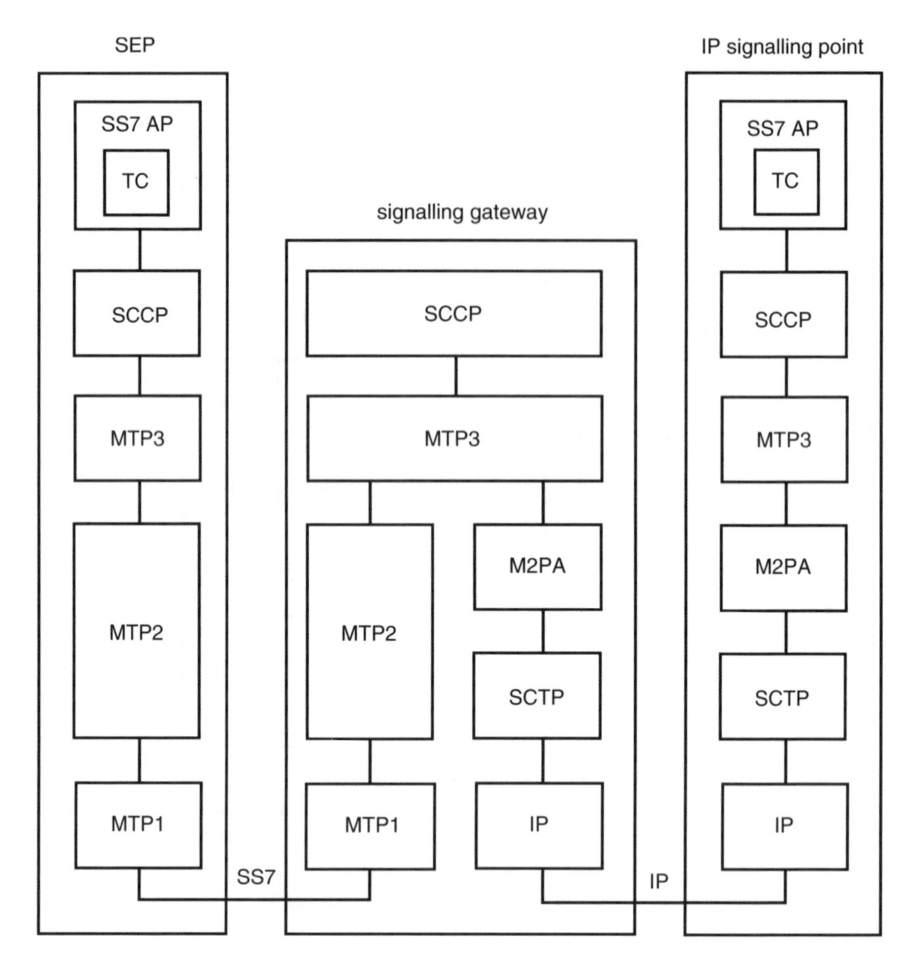

Fig 6.6 Signalling message transport between SG and MGC using MTP3.

The SG sends and receives ISUP/SCCP messages to the PSTN over a standard SS7 network interface, using MTP2. The SS7 signalling link from the SS7 endpoint, or intermediate node, terminates on the SG.

As with M2PA, M2UA runs over SCTP. Within an SCTP association, an agreed number of streams can be opened. These streams can be used to uniquely identify and 'separate' messages relating to individual calls or SS7 links terminated at the SG.

Since no direct SS7 MTP3 functionality is present between the two elements, the SG has a nodal interworking function (NIF) that provides the interworking between the SS7 link and the SCTP stream.

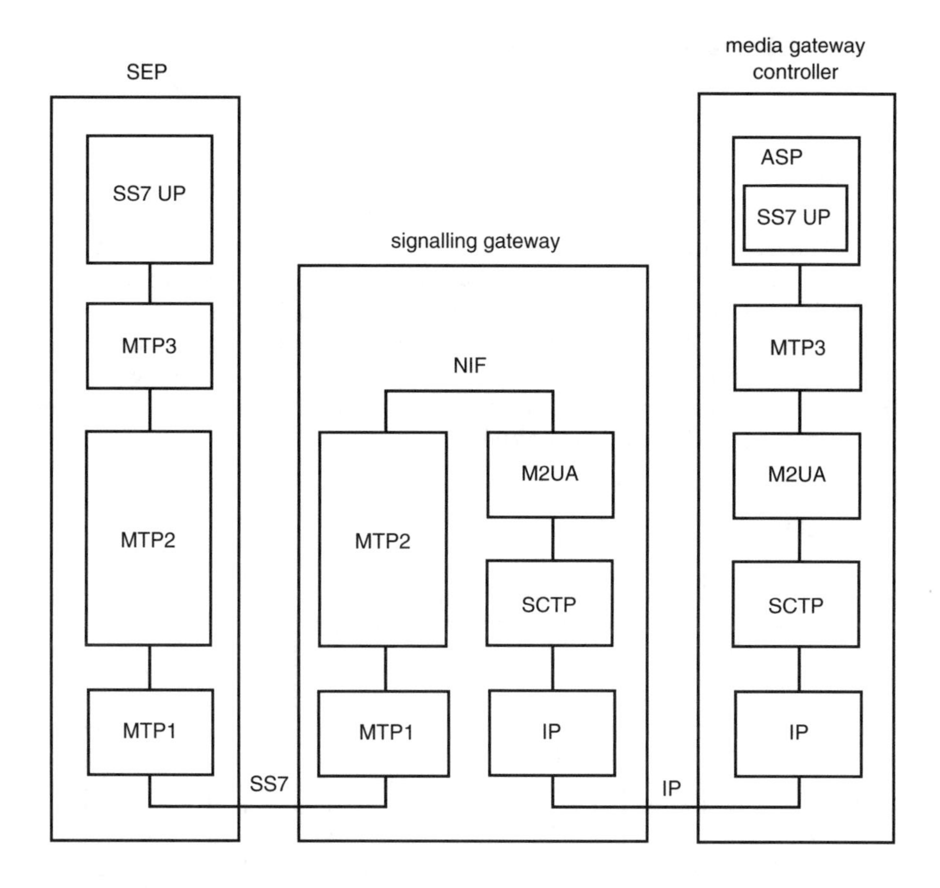

Fig 6.7 Signalling message transport between SG and MGC without MTP3.

This maps MTP3 messages to the SCTP stream and transfers the signalling messages to and from the application server process (ASP). This function is in the MGC and is where the ISUP/SCCP peer protocols layer is located.

M2UA enables access to the SG MTP2 layer services for a remote IP-based application, although it does not provide full MTP2 functions and services. In this scenario, the MGC MTP3 will be unaware that the MTP2 functionality is not being provided locally. In turn, the SG is unaware that the MTP3 is remotely located at the MGC.

As with the M2PA case, since the MGC contains MTP3, an SS7 signalling point code is required as outlined in the MTP specifications.

6.4.3 Signalling Transport using M3UA

6.4.3.1 Transport of ISUP Messages

Other IETF methods suggested for SS7 over IP involve the support of the SS7 user parts directly over the SCTP adaptation layers. Figure 6.8 demonstrates the transport of ISUP messages from an SS7 signalling end-point (SEP) to an ASP located in an MGC. This transfer is via the SG, which provides inter-working between the MTP3 and the SS7 MTP3 user adaptation layer (M3UA) functionality.

In this case, the SG expects the ISUP messages to be received from, and sent to, the PSTN over a standard SS7 interface. The MTP3 is terminated at the SG, and therefore the SG needs to be identified by an SS7 signalling point code. This is not the only option, e.g. IP-based links could be used.

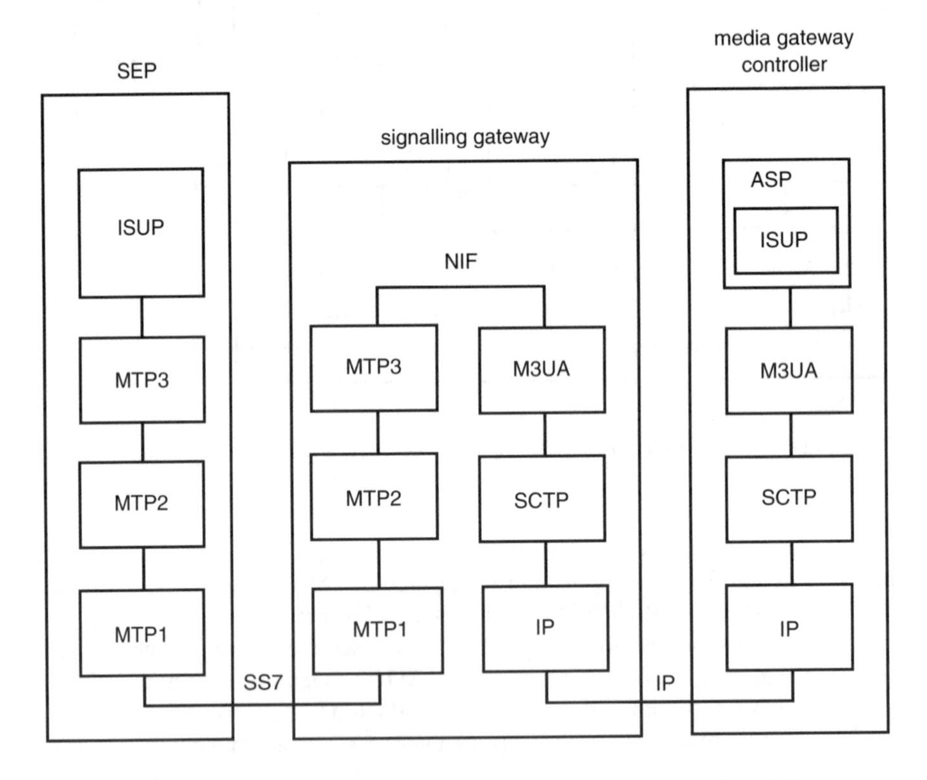

Fig 6.8 ISUP message transport via the SG using MTP3 and M3UA.

In order to enable the SG to transport the received signalling messages between MTP3 and the M3UA, a NIF is provided. As with the other adaptation layers, M3UA uses the functionality provided by SCTP. Just as M2UA provides access to

MTP2 layer services, M3UA enables remote access to the MTP3 layer services located at the SG. However, it must be kept in mind that M3UA does not provide the full MTP3 functionality. It does, however, provide all the primitives required by the user parts. Therefore the ISUP at the ASP should not be aware that the MTP3 functionality is not being provided locally. Similarly the MTP3 at the SG should be unaware that the user parts are not local to it.

In addition to the basic signalling support, MTP3 network management functionality support is also required. The user part located at the ASP therefore needs indications of the availability of the SS7 signalling points, whether the required user parts are available, and the condition of the SS7 network with regard to congestion. With this in mind, the NIF transfers received MTP3 indications, such as MTP-STATUS, MTP-PAUSE and MTP-RESUME, to the M3UA management function. This is, in turn, responsible for the transfer of this information to the MTP3 user lower layer interface at the ASP.

6.4.3.2 Transport of SCCP Messages

SCCP messages can also be transport across SCTP through the use of M3UA since this provides all the primitives required between MTP3 and SCCP (see Fig 6.9).

In the example shown, the SCCP may perform a global title translation at the signalling gateway (SG) on messages received from the SS7 network. The result of this gives an SS7 destination point code (DPC), and may also give an SCCP subsystem number (SSN). These will identify the SCCP peer located in the IP network. The MTP-TRANSFER primitive will then be sent to the SG M3UA, which will provide the required translation and mapping of the network address for onward routing to the required IP destination.

SCCP messages originated in the IP network can also be sent to the SG for a GTT. If this translation results in an address of an SCCP peer located in the SS7 network, then the MTP-TRANSFER request is passed by the NIF to the MTP3 for delivery to the SS7 destination. If not, and the address identifies an SCCP peer in the IP network, then the MTP-TRANSFER request is sent to the M3UA for onward routing.

In order to exchange SCCP user messages between two IPSPs using SCCP (see Fig 6.10), M3UA is used to provide the required primitives between the SCCP and MTP3.

The status of the SCTP association and IP network are indicated through the use of MTP-PAUSE, MTP-RESUME and MTP-STATUS messages passed to the SCCP from the M3UA. In this case, the SCTP association is acting as an MTP signalling relation between the SEPs, where the SCCP peers are located. Due to the direct connection between the IPSPs, when M3UA is used in this way, the procedures to support the MTP3 services are a subset of the MTP3 procedures usually available. Note also that the MTP3 services are not offered remotely.

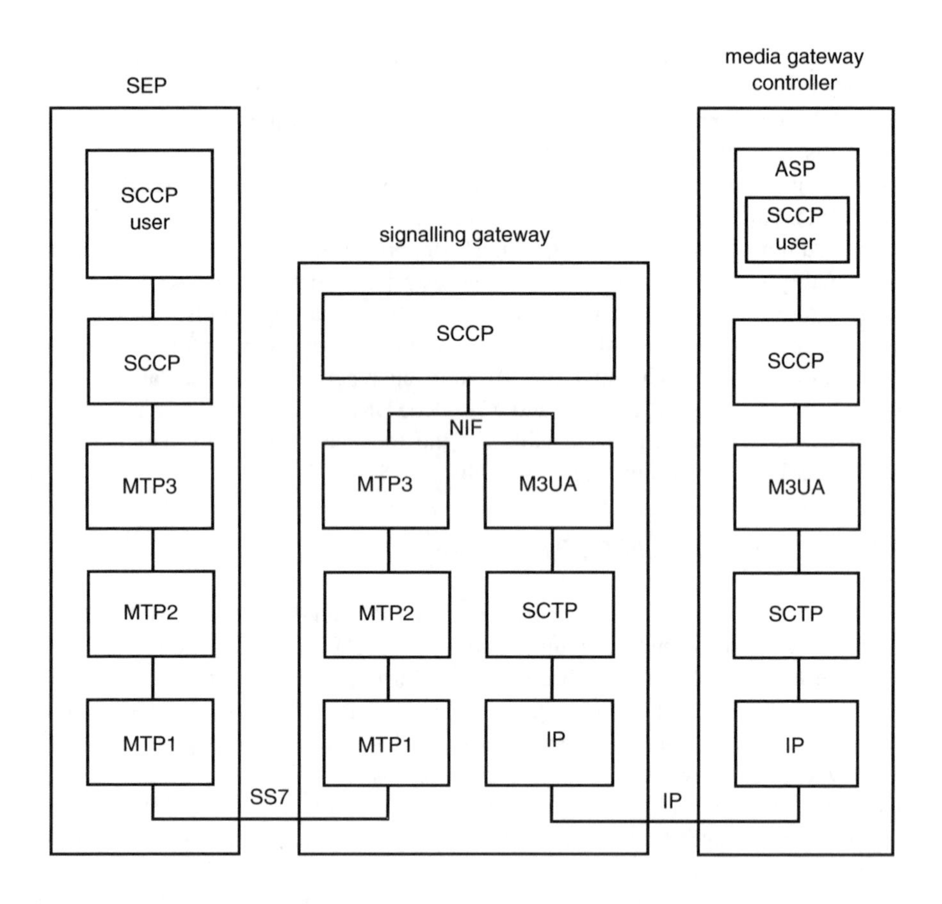

Fig 6.9 Transport of SCCP messages between SG and MGC.

6.4.4 Signalling Transport using SUA

To support the transport of SCCP user messages (e.g. MAP) over an IP network, the IETF has defined the SS7 SCCP user adaptation layer (SUA). This enables transport of these messages between the SG and an IPSP, or between two IPSPs, without the need to carry them over SCCP. Figure 6.11 shows the SCCP user messages being transported from an SS7 SEP, via an SG, to an IPSP.

In this case, the SG expects to receive and transmit SCCP user messages over a standard SS7 network interface. This interface will therefore include SCCP. The SG provides the NIF to ensure transport of the signalling messages between SCCP and the SUA.

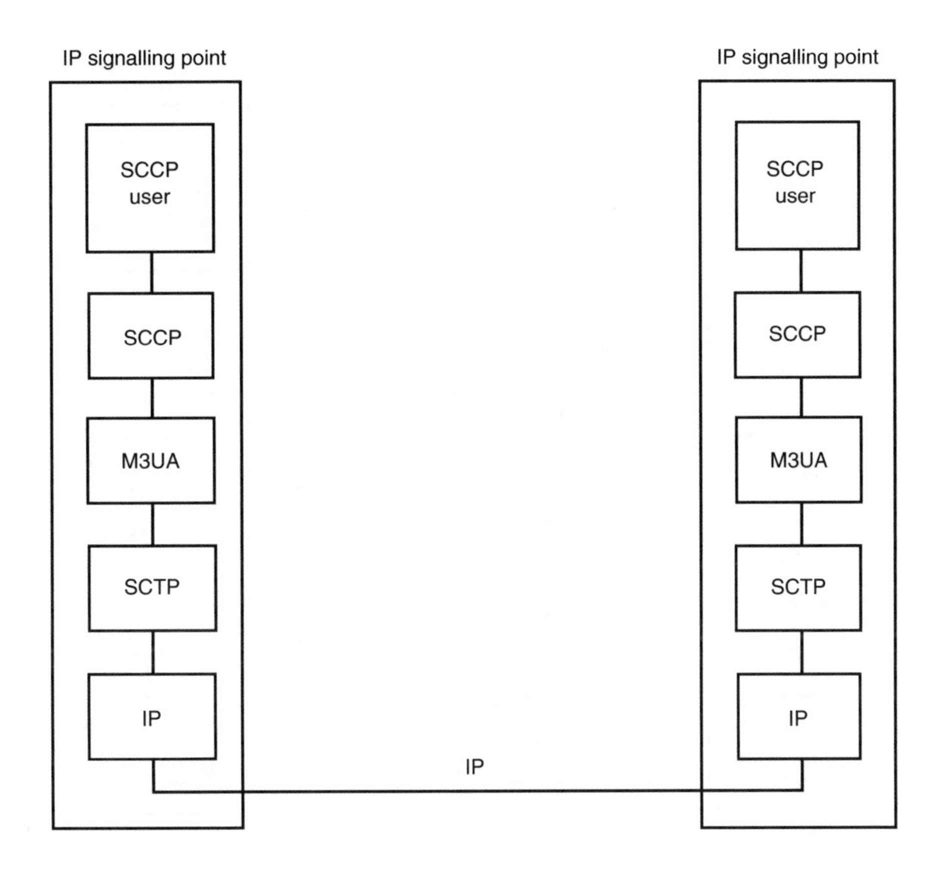

Fig 6.10 Transport of SCCP messages between IPSPs using SCCP.

It also provides interworking of the SCCP management functions between the SG and IPSP.

As with the other adaptation layers, the SUA uses the functionality provided by SCTP and the underlying IP network. It also performs the address mapping of the SS7 addresses to IP addresses. At the IPSP, it provides the required primitive set expected by the SCCP users, which would have been originally provided by the SCCP. The SUA enables remote IP-based applications to access the SCCP layer services at the SG, since M3UA enables access to MTP3 layer services. Therefore, the SCCP user is unaware that the required services are being remotely accessed from the SG.

An alternative scenario could be the direct transport of SCCP user messages, using the SUA, between two IPSPs, as shown in Fig 6.12. In this case, the SCTP association provides the SCCP signalling connection between the two end-points.

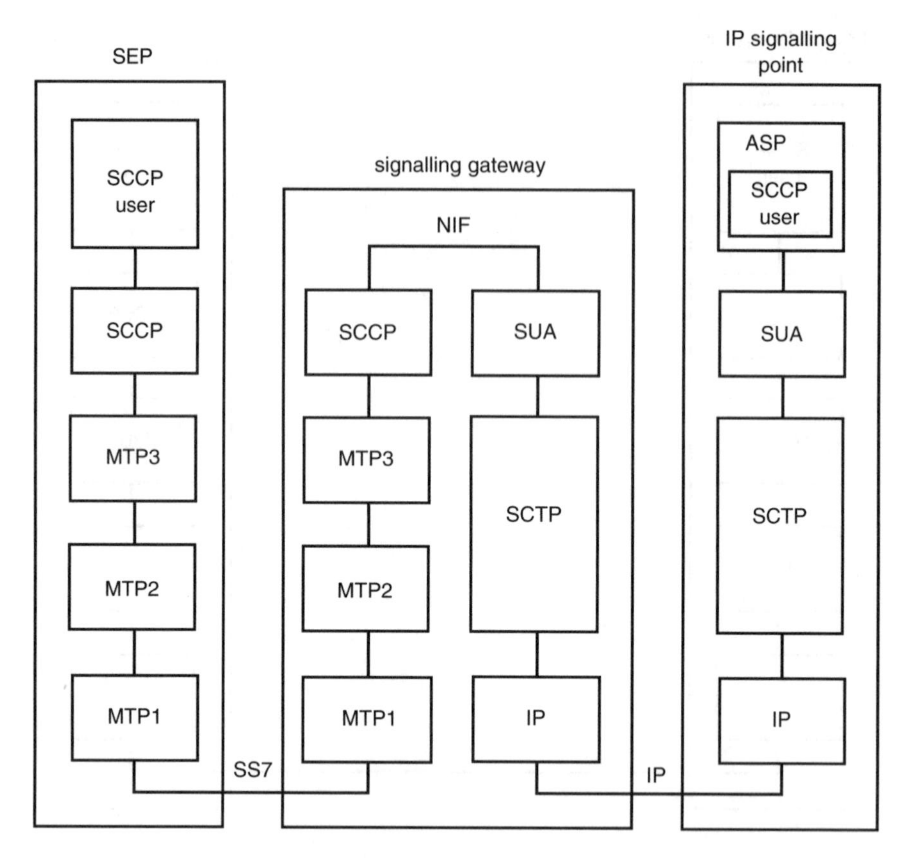

Fig 6.11 SCCP user message transport via the SG using the SCCP and SUA.

6.5 Core Design Issues

A lot of work has been done in order to solve problems with the MTP procedures in the past and these now work in a satisfactory way under a variety of adverse network conditions. However, as is well known, there are always potential problems in real SS7 networks with MTP procedures caused by awkward network structure and poor network planning which can lead to SS7 network outages [12]. Due to the nature of SS7, local problems can rapidly propagate throughout the network and create global problems within the nodes of the network at nearly the same time. Similarly in the case of network interconnection, problems from one network can be transferred to another network via gateway nodes unless great care is taken. It is not unreasonable to conclude that the same problems will be present if SS7 is used over IP in the way described in the previous sections without taking due note of the potential problems that can occur. The following points therefore reflect upon some of the key problems regarding network reliability.

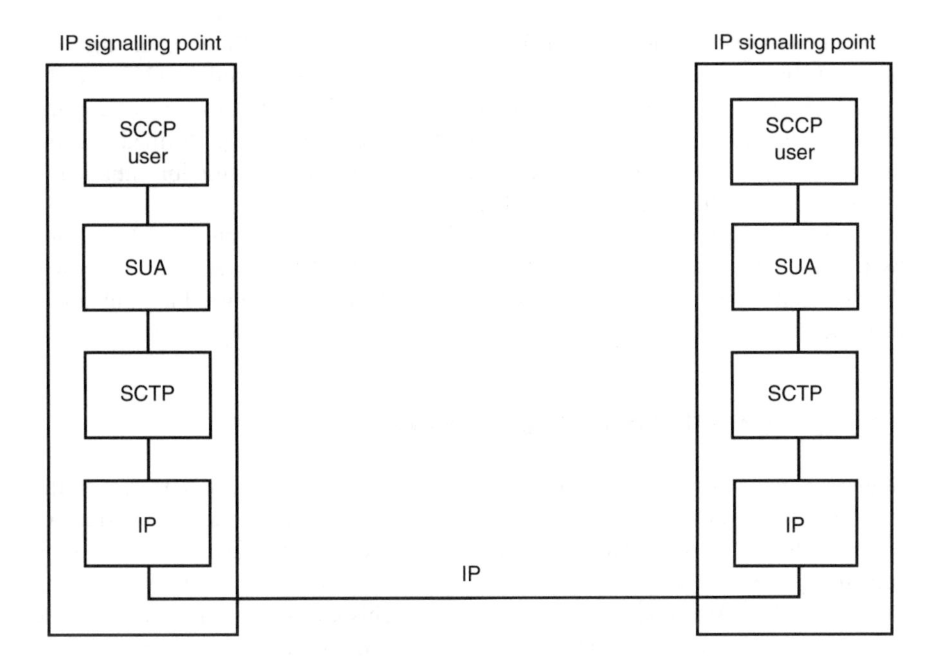

Fig 6.12 Transport of SCCP user messages between IPSPs using SUA.

6.5.1 Congestion Problems and Flow Control

Congestion problems are crucial with respect to network reliability because they can rapidly spread throughout the network. Congestion problems can be caused by failure situations and bottle-necks, in the case of high signalling traffic load, as well as mass problems with signalling route management messages. In order to avoid congestion situations, flow control measures are needed [13]. Since the ideal would be to use the SS7 user parts without change and since they rely on the SS7 flow control, the SS7 flow control must be realised in the SS7 over IP protocols. This, however, has not yet been completed. There is therefore the opportunity to avoid problems with signalling route management messages in real networks with the new protocols, since they offer the possibility to send network management messages for signalling point clusters.

6.5.2 Load Sharing

Load sharing plays an important role regarding network reliability. Load sharing algorithms aim to distribute traffic equally to the available transmission paths in order to avoid bottle-necks and related congestion situations. The SS7 load sharing is based on the assumption that traffic streams, as created by different user parts, are of about the same characteristic. Thus, the destination point code (DPC) and the signalling link selection (SLS) field were considered to be sufficient as routing information. Since user-specific load distribution need not be performed and the message length need not be taken into account, this assumption implied a considerable simplification of the load-sharing mechanism.

However, different traffic streams with different message lengths may be created by new applications and lead to asymmetric load distributions. As is well known, the flow control will not work in this case [13]. Thus, there is a need to think about load sharing in the new SS7 over IP environment.

6.5.3 Restart of Signalling Gateways

Signalling gateways are used to interconnect the SS7 and IP networks. However, the case of a failure of a signalling gateway has not yet been covered in a satisfactory way. For this, a 'co-ordinated restart' of the signalling gateway has to be realised by something like the MTP restart procedure in order to protect the restarting signalling gateway and the interconnected networks [14]. This ensures that problems in one network do not cause problems in the other network. There is therefore still some work to be done to complete the inter-working of network management functions

6.6 Summary

This chapter has described the requirements for reliable and efficient signalling transport for SS7 over IP. The key issue is the maintenance of functionality that will enable the SS7 user parts to continue to function correctly. This has required that the underlying IP transport provide all the facilities and functionality that would have previously been assumed of the SCN SS7 transport layer.

Ongoing work to answer this requirement is taking place in the IETF SIGTRAN Working Group. Specifications are being written that provide a basic reliable transport for IP-based SS7 networks and that meet the key generic transport requirements of the SS7 user parts. In addition to this, adaptation layers are being specified to provide the specific functionality and interfaces required by the different user parts.

At the present time, a number of outstanding key issues remain that will need to be addressed before SS7 over IP can be truly realised at scale. As has been discussed in the previous section, there is still need to complete the interworking of network management functions between the SS7 protocols and the adaptation layers and to correctly realise the SS7 flow control in the IP protocols. Finally, if a network operator wishes to migrate completely to an IP-based network, there would be a strong desire to have the MTP3 functionality available in the protocol stack. Since

M3UA does not itself provide the MTP3 functionality, this may lead to the specification of a new protocol, called for example something like the 'signalling network control protocol' (SNCP). In addition, this would offer the possibility to remove shortcomings with the old MTP3 protocol.

The ongoing work within the IETF and other standards bodies, the resolution of the identified issues, and other further developments are sure to bring us nearer to the goal of SS7 over IP. However, several of the SIGTRAN protocols will not, in themselves, meet the performance and reliability requirements of SS7. Therefore, it is necessary to specify a physical network architecture that will enable these stringent requirements to be met, which is discussed in some detail in Rufa [2].

References

1 Dufour, I. G.: '*Network Intelligence*', Chapman & Hall, London (1997).

2 Rufa, G.: '*SS7 over IP*', Baseline Document, Mannheim University (September 2000).

3 Manterfield, R. J.: '*Common-channel signalling*', Institution of Electrical Engineers, London (1991).

4 ITU-T Recommendation Q.700: '*Introduction to CCITT Signalling System No.7*', (March 1993).

5 IETF Internet Draft: '*Telephony signalling transport over SCTP — applicability statement*', (December 2000).

6 IETF RFC2960: '*Stream control transmission protocol*', (October 2000).

7 Draft ITU-T Recommendation Q.2111: '*Service specific connection oriented protocol in a multi-link and connectionless environment (SSCOPMCE)*', (1999).

8 IETF Internet Draft: '*SS7 MTP2-user peer-to-peer adaptation layer*', (March 2001).

9 IETF Internet Draft: '*SS7 MTP2-user adaptation layer*', (February 2001).

10 IETF Internet Draft: '*SS7 MTP3-user adaptation layer*', (February 2001).

11 IETF Internet Draft: '*SS7 SCCP-user adaptation layer*', (February 2001).

12 Rufa, G.: '*MTP Procedures, Network Structure and SS7 Network Outages*', Mannheim University (October 1999).

13 Rufa, G.: '*SS7 Flow Control*', Mannheim University (March 2000).

14 Rufa, G.: '*MTP Restart Procedure*', Mannheim University (September 2000).

7

VoIP GATEWAYS AND THE MEGACO ARCHITECTURE

B Rosen

7.1 Introduction and Background

7.1.1 Gateways Defined

Generally, a device that sits between two different types of network technology is called a 'gateway'. As identified in Chapter 1, a common gateway for telephony sits between the current TDM domain and the emerging packet domain, such as an IP network. For example, in H.323, a gateway is a device that connects conventional telephones to an H.323 network. In principle, gateways could connect directly either to an analogue telephone, to a PBX (Fig 7.1), or to a TDM telephony switch on one side, and to an ATM, IP or MPLS network on the other side.

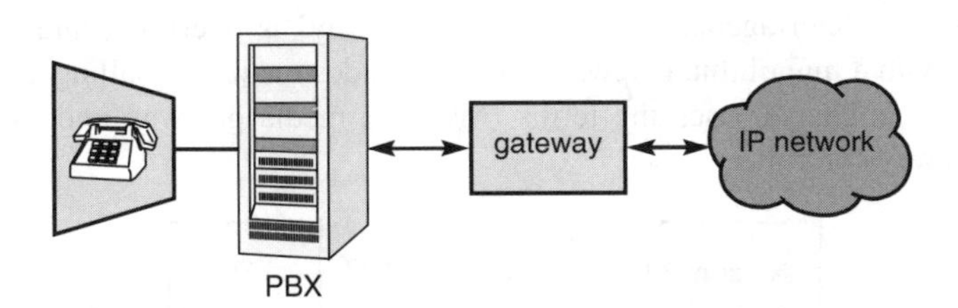

Fig 7.1 PBX-to-packet gateway.

Gateways can also connect two packet domains (Fig 7.2). For example, a device could be built that takes ATM AAL2 voice on one side and produces VoIP (RTP) on the other side. Even a device that provided transcoding between two compression algorithms, but otherwise maintained the same transport, could be considered to be a gateway.

In the context of the media gateway control protocols discussed in this chapter, gateways may also include devices that provide services to audiovisual conferences, such as MCUs and IVR boxes.

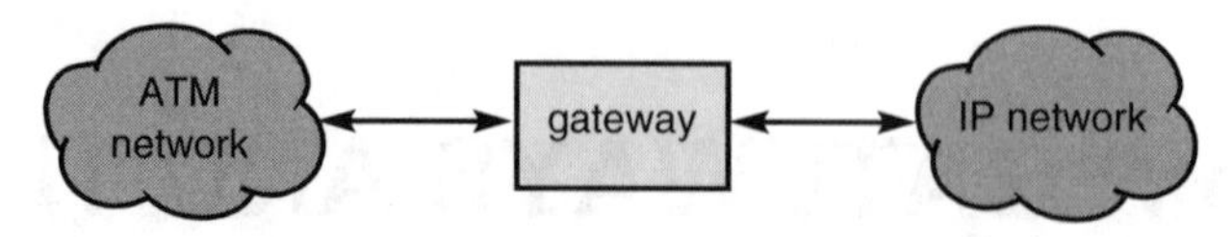

Fig 7.2 Packet-to-packet gateway.

7.1.2 Decomposed Gateways

As discussed in Chapter 1, in a conventional gateway, the same device that provides the media conversion from TDM to packet also provides the signalling interworking (Fig 7.3).

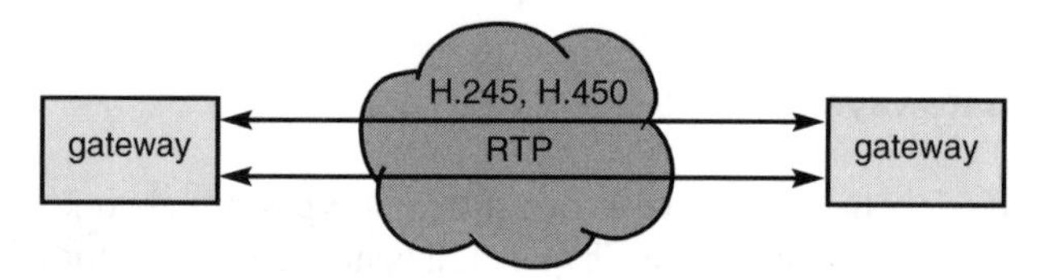

Fig 7.3 Conventional gateway.

A new class of gateway has been proposed that splits the media-handling function from the signalling function. These 'decomposed' gateways assign the media functions to a 'media gateway' and the control functions to a 'media gateway controller' (Fig 7.4). In earlier protocols, the two components were referred to as 'gateways' and 'call agents', but the term 'gateway' is overused, and there was confusion with a monolithic gateway, which includes media, signalling and control functions. This chapter uses the terms 'MG' for media gateway and 'MGC' for media gateway controller.

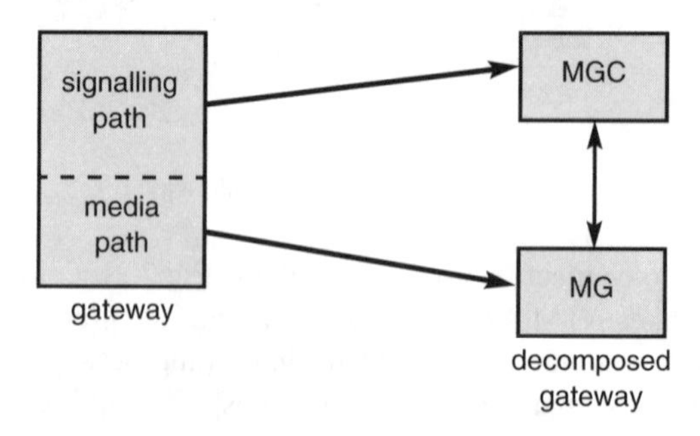

Fig 7.4 Decomposing gateways.

As the names imply, there is an inherent master/slave relationship between the MG and an associated MGC. Further, when engaged in a conference between two similar sets of devices (Fig 7.5), there is a peer relationship between the MGs (the bearer path) and another peer relationship between the MGCs (the signalling path).

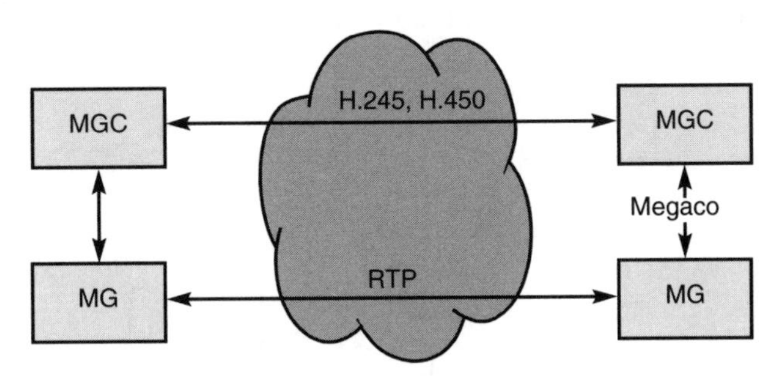

Fig 7.5 Decomposed gateway.

In principle, the protocols between the MGs could be any of the packet formats for real-time audiovisual media, most commonly RTP for IP networks, AAL1/AAL2/AAL5 for ATM-connected networks, and FRF.11 for frame relay connected networks. Similarly, the signalling protocols between the MGCs might be H.323 (see Chapter 10), SIP (see Chapter 11) or BICC (see Chapter 8) among other possibilities. However, the protocol between the MG and the MGC (as discussed in Chapter 6) is the subject of this chapter, and today would most typically be the media gateway control protocol (MGCP) or Megaco (H.248).

The motivations to decompose gateways are varied. In large 'trunk' gateways that have a very large number of TDM ports in and out, the processing requirements of the signalling often require too much physical space and power to efficiently collocate it. Further, the processing is code-intensive, rather than I/O intensive. It is therefore much more efficient to move the code to a low-cost 'COTS' computer and let the special-purpose hardware on the gateway handle the media.

In residential gateways, such as those employed in voice-over-cable, or VoDSL, there is a valid concern on the part of network operators that extending network system signalling out to a device inside the home, not controlled by the operator, is not conducive to maintaining a highly reliable system. In these cases, the device in the home is an MG, and the MGC is a computer within the operator's network.

In access gateways, regulators may require that competitive carriers receive access to ports on the gateway. In these cases, multiple MGCs may control one MG. The most recent GCPs permit such sharing. Also, the MGC may be specialised by features and services provided. Different customers may have different service sets, which may require that their ports be controlled by specialised MGCs, while other ports in the same physical device are controlled by other MGCs. Trying to run multiple signalling entities within a single device is more difficult than excising the signalling into multiple MGCs controlling a single MG.

7.1.3 History of GCPs

There have been a number of gateway control protocols proposed and implemented to provide the MG-MGC interface. The primary ones include SGCP, IPDC, MGCP, MDCP and most recently Megaco, as shown in Fig 7.6.

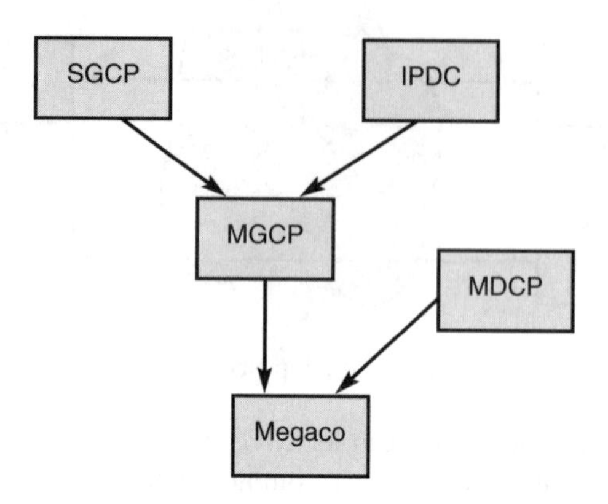

Fig 7.6 Gateway control protocol history.

7.1.3.1 IPDC

In 1998 the then new competitive carrier Level(3) assembled a group of vendors and other interested parties into a 'technical advisory committee' that designed the 'Internet protocol device control' (IPDC) protocol. IPDC has been implemented and is still in service within the Level(3) and other networks at the time of writing.

7.1.3.2 SGCP

Independently of the Level(3) TAC effort, researchers at Telcordia (formerly Bellcore) designed a similar protocol they dubbed 'simple gateway control protocol' (SGCP) which had similar goals. SGCP was implemented in research equipment, but had limited commercial success. However, the Telcordia effort was a significant development by a highly respected team.

7.1.3.3 MGCP

The principals behind IPDC and SGCP came together and devised a new protocol combining the best concepts of the two constituents. All parties viewed the media gateway control protocol as the successor to both the original gateway control protocols. The MGCP effort was still essentially a private effort of a closed group of companies and researchers. The MGCP proponents took their protocol to the

Internet Engineering Task Force (IETF) intending to standardise MGCP. Eventually, the IETF agreed to publish a version of MGCP as an 'informational RFC', which is not a standard, but publicly documents a *de facto* standard (see Chapter 5). There are a number of MGCP implementations completed, and one major organisation, CableLabs [1], has adopted MGCP as its protocol for the packet over cable effort that is part of PacketCable (see alo Chapter 5, para 5.8.2).

7.1.3.4 Megaco

Independently of MGCP, work was started in the International Telecommunication Union (ITU-T) to create a similar protocol. One of the contributors to ITU-T and IETF, Lucent Technologies, brought yet another protocol to the table, the media device control protocol (MDCP).

In a historic agreement, the ITU [2] and the IETF [3] agreed to work together on a unified effort to define a single, worldwide protocol for media device control. Using MGCP and MDCP as the starting point, the MeGaCo Working Group of the IETF and Study Group 16 of the ITU-T worked to define a new protocol. It was not politically possible to get all parties to agree to base the effort on MGCP. While many would have preferred that the result be very closely aligned with MGCP, and in fact be accepted as simply the next version of MGCP, there were many more parties at the table than at the original MGCP effort, and the result was a new protocol which is published both as a standards-track RFC (RFC 3015), and as an ITU-T Recommendation (H.248).

The texts of RFC 3015, and H.248 are identical except for the organisational boilerplate wording. The section headers and numbering are the same, and thus references to sections are valid for either document. There is no difference at all in the defined protocol, nor in the original annexes (Annex A through Annex G, plus Appendix A). Megaco is very similar to, and shares many concepts with, MGCP, but the basic model, commands and constructs are different.

7.2 Megaco Protocol

7.2.1 Protocol Elements

Details about each of the five protocol elements are given below.

- Terminations

  — represent sources and sinks of media flow;

  — are abstractions of the protocol that represent ports connected to the gateway;

  — are said to be 'physical' when they represent external interfaces such as analogue lines or digital trunks, and are instantiated at boot time;

— are said to be 'ephemeral' when they represent media flows on the network, and are created as needed;

— are created by the MG and are given a name (TerminationId) by the MG;

— can be multimedia, and thus represent a set of related media flows (audio, video), each of these flows being called a 'stream' and each stream given a StreamId by the MG;

— have properties (parameters) that can be set or examined by the MGC — these properties have names (PropertyIds).

- Contexts:

— are associations between groups of terminations;

— connect each of the streams of the terminations in the Context together, by default in a 'star' connection in such a way that whatever is received by any Termination is transmitted by all other Terminations in the Context (Fig 7.7);

— are multimedia, and the mixing of Streams in the Context is by StreamId (all streams with StreamId=1 are mixed, but are separate from all Streams with StreamId=2);

— are created by the MG under the direction of the MGC and are assigned a ContextId by the MG when it is created;

— as the MGC processes calls, it creates active Contexts and moves Terminations into, and out of, the Contexts, a Context being created when the first Termination is Added to it, and destroyed when the last Termination is removed (Subtracted) from it;

— a special Context called 'Null' holds all the physical Terminations that are not in an active Context — upon boot, all physical Terminations will be in the Null Context, and, when a physical Termination is subtracted from a Context, it returns to the Null Context, but, when an ephemeral Termination is subtracted from an active Context, it is destroyed.

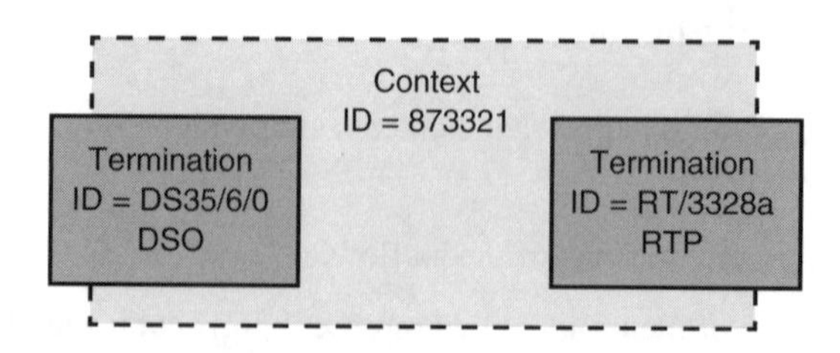

Fig 7.7 Protocol elements.

- Events:

  — the MG may detect asynchronous conditions (Events) the MGC may wish to know about;

  — the MGC may specify a list of such Events, detection of any one of which will cause the MG to notify the MGC of its occurrence;

  — off-hook on an analogue line is an example of such an Event;

  — detection of a DTMF digit is another common Event, each of the DTMF digits being a separate Event.

- Signals:

  — Terminations may have Signals applied to them by the MG under the direction of the MGC, and these Signals may be audible tones, such as dialtone, or mechanisms such as power ring;

  — Signals are normally applied from the Termination out to the exterior of the MG;

  — there are three types of Signal:

    brief Signals last for a short period of time and then stop on their own (an example being the 'call waiting' alert tone) ;

    timeout Signals have a duration property whereby the Signal will stop on its own after the duration but will also stop upon detection of an Event (a dialtone Signal would normally be played for a period of time, but detection of a DTMF digit would stop the dialtone Signal — if no digits were detected for a 'long' time, an error would be declared, dialtone thus being an example of a timeout Signal);

    on-off Signals which remain playing until they are explicitly turned off by the MGC;

  — the MGC may specify that a single Signal be applied to a Termination, or more specifically, to a specified Stream of a Termination — alternatively, the MGC may specify a Signal List, which plays each Signal on the list sequentially.

- DigitMaps:

  — the collection of a dialled number is a common task where the use of a separate event notification for each dialled digit puts a response load on the MGC, a DigitMap being a pattern recognition feature of Megaco that allows the MGC to instruct the MG to look for a complete dialled number — of course, the definition of 'complete' depends on the dial plan under which the MG is operating, and thus the DigitMap is an expression of a dialling plan sufficient to determine the terminal digit);

— DigitMaps consist of a number of patterns, expressed in a simple string notation, the patterns including tokens for specific digits, ranges of digits, and timers (for indeterminate length number sequences such as international numbers);

— the state machine applies each of the patterns sequentially to the current detected digit string, looking for a complete match, and, when a complete match is detected, the MG sends the MGC a single Event notification, including the entire dialled number in the notification.

7.2.2 Commands

The protocol messages are expressed as a series of commands that operate upon the Terminations and Contexts. Each command includes the name of the command to be executed, the TerminationId(s) on which the command is to be executed, and a series of parameters called 'Descriptors'. Commands are grouped into 'actions', each one of which specifies the Context(s) that will be affected by the commands in the action.

- Add

  Add is used to add a termination to a Context. The MGC may specify that a new Context be created to hold the first Termination Added to the Context.

  The Add command may specify new values for the properties of the termination. It may also specify events to be detected on it, signals to be applied to it, DigitMaps to be used on it, etc. Each of these specifications is described in a Descriptor.

- Subtract

  A Termination is removed from a Context with a Subtract command. Ephemeral Terminations are destroyed upon being Subtracted. If the last Termination is Subtracted, the Context is destroyed. When a Termination is Subtracted, by default a series of statistics are sent by the MG for the Termination.

- Modify

  The Properties, Events, Signals, DigitMaps and other parameters of a Termination may be changed with the Modify command.

- Move

  A Termination may be moved from one Context to another atomically with the Move command. Move may also change Properties, Events, Signals, DigitMaps, etc, as in an Add command.

- Notify

  The Notify command is used by the MG to notify the MGC of one of the Events it was asked to detect. The Notify command includes the Event detected, a timestamp, and may have parameters. A RequestId is specified by the MGC when it asks for Events to be detected on a specific termination, and that RequestId is returned in the Notify.

- AuditValue

  The current values of all Properties, list of Events, current Signals, defined DigitMaps and other state of terminations, as well as the TerminationIds and the ContextIds to which they belong, may be read by the MGC using the AuditValue command.

- AuditCapability

  The MGC may discover the possible values that any given Property, Event, or Signal may have by use of the AuditCapability command. Where AuditValue reports the current value, AuditCapability reports all the possible values AuditValue might return. Using AuditCapability, especially in the beginning of a control association between an MG and its MGC, allows the MGC to discover what the MG can do.

- ServiceChange

  The MG and the MGC report major state change to each other with the ServiceChange command. The MG also uses ServiceChange to create a control association between itself and its controlling MGC. Either the MG or the MGC can put terminations out-of-service with this command. The MG or the MGC can notify its partner of redundant box fail-over with ServiceChange.

7.2.3 Descriptors

Each of the commands has a set of parameters. Parameters are specified with Descriptors, which are common across most commands. Each Descriptor has a name, often an Id (e.g. StreamId) and usually a list of sub-parameters expressed as a 'property = value' construct.

- Media

  The Media Descriptor (Fig 7.8) is a container enclosing descriptors that specify properties of the termination:

  — TerminationState: properties of a termination that are not Stream specific are specified in the TerminationState descriptor;

— Stream: the Stream Descriptor is a container for enclosing descriptors that specify properties for a specified stream, including the StreamId for which the descriptor specifies properties:

'LocalControl' — properties of a termination that are stream specific are specified in LocalControl;

'Local' contains all the properties of streams received by the MG from a remote entity, and has two forms, corresponding to two possible encodings for the protocol — for text encoding (like HTTP), Local is specified using the session description protocol (SDP) (RFC 2327), and for binary encoding (BER, like H.323) Local uses a typed parameter list with properties equivalent to those expressible with SDP contained in Annex C of the specification;

'Remote' contains all of the properties of media streams sent to a remote entity (like Local, Remote is described by SDP or Annex C).

```
Media {
    Termination State
    Stream {
        LocalControl
        Local
        Remote
    }
}
```

Fig 7.8 Relationship between Media, Stream and Local/Remote.

- Events

  This descriptor contains the list of events the MGC requests the MG to detect if and when they occur on the termination. Each Event in the EventDescriptor may have parameters. A RequestId included in the EventDescriptor is returned in a corresponding Notify.

  Each Event in the EventDescriptor can include an 'Embedded Event' or an 'Embedded Signal' descriptor. These look like an EventDescriptor or SignalDescriptor and become the actual Event/SignalDescriptor when an Event in the original EventDescriptor is detected. These embedded descriptors provide a fast turnaround response to detection of the event. An Embedded Event or Embedded Signal Descriptor cannot contain another Embedded Descriptor, so this capability extends only one level.

  Each Event can also specify a DigitMap to be used with the Event. The DigitMap can be defined in the EventDescriptor, or it can refer to a named DigitMap previously defined.

- Signals

  The Signals Descriptor contains a list of signals to be applied to the termination. If there is more than one signal in the Signals Descriptor, they are applied simultaneously. The 'SignalList' construct can also be in the list, where each of the signals in SignalList is played sequentially. Each signal may have parameters, one of which can be the type of signal (Brief, Timeout, On/Off) and another of which could be duration. The duration in the Descriptor overrides the default signal type in the definition of the Signal.

- DigitMap

  To define a new DigitMap, the DigitMap Descriptor contains a list of patterns. Each pattern is a string of tokens, with tokens representing each of the DTMF tones, ranges of digits, wildcards, and one of three timers that can be used to define a complete dialling plan. DigitMaps can be named, which allows the definition to be reused. When the terminal digit in any of the patterns is found, the DigitMapComplete event is raised, which returns the digit string. The timers can be used to terminate a pattern.

- ObservedEvents

  ObservedEvents is the Descriptor returned with a Notify, which contains the Event detected, a timestamp, any parameters defined by the Event and the RequestId of the EventDescriptor that listed the Event.

- Audit

  The Audit Descriptor is used with the AuditCapability and AuditValue commands to describe which Descriptors are desired in the reply for the Audit command. Each of the Descriptors has a token that can be included in the Audit Descriptor. The reply to the Audit command contains at least one descriptor of that type.

  When using AuditCapabilities, more than one descriptor of a given type mentioned in the Audit Descriptor can be returned which is just one of the mechanisms that are used to indicate alternative values.

- Service

  The parameter to the ServiceChange command is the Service Descriptor. It contains a number of properties that indicate what is happening. Method and Reason describe the conditions that prompted the ServiceChange command. Address supplies the desired address to be used by the partner for subsequent commands. Version indicates the version of Megaco the entity can support. The Delay parameter is used with the 'graceful' method to indicate a time period during which the entity will change the status of terminations (generally to 'out of service'). MgcId is used on a registration (creation of a control association between an MG and an MGC) to indicate an alternative MGC that may offer service to the MG. Profile supplies the name of the Profile that the entity implements. Profiles are documents describing selections of options in the protocol. Generally, it is a shorthand way of determining gross MG characteristics that the MGC uses to determine if it has the capability to provide service to an MG.

7.2.4 Transactions

Commands are grouped into 'Actions' that operate on a Context or set of Contexts. Within an Action, Commands are executed sequentially in order. Each Command operates on a Termination or set of Terminations. Actions are grouped into Transactions that are sent as a unit to the corresponding entity. Within a Transaction, Actions are executed sequentially in order. The sender assigns a TransactionId to each Transaction (Fig 7.9).

```
Transaction = transId {
    Context = cntxId {
        Add = termId {
            Event = RqstId {. . .}
        }
    }
}
```

Fig 7.9 Relationship between Transaction, Action, Command and Descriptor.

The recipient acknowledges transactions when they are executed. The reply includes the TransactionId specified in the request, and for each action, and each command within the action, a reply is normally generated.

For each action, the response includes the ContextId against which the action was performed. In the request, ContextIds for each action may be fully specified, which would mean a specific, existing ContextId (or Null), or the ContextId can be underspecified or overspecified. An underspecified ContextId uses the 'CHOOSE' construct, which indicates that a new Context is to be created. CHOOSE is specified as the ContextId in the Action on the request, and the actual new ContextId is returned in the reply. An overspecified ContextId refers to the use of a wildcard, which might evaluate to zero, one, or more than one actual ContextIds. Overspecifying a ContextId in the request generates a reply for each matching ContextId. The effect is as if the action was repeated in the request with a fully specified ContextId in each repetition.

For each command, the response includes the command that was executed, and the TerminationId against which the command was performed.

The request can fully specify, underspecify, or overspecify the TerminationId. A fully specified TerminationId would be an exact, existing TerminationId. An underspecified TerminationId, using the CHOOSE construct, would specify that the MG is to choose a TerminationId that is appropriate to the command, and the parameters specified for that command. In most cases, CHOOSE is used to create a new ephemeral Termination, but it can also be used to ask the MG to select a termination from a group of terminations that have common characteristics. An overspecified TerminationId uses the wildcard construct which can match zero, one or more than one actual TerminationIds. The reply for an overspecified TerminationId would consist of one command for each matching TerminationId, as if the request repeated the command with the fully specified TerminationIds.

For each command, the response also may include Descriptors as defined above. The Descriptors that are returned depend on what the command was, and how the parameters in the Descriptors were specified.

Some commands force Descriptors to be returned. For example, Subtract normally forces the Statistics Descriptor to be returned. The Audit command causes each of the descriptors whose tokens appear in the Audit Descriptor to be returned.

Parameters to Descriptors can be fully specified or underspecified. If any of the parameters in a descriptor were underspecified (using the CHOOSE construct), the Descriptor with a value chosen by the responder is returned.

If an error is encountered, the reply stops at the point (Transaction, Action, Command, Descriptor) where the error was detected. An Error Descriptor is sent in place of the Action/Command/Descriptor. In some cases it is desirable to allow processing of commands, and subsequent actions, even though an error was detected. For this reason, it is possible to mark a command such that processing of the transaction is not halted by detection of an error, but, if the MG is able to do so, processing continues.

In some cases, use of wildcarded TerminationIds could lead to a very large amount of data being returned which could be redundant or not useful. For this reason, it is possible to mark Commands that use the wildcard construct as 'wildcard-response'. In this case, a single composite response will be returned representing the union of all the replies that otherwise would have been returned individually.

Some of the Megaco protocol transports are not reliable. For example, UDP is commonly used as the transport for Megaco. With such transports a Transaction Ack is sent by the sender to the recipient to indicate that it successfully received the reply for a transaction. This three-way handshake mechanism can be used with retry timers to provide reliable transaction processing.

There is no requirement that only one transaction be outstanding at a time. If a recipient has more than one transaction, it is free to execute transactions in any order.

7.2.5 Encoding

The protocol includes both a binary and a textual encoding. Although there were major efforts during the development of the protocol to try to come to a compromise on a single encoding, it was ultimately not possible to agree, and thus there are two. The binary encoding is based on a BER encoding of an ASN.1 description of the protocol. The text encoding is based on an ABNF description. The two descriptions are identical except for the use of SDP in the text encoding of Local and Remote, with Annex C typed values in the binary encoding. The encoding method can be determined by the port the first (registration) Command sent and entities can implement text, binary or both encodings. At present, there appear to be many more text implementations than binary, but this could change. To the dismay of binary supporters, some preliminary results on implementations that include both binary and text encodings show that the text encoding is both shorter and faster to encode/decode than the exact equivalent binary encoding.

7.2.6 Transport

The base protocol defines UDP and TCP transport. Several ATM transports are being standardised, and SCTP transport is also being standardised for Megaco. UDP is currently the preferred transport for Megaco when on an IP network.

7.2.7 Security

IPsec is specified as the security mechanism for Megaco. At the time the protocol was developed, many embedded operating systems on which media gateways were being developed did not support IPsec. The Megaco protocol document includes a definition of a so-called interim-AH security header, to be included in a transaction header that mimics the functions of the IPsec AH header. At the present time, most of these embedded OSs now have IPsec available, and the interim-AH scheme may not see much deployment.

7.3 Packages

7.3.1 Dealing with a Variety of Different Gateways

Megaco is designed to be used with a very wide range of gateways. Each type of gateway has different requirements, specifically on the characteristics of the terminations it implements. To cater for this diversity, Megaco has an extension mechanism called a 'package'. A package is described in a document, which may come from a standards organisation, a consortium or other trade group, a vendor or even an operator. Packages include definitions of Properties, Events, Signals, and Statistics that are related. An MG implements sets of terminations. Each termination realises a set of packages that specialise that termination to the function it performs. For example, there is an analogue line package that would be implemented on terminations that represent POTS lines.

To ensure the integrity of the solution, each package has a unique name which is registered with IANA. This registration procedure ensures that regardless of the source specifying the package, no two packages will have the same name. The document defining a package includes identifiers for the ASN.1 definitions of the objects defined by the package as well as the text names employed. For example, Properties, Events, Signals and Statistics may be defined in a package and each of these is given a name that is local to the package. So when a given Property, Event, Signal or Statistic is specified in the protocol, the name used (PropertyID, EventID, etc) is the combination of the name of the package and the name of the item as defined in the package.

A package can be defined to extend another package. Extended packages inherit all of the elements of the base package, and the extended package may define new Properties, Events, Signals and Statistics. Extensions may not delete or redefine existing definitions, but may only add new definitions.

The Audit commands can determine which packages are implemented on a given termination.

7.3.2 Existing Packages

The base protocol document includes in its Annex E a number of packages. These packages do not have to be implemented in every gateway, but they provide basic capability expected to be widely deployed.

7.3.2.1 Generic

Generic provides two generic events that any termination could use. The generic error event will notify the MGC when any asynchronous error condition arises. It also can be used to detect dropping of a connection (for connection-oriented networks). The signal completion event can be used to notify the MGC of the completion of a signal application on a termination. The signal completion event includes a parameter that specifies why the signal stopped, including timeout, interruption by an event, or loading a new signal. The Signals Descriptor includes an enable parameter to allow this event to be used selectively.

7.3.2.2 Root

The Root package is defined on Root, that is, the gateway as a device rather than a specific termination. Root includes properties defining the maximum number of contexts, and the maximum number of terminations per context. It also contains properties defining the normal MG execution time, normal MGC execution time, and provisional response time. These are used to estimate, and control transaction transmission and acknowledgement.

7.3.2.3 Tone Generation (Tonegen)

Tone Generation is a common MG function. The Tonegen package defines the basic mechanism to play tones on terminations. The package defines two mechanisms — one defines a signal for each tone, the other defines a single signal that takes a list of parameters that have the names of the tones. The latter includes an inter-tone delay, for use with typical 'outpulse' applications.

The package does not actually define any tones — they are defined in packages that extend Tonegen.

7.3.2.4 Tone Detection

Like Tone Generation, Tone Detection provides the basic mechanism to detect tones on an MG. As with Tonegen, a basic mechanism is defined, and other packages define the 'toneids' for specific tones. Tone Detection defines three events, one that detects the start of a tone, one that detects the end of a tone, and one that detects a 'long tone', that is, one that lasts longer than a parameter-defined length. The start tone detection can be paired with the 'long' tone detection to provide a one-event-per-tone, with an extra event for tones longer than a defined limit (such as a long hash, used by some calling card applications).

7.3.2.5 DTMF Generate

DTMF Generate extends the basic Tone Generation package and defines the usual DTMF tones.

7.3.2.6 DTMF Detect

DTMF Detect extends the basic Tone Detection package and defines the usual DTMF tones. DTMF Detect also defines the DigitMap completion event.

7.3.2.7 Call Progress Tone Generate

Like DTMF Generate, Call Progress Tone Generate extends Tonegen and defines typical call progress tones. This package uses the definition of tones in ITU-T E.180, but the package does not define all of the tones that are described in E.180.

7.3.2.8 Call Progress Tone Detect

This package provides the analogous detection for the call progress tones defined in the Call Progress Tone Generation package.

7.3.2.9 Analogue Line Supervision

This package provides basic line supervisory functions for analogue lines. It includes Events for onhook, offhook and flash detection, as well as a Signal for power ring.

7.3.2.10 Basic Continuity

The Basic Continuity package provides the mechanism for basic loopback testing, typical of an analogue trunk implementation. It includes a Signal to initiate a continuity test, and an Event to detect completion of it.

7.3.2.11 Network

The Network package provides a base for most media flow termination, other than analogue lines or trunks, especially for ephemeral (network) streams such as RTP. It includes a parameter to define the jitter buffer size, an Event to detect network failure, an Event to detect the failure of the network to deliver agreed quality of service, and statistics for duration of the connection as well as octets sent and received.

7.3.2.12 RTP

The RTP package extends the Network package and provides an Event to detect transitions from one payload type to another, as well as statistics for packets sent and received, packet-loss percentage, jitter and delay.

7.3.2.13 TDM Circuit

The TDM Circuit package also extends the Network package, and provides properties to enable echo cancellation and to set gain.

7.3.3 Creating a New Package

As stated previously, any organisation can create a new package. The protocol specification defines the procedure concerning how new packages are defined. The requirements include creating a document describing the name of the package, its version and which package it extends, if any. The document then describes each Property, Event, Signal and Statistic, together with any parameters and enumerated values.

While there are efforts under way in both the IETF and ITU to standardise additional packages (which will be additional annexes to Recommendation H.248 and separate RFCs in the IETF), there is no requirement to standardise a package in order to use it. Of course, both the MG and MGC need to implement a package before a specific MG implementing it could be controlled by a corresponding MGC.

7.4 Profiles

7.4.1 Coping with the Explosion of Packages

There are expected to be quite a few packages defined by a variety of organisations. Since an MGC has to implement the union of the packages implemented by the set of MGs it controls, there are a myriad of interoperability issues raised. To ease the burden of defining compatible MGs and MGCs, a mechanism is defined that summarises the set of packages implemented. A profile is a document described by an organisation that includes, among other things, a list of packages to be implemented. The profile implemented by an entity is included in the registration ServiceChange command. The MGC can verify that it is compatible with a registering MG by looking at the profile.

7.4.2 Definition of a Profile

A profile is a document, typically not defined by the IETF or the ITU. It contains, at minimum, a name and a list of packages. It might also contain additional statements which limit the other options that the Megaco/H.248 protocol defines. For example, a profile might define a convention for TerminationIds and ContextIds. It might define encoding, and possibly transport. A good profile would, specifically, define a minimum implementation for each option presented by the protocol.

An MG that implements a profile should be compatible, in every respect, with an MGC that implements the same profile. A profile might include options, but compatibility between MGs and MGCs should not be imperilled due to either side implementing, or not implementing, an option. Profiles have versions, which are significant (but backwards compatible).

In general, an MG implements one profile. In some cases, an MG might implement a profile, and several less-stringent related profiles. Since the ServiceChange only allows a single profile in the Service Descriptor, an MG would attempt to register using its most complex profile. If it could not find an MGC which would service this profile, it might try again with a less complex profile.

An MGC on the other hand, often implements multiple profiles. The MGC would compare the profile of a registering MG with its list of profiles. If it implemented the profile offered by the MG, it might agree to serve the MG, returning the corresponding profile in the reply to ServiceChange.

7.4.3 Example — the MSF Trunk Gateway Profile

The Multiservice Switching Forum [4] is defining a set of Implementation Agreements which contains profiles for several types of media gateway, initially on ATM networks. These profiles provide an excellent blueprint for what might be in a profile. At the time of writing, these profiles have not been finalised, and therefore specific choices mentioned here may not be the final ones.

The MSF profiles each follow a pattern. After the introductory text explaining the definition of the gateway defined, and the relationship of the profiles to the MSF architecture, the profiles consist of:

- the name of the package;
- a list of required, and optional packages — the optional packages are used for specific exterior interface types (not all access gateways have all interfaces), with the package definitions including the package name, where it is defined, and the version number;
- naming conventions for the MG, MGC and TerminationIds on both physical trunks and ATM network connections;
- a statement that the Topology descriptor is not required — implementation of Topology is optional in the protocol;
- a statement specifying the minimum number of provisionable MGCs in the MG used for registration;
- the transports to be supported;
- the encoding to be supported;
- a statement that real IPsec must be supported.

7.5 Current Work

Megaco is not a static work product. Many implementations are under way, and, as these mature, it is found that additional work is required, to correct errors in the text, explain unclear wording, and to extend the protocol to cover unforeseen circumstances.

7.5.1 Ongoing Effort in the ITU

Study Group 16 has a large number of additional packages in development, has defined an additional set of transports (including SCTP and several ATM transports), and is actively creating what it calls an 'Implementor's Guide' which consist of errata, additional text, and guidance information.

SG16 is considering starting work on a Version 2 of the protocol, which would extend it in several ways. These efforts are still in early discussion.

7.5.2 Ongoing Effort in the IETF

The MeGaCo Working Group also has several packages in development. It has agreed with SG16 to work together on the packages in development at both groups. The IETF mailing list is the central place for discussion on the packages and other annexes under development.

7.5.3 Interoperability Testing

The Multiservice Switching Forum and the International Softswitch Consortium have co-sponsored Megaco/H.248 interoperability testing events at the University of New Hampshire's InterOperability Laboratory [5]. These events draw engineers from around the world who bring their implementations to test against others. At the events, MGs are paired with MGCs and a series of compatibility tests are conducted. Additional testing with multiple implementations is also conducted, and even in the first event, held in August 2000, complete basic call tests between a telephone connected to one MG implementation successfully completed, including media flow to another MG implementation under control of an MGC from a third implementor. At this time, there are at least 30 Megaco implementations, all with independently developed code bases, in various stages of maturity.

7.5.4 Adoption in the Industry

A number of organisations are now specifying Megaco/H.248 in their work. The MSF, as mentioned above, is specifying Megaco as part of their implementation agreements. The ATM Forum is extending its broadband loop emulation standard to use H.248. The ITU SG11 is using it in its definition of bearer-independent call control (BICC-Q.1901.x) (see Chapter 8). 3GPP has specified Megaco for use within 3rd generation mobile networks. A number of prominent network operators are now telling vendors that Megaco/H.248 is required for future acquisitions.

In response, most media gateway and softswitch vendors are implementing Megaco in their products. In some cases, they are using stacks from one of at least six vendors who have implemented the protocol for licensing to others. Products incorporating the protocol have been announced, and some laboratory trials are known to be under way, but as of this writing, no production shipments have been made.

7.6 Whither MGCP?

While IPDC, SGCP and MDCP have faded, at least for new designs, MGCP is still a viable protocol. The PacketCable effort at CableLabs is a prominent supporter of MGCP, and that alone has assured vendors that MGCP has some future. It is not clear yet if the on-the-ground reality of MGCP, which has quite a few shipping products supporting it, will trump the efforts of the IETF and ITU to standardise Megaco/H.248.

There are several indications that suggest that MGCP may not survive, with the possible exception of the PacketCable efforts. One is that, while there are operators who are specifying Megaco/H.248 for future acquisitions, the author is not aware of similar operators with a long-term specification for MGCP. Most operators, to be sure, are in the 'undecided' camp.

Another indicator is the number of participants in continuing work. There is a small group in the Softswitch Consortium working on continuing the development of MGCP, and of course PacketCable continues to use a profile of MGCP in its specifications. There is a much larger set of contributors to the work in the ITU and the IETF.

Finally, further standards developers who need a media gateway control protocol have been selecting Megaco/H.248 in preference to MGCP, as indicated above. The BICC, 3GPP and BLES efforts are indicative that the imprimatur and standards-track status of Megaco/H.248, plus its technical merit, are propelling it to the leadership position.

Nevertheless, prudent vendors are implementing both protocols in near-term products. The final chapter is definitely not written.

References

1 CableLabs — http://www.cablelabs.org/

2 ITU-T — http://www.itu.int/

3 IETF — http://www.ietf.org/

4 Multiservice Switching Forum — http://www.msforum.org/

5 University of New Hampshire, InterOperability Lab — http://www.iol.unh.edu/

8

BEARER-INDEPENDENT CALL CONTROL

R R Knight, S E Norreys and J R Harrison

8.1 Introduction

The patterns of telephony traffic in the networks run by public network operators (PNOs) have changed drastically over the last five years. Communications methods have changed and the demand for access to the Internet seemingly grows daily. Throughout the year 2000 many PNOs have reported that the Internet and related IP traffic has surpassed, in simple bandwidth, the demand for traditional telephony traffic.

It is easy to forget, however, that the volume of non-Internet telephony traffic also continues to grow. Better marketing information enabling personalised direct telesales, and the enormous growth in mobile telephone sales both contribute to the demand for telephony-type services. The telephony growth, however, is small (about 3 - 5% per annum, depending on statistics and network operator) compared to the phenomenal growth in Internet and IP-related traffic (anything from 100% to wildly imaginative figures exceeding the world's production capabilities).

Recently, PNOs have faced a choice in the development of networks — to continue the development of existing, regulated switched-circuit networks (SCNs), also known as time-division multiplex (TDM) networks, or to leave the existing infrastructure frozen in its current state, and start to deploy a new network, better suited to IP-based terminals, outside the constraints of regulation and offering the potential of new services with faster deployment, and at lower costs. If the demand for telephony was not also growing, this strategy might be extremely attractive; however, many TDM networks are running out of switching and transmission capacity.

First conceived in the early 1970s, integrating voice and data to provide a single, comprehensive, multi-purpose communications network [1] offered a solution for PNOs. Reduced operating costs from simpler network and service management still provide the rationale for integration today.

The Internet Engineering Task Force (IETF) is the standardisation body charged with developing protocols for the Internet (see Chapter 5 for further information on the work of the standards bodies). They have made considerable progress in specifying protocols to enable the packetisation of speech and the provision of multimedia applications, for example, through the development of the session initiation protocol (SIP) which is discussed in Chapter 11. The IETF approach has also been to specify enhancements to SIP to support the basic call and a few selected supplementary services. However, the IETF began work on VoIP as a means of supporting just another media type to be transferred using the Internet. As considered in Chapter 6, adding the emulation of existing public switched telephone networks (PSTNs) or integrated services digital networks (ISDNs) has been a more recent development. However, a major factor in the development of these protocols is the commercial opportunities offered by new services integrating real-time media, including voice, and data.

8.2 Overview and Scope of Bearer-Independent Call Control

While the IETF are the prime standardisation body for the Internet and general IP-related protocols, the international standards organisation with the greatest expertise and experience in telephony is the International Telecommunication Union — Telecommunication Standardisation Sector (ITU-T). It was the ITU-T (formerly the CCITT) that standardised the main signalling protocol for telephony networks — the integrated services user part (ISUP) of Signalling System No 7 (SS7).

The ITU-T looked at the work that the IETF were undertaking on VoIP and took a very simplistic view on PSTN/ISDN replacement — if you try to emulate existing services using a new protocol, the result will not be an identical service. Even more drastic was their opinion that nothing less than 100% of the existing services should be supported in any new technology that intends to replace the TDM network and support existing customers. This strategy has considerable merit, since it does not seem a wise strategy to take services away from customers, only to re-instate them again at some point in the future. Telecommunications customers purchase service, not technology, from their network operator. The service that they purchase is very much richer than simple voice communication — it is an intricate set of complex services available for a very simple terminal that operates to provide access to global communications.

Even within the existing TDM network, there are instances of service interactions that fail due to incompatibilities between implementations of ISUP, and the success of emulating all of the services, at day one, in a different technology and using a different signalling system was seen as doubtful by many network operators active in the ITU-T.

This simple view of replacing the technology underlying the existing telecommunications networks was given further impetus as TDM networks began to reach their design capacity, and network operators were unwilling to add further investment in TDM-based technology. As the demand for new TDM switches (exchanges), and the need for new services to be added to existing TDM networks, fell, the cost of TDM switches grew. The immediate issue of relieving the capacity problem needed a solution that would be compatible with the longer term aims of the network operators. Another simple solution was arrived at. Rather than develop an ISUP derivative for the IP world, and another as a stop gap for the capacity problem, the ITU-T began work on defining an ISUP derivative that would be 100% compatible with the existing networks, but could operate in any environment that could transport voice with reasonable quality. The term that they coined for this protocol was bearer-independent call control — BICC.

The scope of this new work was severely limited — it had to:

- be based on ISUP, to be fully compatible with existing services;
- be independent of the underlying technology;
- reuse existing signalling protocols to establish the communications within networks.

These restrictions positioned the standardisation work independent from other standards organisations and with the single simple goal of PSTN/ISDN replacement. Many standards bodies have begun work with a very simple goal only to be distracted along the way with new ideas and aspirations. The success of BICC has been a dogged determination to remain within the existing scope and not to extend it to cover such exciting features as multimedia support. One single enhancement, beyond existing ISUP capabilities, was added — better support for mobile calls. Many of the world's fixed networks provide the interconnect for mobile calls, and one of the main causes of loss of speech quality for mobile-to-mobile calls is the conversion to the fixed network voice coding schemes and back again (transcoding). The ability to signal codec usage, and to negotiate the best codec to use for a particular call could easily be added without compromising the interworking with ISUP, and therefore the backwards compatibility.

The goal of BICC was simple, and the starting point was to take ISUP and determine which parts were not applicable to alternative technologies, and to add to ISUP only features to enable connections to be established by the new technology. Three technology types were identified for voice transport — ATM AAL1, ATM AAL2 and IP. Since many network operators had already deployed networks with these technologies, the signalling systems that could establish connections across these technologies were identified as being:

- the ATM Forum UNI4.0 and ITU-T DSS2 for access signalling in ATM;
- ITU-T B-ISUP, ATM Forum PNNI and AINI for core network signalling in ATM;

- AAL Type 2 signalling [2] as a true bearer control protocol in ATM;
- SDP [3] for IP networks.

This was felt to be a considerable task to achieve in one go, and, with the pressure from network operators for a short-term solution to the capacity shortage problem, the standardisation work was split into two capabilities — ATM for the short-term problem and the addition of IP for the longer term. It should be noted that IP was always seen as the target, and that the ATM work would merely be a step along the way.

8.3 BICC Architecture

The starting point for the BICC architecture was to assume that calls would enter (ingress) and exit (egress) the new network technology through interface serving nodes (ISNs). More generally, a serving node (SN) is a point in the network that provides the functionality for existing PSTN/ISDN services. Networks tend to grow from the centre outwards, since it is pointless to provide functionality at one access gateway that cannot be supported elsewhere. The ISN initially had to provide a signalling interface between the narrowband ISUP and a peer ISN, as shown in Fig 8.1.

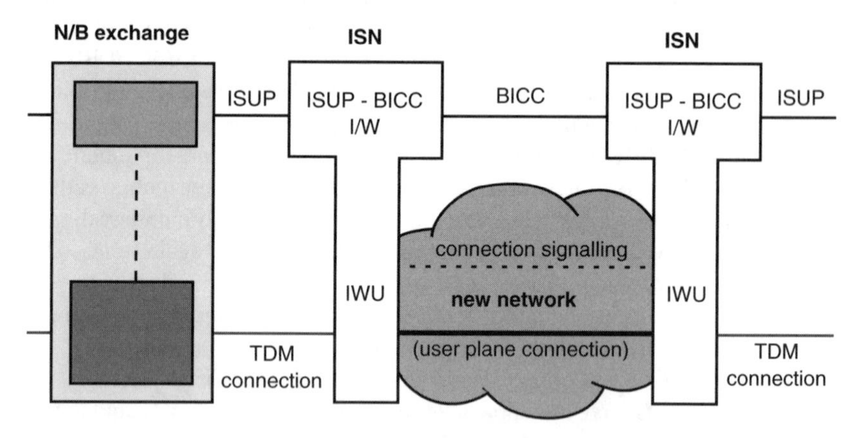

Fig 8.1 Initial BICC architecture.

This simple architecture, although realistic, does not provide any flexibility. In a large network, the interconnections are much more flexible, with core network nodes to spread traffic evenly across the network. This initial scenario also would not prove that a robust and generic standard was being produced, since BICC would only interface between ISUP and itself. A more realistic scenario allows the interconnection with other licensed operators (OLOs), which is conceptually performed at a PSTN gateway. Retaining the PSTN naming conventions, this new node was named a gateway serving node (GSN). It provides connectivity between two BICC networks and interconnection consists of a pair of GSNs communicating via a simple transmission link. Although this is not necessarily exactly how network operators perform network interconnection, it is a simple model that is sufficient for standardisation.

If two network operators can interact, then it should also be possible for a single network operator to provide PSTN/ISDN services at a node within a network, as a point of flexibility. This node is similar in function to a transit exchange, and was named the transit serving node (TSN). Figure 8.2 shows the ISN, TSN and GSN in a simple configuration.

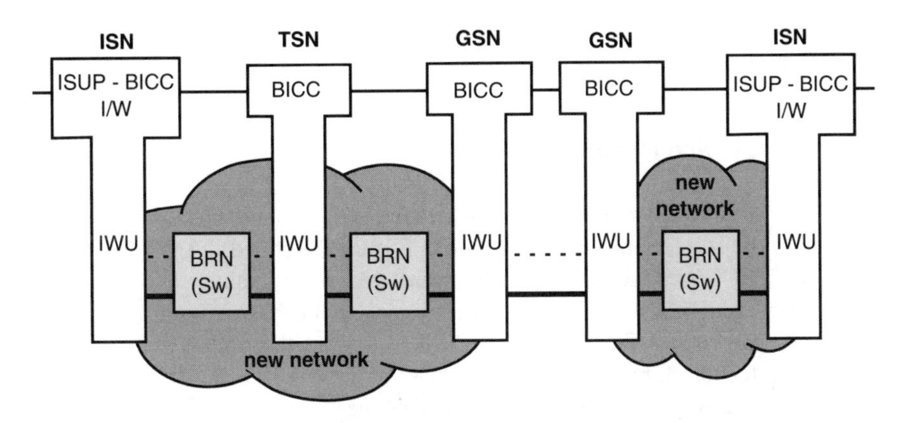

Fig 8.2 BICC serving nodes.

Figure 8.2 also shows that account must be taken of all types of network technology. ATM, recognised as the short-term solution for relieving capacity, uses intermediate switches set up by signalling. These nodes are termed bearer relay nodes (BRN), and not all network technologies need these nodes.

One of the key requirements of BICC was to fully support 100% of the narrowband services on first deployment, and this clearly includes the services derived from intelligent networks. Intelligent network services are often provided by a transit exchange, on behalf of a local exchange. In some smaller networks, it even makes sense to provide a single node that offers access to IN. Additionally, in IP networks, it is not necessarily very efficient to provide, at a TSN offering IN access, conversion of the user plane packets through the TSN. This type of scenario occurs, for example, where charge card services are provided by IN. The initial number dialled takes the call as far as an IN special resource (located at a TSN), which then collects further digits providing charge card authorisation and destination number. The call is then routed from the TSN — but optimisation could occur if the TSN was allowed to migrate to a node that supported the BICC signalling, but no longer had the user plane present. For these reasons, a special call mediation node (CMN) was introduced. This is really just a special case of a TSN without a user plane, and so does not serve the user plane (hence it was not labelled a call serving node). This is covered in more detail in section 8.4.2.

8.4 Requirements

The requirements for the signalling protocols are derived from the services to be supported and the functional modelling. Functional modelling decomposes network functionality to logical entities and the information to be exchanged by the entities to achieve the services. The functional modelling in BICC builds upon the network architecture to produce a conceptual model and the interactions between the functions in the model.

8.4.1 Service Requirements

Since BICC must support all existing narrowband calls, the service requirements are based upon the services that ISUP supports. These are defined in Q.761 in tabular form. The BICC service requirements were derived by taking this table and determining which ISUP services were not applicable in non-narrowband technology networks. This removes those services that relate more to the establishment, control, maintenance and release of the circuit within TDM switches, and the TDM transmission path between the switches (see Table 8.1).

8.4.2 Functional Model

The architecture for managed IP networks supporting real-time services is built around the concept of media gateways (MGs) that provide conversion of user plane communications from one format (e.g. TDM) to another (e.g. IP). The control of these MGs is then performed by media gateway controllers, providing a split of responsibilities between two separate physical units. Since it is highly desirable that BICC can co-exist in the same environment, the functional model first separates the generalised serving node into portions that can be mapped on to an MG/MGC architecture. Requirements for VoIP networks have been derived in the ETSI Project 'TIPHON' (see Chapters 5 and 16), and other work has been performed in the Multiservice Switching Forum. The work of both of these groups was taken into account in deriving the functional model, which allows a physical split between the call and bearer control, as well as the more general split between control functionality and the media gateway. Figure 8.3 shows the basic split of functionality.

Table 8.1 ISUP services supported by BICC.

ITU-T ISUP 2000 Function/service	Applicability to BICC
Basic call	
Speech/3.1 kHz audio	Required
64 kbit/s unrestricted	Required
Multi-rate connection types	Required
N X 64 kbit/s connection types *en bloc* address signalling	Required
Overlap address signalling	Required
Transit network selection	National Option
Continuity check	Not Required
Forward transfer	Required
Simple segmentation	Required
Tones and announcements	Required
Access delivery information	Required
Transportation of user teleservice information	Required
Suspend and resume	Required
Signalling procedures for connection type allowing fallback capability	Required
Propagation delay determination procedure	Required
Enhanced echo control signalling procedures	Not required
Simplified echo control signalling procedures	Required
Automatic repeat attempt	Required
Blocking and unblocking of circuits and circuit groups (in Q.BICC, circuits = CIC which is equal to the CCA-ID)	Required
CIC group query (in Q.BICC, CIC = CCA-ID)	National option
Dual seizure (in Q.BICC, dual seizure applies to CIC = CCA-ID and does not refer to circuits)	Required
Transmission alarm handling for digital inter-exchange circuits	Not required
Reset of circuits and circuit groups (in Q.BICC, circuits = CIC which is equal to the CCA-ID)	Required
Receipt of unreasonable signalling information	Required
Compatibility procedure	Required
Temporary trunk blocking	Not Required
ISDN user part signalling congestion control	Required
Automatic congestion control	Required
Interaction between N-ISDN and INAP	Required
Unequipped circuit identification code (in Q.BICC, CIC = CCA-ID)	National option
ISDN user part availability control	Not required
MTP pause and resume	Required
Over length messages	Required
Temporary alternative routing (TAR)	Required
Hop counter procedure	Required
Hard-to-reach	Required
Calling geodetic location procedure	Required
Generic signalling procedures	
End-to-end signalling — pass along method	Required
End-to-end signalling — SCCP connection orientated	Required
End-to-end signalling — SCCP connectionless	Required
Generic number transfer	Required
Generic digital transfer	Required
Generic notification procedure	Required
Service activation	Required
Remote operations service (ROSE) capability	Required
Network specific facilities	Required
Pre-release information transport	Required
Application transport mechanism (APM)	Required
Redirection	Required
Pivot Routing	Required
Supplementary services	
Direct-dialling-in (DDI)	Required
Multiple subscriber number (MSN)	Required
Calling line identification presentation (CLIP)	Required
Calling line identification restriction (CLIR)	Required
Connected line identification presentation (COLP)	Required
Connected line identification restriction (COLR)	Required
Malicious call identification (MCID)	Required
Sub-addressing (SUB)	Required
Call forwarding busy (CFB)	Required
Call forwarding no reply (CFNR)	Required
Call forwarding unconditional (CFU)	Required
Call deflection (CD)	Required
Explicit call transfer (ECT)	Required
Call waiting (CW)	Required
Call hold (HOLD)	Required
Completion of calls to busy subscriber (CCBS)	Required
Completion of calls on no reply (CCNR)	Required
Terminal portability (TP)	Required
Conference calling (CONF)	Required
Three-party service (3 PTY)	Required
Closed user group (CUG)	Required
Multi-level precendence and pre-emption (MLPP) [Note: only transiting of MLPP information is supported]	See note
Global virtual network service (GVNS)	Required
International telecommunication charge card (ITCC)	Required
Reverse charging (REV)	Required
User-to-user signalling (UUS)	Required
Additional functions/services	
Support of VPN applications with PSS1 information flows	Required
Support of number portability (NP)	Required

The call serving function (CSF) provides the BICC procedures on a per-call basis. The bearer control function (BCF) provides a point of flexibility, and could be considered to be drawing upon functionality in more general-purpose media gateway controllers. The simple media gateway functions (without any control functionality) is provided by the media mapping and switching function (MMSF).

Since the BICC standardisation concentrates on the requirements to support existing narrowband services utilising broadband technology, the BCF and MMSF are logically grouped together to form a bearer interworking function (BIWF). The

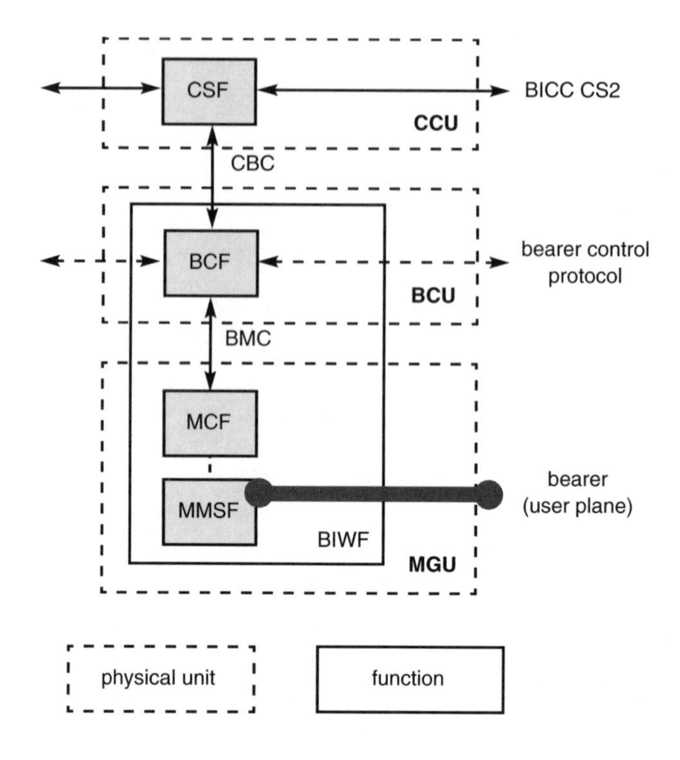

Fig 8.3 Decomposed CS2 serving node.

BICC architecture allows for the interface between the BCF and MCF to be open (the 'BMC' interface), but the split of functionality between the BCF and MMSF is not considered to be the responsibility of the BICC standardisation group.

A more general architecture allows a single call server to control multiple media gateways (MGs), and this is reflected in the BICC functional model as shown in Fig 8.4.

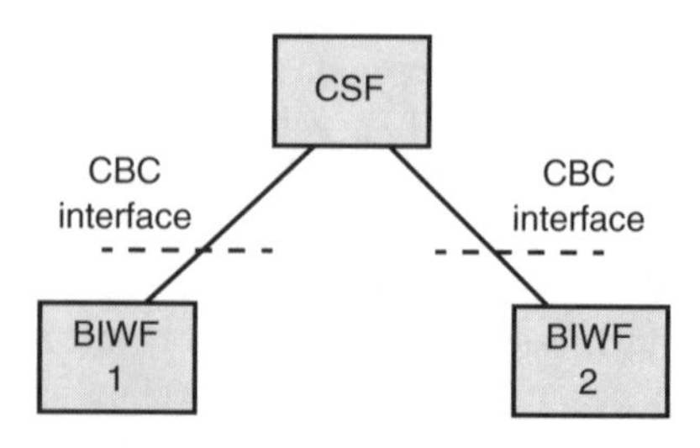

Fig 8.4 Control of multiple BIWFs.

The requirements for supporting mobile applications have also been incorporated, and this requires multiple CSFs to control a single BIWF as shown in Fig 8.5. The reasoning for this is to enable a call and bearer to be progressed to the mobile network, and the call only to be progressed within the mobile network to the terminal. At this point, tones may need to be applied, under the control of the call server in the mobile network, but implemented by the existing BIWF.

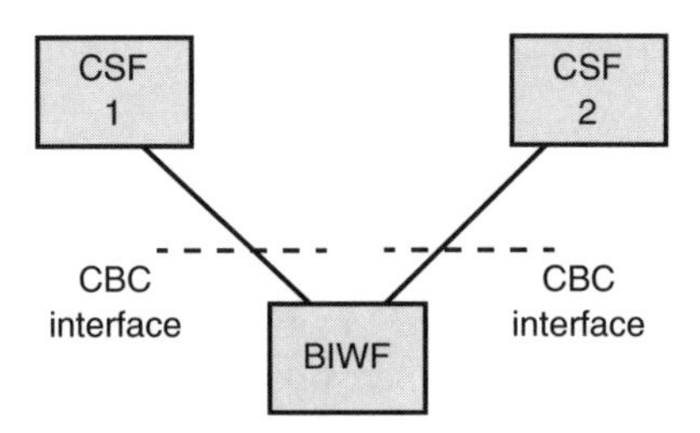

Fig 8.5 Control of BIWF by multiple CSFs.

Figure 8.6 shows a simplified functional model. It should be noted that BICC CS2 also supports all of the functionality of CS1, and therefore a TSN is also supported, but this is not shown. Also not shown is the support for access networks, which is an integral part of the ISUP-DSS1 interworking, and the IN functionality.

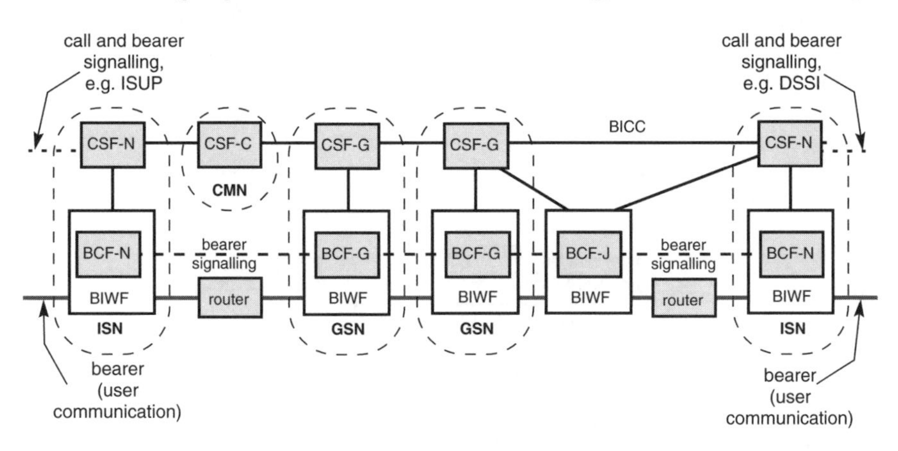

Fig 8.6 BICC CS2 simplified functional model.

8.5 BICC CS1

During the second half of 1999 and the early part of 2000, ITU-T Study Group 11 conducted intensive work on the initial capability set of BICC (BICC CS1). Although it has always been recognised by the designers of BICC that it should be both signalling transport and bearer transport independent, in order to provide a meaningful and useful CS1 package, in the time-scales allowed, the initial capability set work focused on a subset of BICC requirements. This subset was selected by reference to the immediate requirements of certain network operators and equipment implementors, but care was taken to prevent short-term expediency limiting the scope of future BICC capabilities.

In February 2000, BICC CS1 work was completed and formal ITU-T approval was given in June 2000.

8.5.1 What Does BICC CS1 Offer?

BICC CS1 is designed to allow a large incumbent ISUP-based network operator to migrate from using MTP3 signalling networks and TDM bearer networks towards alternative packet-based technologies. The BICC CS1 model allows an ATM segment to be inserted into an existing ISUP narrowband network without loss of ISUP or IN features and services (see Fig 8.7).

The BICC protocol is very closely based on the ISUP protocol. It has been designed to interwork 'naturally' with narrowband ISUP taking full benefit from its strong forward and backward compatibility features. This means that it has inherited all the narrowband ISUP features and services that are applicable in a separated bearer environment. Furthermore, non-BICC relevant ISUP signals can be transferred transparently between two PSTN/ISDNs connected by a BICC segment without loss of services offered.

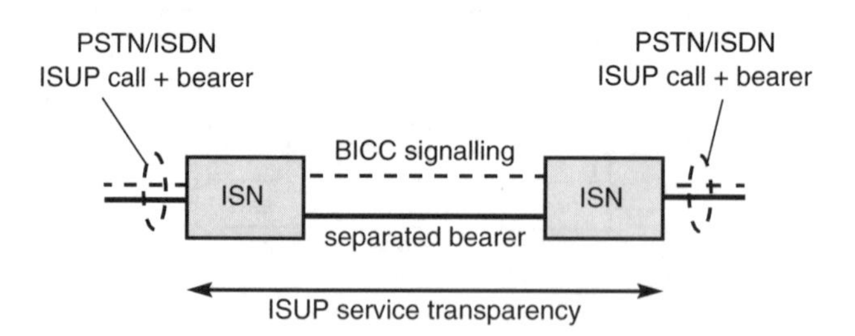

Fig 8.7 BICC segment inserted into an existing PSTN/ISDN.

BICC CS1 has also introduced new optional 'codec negotiation' and 'codec modification' capabilities not defined in narrowband ISUP. These will allow BICC to offer transcoder-free operation in core networks supporting, for example, mobile services. Transcoder-free operation improves speech quality by avoiding unnecessary transcoding between different speech encoding/compression schemes within the network.

8.5.2 The BICC CS1 Specification Model

The model of the interface serving node (ISN) used for BICC CS1 is shown in Fig 8.8. It can be seen that this is a subset of the comprehensive model used for BICC CS2 shown in Fig 8.6.

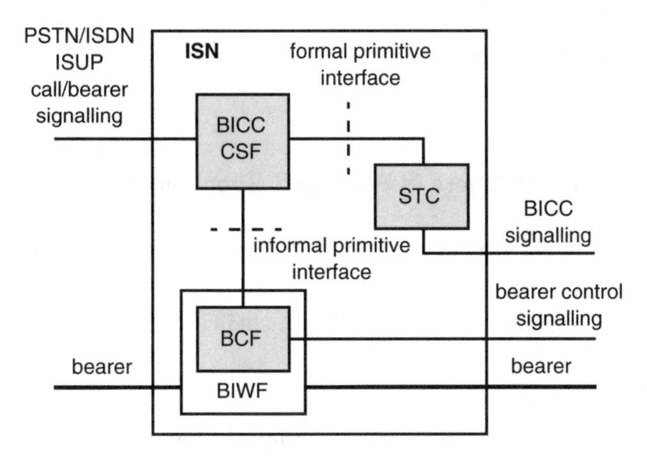

Fig 8.8 Functional components of the ISN as used for BICC CS1.

The call service function (CSF) performs three functions:

- interfaces to the traditional ISUP signalling in the TDM domain;
- provides a generic interface to the signalling transport-specific converter (STC);
- provides a generic (but loosely defined) interface to the bearer interworking function/bearer control function (BIWF/BCF).

The signalling transport converter maps the generic BICC on to a specific signalling transport. For BICC CS1, STCs are defined by ITU-T for the following signalling transports:

- MTP3/3B;
- SSCOP;
- SSCOPMCE.

The BIWF provides interworking of different bearer technologies (e.g. TDM to ATM). The BCF provides bearer-specific signalling to establish new bearers for use by BICC calls. A BICC CS1 CSF controls the BIWF/BCF through a generic, but loosely defined, primitive interface. To allow for adequate interoperability between the ISN implementations of different manufacturers, it is necessary to describe BICC mapping to/from specific bearer network technologies. For BICC CS1, mappings on to the following bearer network interfaces have been defined by ITU-T:

- AAL1 using DSS2 signalling;
- AAL1 using B-ISUP signalling;
- AAL2 using AAL2 (CS1) signalling.

Furthermore, the ATM Forum (ATM-F) has defined mapping of BICC on to:

- AAL1 using ATM-F UNI signalling;
- AAL1 using ATM-F PNNI signalling;
- AAL1 using ATM-F AINI signalling.

8.5.3 How and Where is BICC CS1 Specified?

BICC CS1 is specified as a 'delta' to ISUP2000. The ISUP2000 base text is defined by:

- ITU-T Recommendations Q.761 to 764 (including an Addendum to both Q.762 and Q.763);
- ITU-T Recommendation Q.765.

The BICC 'delta' is defined in ITU-T Recommendation Q.1901.

Annexes to ITU-T Recommendation Q.1901 describe interworking of ISUP with BICC at an ISN and the requirements for specific signalling transport converters (see section 8.7.4).

The specific application of the BICC/ISUP application transport mechanism (APM) used in BICC to support the separated bearer procedures is defined in ITU-T Recommendation Q.765.5 (2000).

Descriptions of BICC operating with ITU-T bearer control signalling systems are in ITU-T supplements:

- TRQ.3000 (operation of BICC with DSS2);
- TRQ.3010 (operation of BICC with AAL2 CS1);
- TRQ.3020 (operation of BICC with B-ISUP for AAL1).

The description of BICC operating with ATM-F bearer control signalling systems is in AF-CS-VMOA-0146.000 (operation of BICC with SIG4.0/PNNI1.0/AINI).

8.6 Call Flows

Figure 8.9 shows a simplified version of the call flows and the interactions between the vertical and horizontal interfaces. They also show how the BICC environment interacts with the PSTN/ISDN environment.

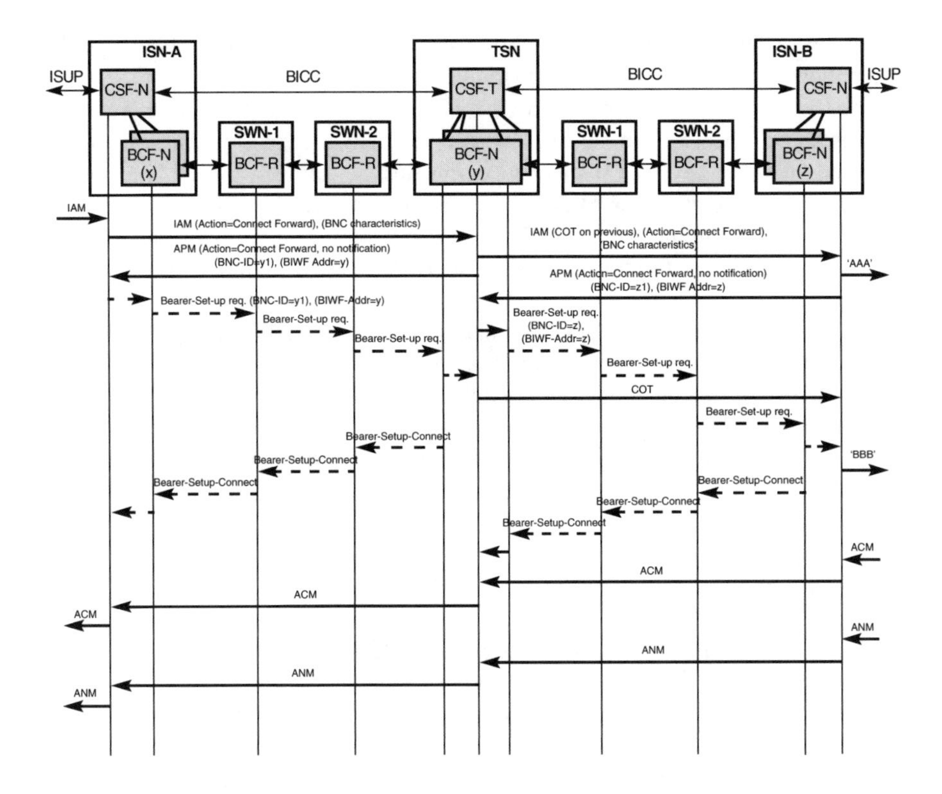

Fig 8.9 Call flow for forward bearer establishment of the backbone network.

The incoming IAM message is received at the originating ISN and the outgoing IAM is generated and sent out to the succeeding node with the BNC characteristics (identifying the connection type — AAL1, structured AAL1, AAL2, etc) encapsulated in an APP parameter.

Upon receipt of an APM message with the bearer addressing information, the CSF will then notify the BIWF of the request for a bearer and the bearer-addressing information.

The actions at the destination CSF are to notify the preceding node, in an APM message, of the addressing information required to set up a bearer. If this node is a TSN it will then generate and send out an IAM with a 'COT on previous' indication plus its BNC characteristics encapsulated in an APP parameter and await an APM with the bearer-addressing information. If this node is the destination ISN, the same action to send back an APM message applies, but the outgoing IAM with 'COT on previous' indication is sent to the ISUP call control in the TDM network. If the 'COT on previous' indication is not used then the outgoing IAM at the destination CSF will be delayed until the 'BBB' point.

Further APM message exchanges are optional depending on whether codec negotiation or tunnelled bearer information is deployed.

The COT message is then sent by the TSN once the bearer requirements have been met, and a bearer path has been established between the SNs. The destination ISN sending out an ISUP COT message completes the call.

The message flows for ACM and answer are the same as for ISUP.

The release signalling sequences, which are not shown, are also unchanged from ISUP, although an instruction from CSF to BIWF is also required to release the bearer.

The recommendations describe a number of options for establishing bearers, e.g. forward and backward bearer set-up, reuse of an idle bearer and use of a BICC tunnel to carry bearer signalling. Figure 8.9 shows a simplified version of a call flow (forward bearer set-up). The backward bearer set-up uses the same message flows except that the bearer characteristics and tunnelling information are generated at the destination BIWF and passed back between CSF in the first backward APM.

8.7 BICC Protocols

8.7.1 BICC Protocol in IP Networks (CS2)

BICC CS2 call control protocol builds on the work in CS1 in that the protocol is based upon ISUP; it includes most of the services supported by ISUP as the architecture now includes local exchange functionality. The impact of this has required that the basic call recommendation, developed in the ITU-T SG11, is now complete text instead of a delta on the ISUP recommendation as published in CS1. The recommendations in the basic call series are shown in Table 8.2.

Table 8.2 BICC basic call recommendation series.

BICC recommendation	Equivalent ISUP recommendation	Describes	Comments
Q.1902.1	Q.761	Scope	A complete stand-alone text
Q.1902.2	Q.762	Definitions	Shared with ISUP
Q.1902.3	Q.763	Formats and codes	Shared with ISUP
Q.1902.4	Q.764	Procedures	A complete stand-alone text

Any future developments of ISUP will now include new versions of Q.761 and Q.764 with the definition and layout of any new parameters and messages documented in Q.1902.2 and Q.1902.3. A complete listing of the various BICC documents can be found in the Appendix.

As BICC CS2 includes the local exchange functionality, it has to support the ISUP supplementary services as well as the originating and destination exchange functions, e.g. building the IAM and supporting call diversion. This is done by interworking the BICC functionality with the ISUP functionality as illustrated in Fig 8.10.

Some of the ISUP protocol is not required for the BICC protocol, either because it is a pure ISUP TDM circuit-control functionality or the service is not supported. So, for example, the circuit identification code (CIC) has changed its meaning in a BICC environment to call instance code and is larger than the ISUP CIC, thereby increasing the number of simultaneous calls that can be supported by the protocol (although exchange performance is more likely to be the limit on the number of calls supported).

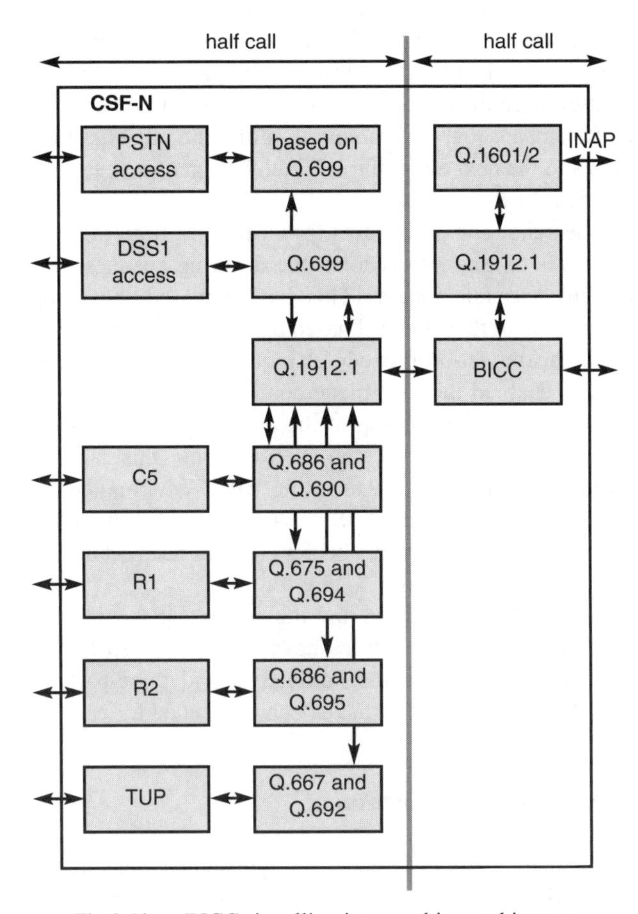

Fig 8.10 BICC signalling interworking architecture.

The ISUP local exchanges are the nodes which support the majority of the supplementary services functionality and as such the BICC local SN will have to emulate this to support the full PSTN/ISDN services. This has had an impact on some of the supplementary services, e.g. the call diversion supplementary service which could take account of the global call reference and therefore change the service. Other supplementary services such as conference and three-party calls need to take into account that more than one bearer could be involved — as this has not yet been resolved, future work may be required.

A new generic service, bearer redirection, has been developed in which the bearer is redirected to a new destination SN but the call is not. This could be deployed in the BICC network for interworking to an IN service such as a charge card service where the IN needs to be involved in the call but the bearer does not, and the call diversion services could also use this functionality.

The other important aspect of BICC CS2 is that it is supported on an IP bearer and that the bearer control and call control are separated. The model also allows for one CSF to control more than one BIWF and for a BIWF to be controlled by more than one CSF. This means that the amount of information that has to be carried from the originating SN to the next SN in the call path has increased; this is for call/bearer correlation and identification purposes.

A number of bearer set-up procedures for a call have been included in the BICC signalling procedures to take into account the differing bearer set-up methods, i.e. forward, backward and idle bearer. This has increased the complexity of the signalling procedures, but they are still based upon the ISUP signalling.

As some of the bearer set-up procedures require a confirmation that each SN in the bearer path has enabled the user plane communication, this could result in the destination SN holding up the sending of the outgoing IAM. This could lead to an unacceptable delay in setting up a call; to overcome this the ISUP continuity signalling procedure has been included in the BICC environment. The continuity 'test' of the connection as defined for ISUP is not included in BICC, but the COT signal has been reused to indicate that a bearer path has been set up and that the IAM is allowed to go forward from the BICC environment.

To allow for network optimisation and the support of mobile telephony, codec negotiation and modification procedures have been added to the BICC protocol. This could have an impact upon the conference and three-party supplementary services where a different codec may have been negotiated for each call leg.

8.7.2 IP Bearer Control Protocol

IP networks are fundamentally different from connection-oriented networks, such as those based on TDM and ATM, in that signalling messages do not need to be acted upon by all network elements in order for a user plane communication to take place. If the gateways or interworking points can exchange IP addresses (including port numbers), then the IP network can route the user plane packets between the two gateways. This is sufficient for a best-effort service; however, an IP network that provides a level of guaranteed service must act upon any request for a connection. Since BICC is only concerned with reusing existing protocols within a network based on IP technologies, an assumption has been made concerning the provision of quality of service. A request for a connection simply allocates the bandwidth at the gateway that receives the request. Also, IP networks do not automatically set up bi-directional communications and to provide this (vital for conversational speech) a flow of information in each direction is required. Fortunately, this exchange of information can be very simple — the exchange of the IP addresses and port numbers, plus some information on the type of media required, is all that is needed.

In IP networks utilising SIP [4], the information concerning the media is defined separately in the session description protocol (SDP) [3]. Although this is referred to as a protocol, it is really only the definition of the format of the data (and its meaning) that can be exchanged. SDP specifies no messages and is therefore carried by SIP. In order to make BICC as compatible as possible with SIP-based networks, this approach was also adopted. The BICC call control messages have been extended to carry (optionally) the SDP media information. This information is derived at the MMSF (see section 8.4.2) and must be carried by the CBC (H.248) interface vertically to the BICC call control and then is passed horizontally as information tunnelled within BICC. This feature allows SDP information to be exchanged to 'open up' the user plane communication, without the BICC recommendation needing to redefine all the information specified in SDP. It is referred to as a tunnel, because the BICC protocol does not inspect, check or understand the information it passes.

To allow the tunnelling mechanism to be generic, ITU-T has produced a separate recommendation defining a header at the start of the tunnelled information to identify the protocol and a version number. The version number allows future generic capabilities to be included (such as segmentation and re-assembly).

8.7.3 BICC Use of H.248 and the Megaco Protocol

The original intention of the BICC CS2 work was to reuse existing signalling protocols. When considering the CBC, the vertical interface, the obvious candidate was H.248 and the Megaco protocol. A fundamental problem was identified — the need for BICC information to be exchanged between the CSF and BCF. This information included the tunnelling information as well as some BICC-specific information. The solution was to use the generic procedures of H.248 or Megaco as far as possible and then to extend the functionality by defining a new package. One of the more interesting aspects of this work has been to define a package that is initiated solely by the BCF. Further information on H.248 and Megaco can be found in Chapter 7.

8.7.4 Signalling Transport

As discussed in Chapter 6, an important aspect of any architecture is the ability to transport signalling messages between peer entities. The goal of BICC was to use whatever already exists, allowing an adapted ISUP to be 'grafted' on to new network technology. The signalling transport therefore had to provide BICC with exactly the same services that ISUP receives from the lower layer signalling transport, i.e. SS7 message transfer part level 3 (MTP3) primitives. On the other hand, it must be deployed in networks that do not implement SS7 (or only adaptations and emulation of SS7). Reconciling these two apparently disparate requirements, without specifying new signalling transport mechanisms for each network technology to be supported, was achieved using signalling transport converters as shown in Fig 8.11.

The signalling transport converters document the procedures and mapping to adapt the generic primitives defined between ISUP and MTP3 on to specific primitives for the particular signalling transport.

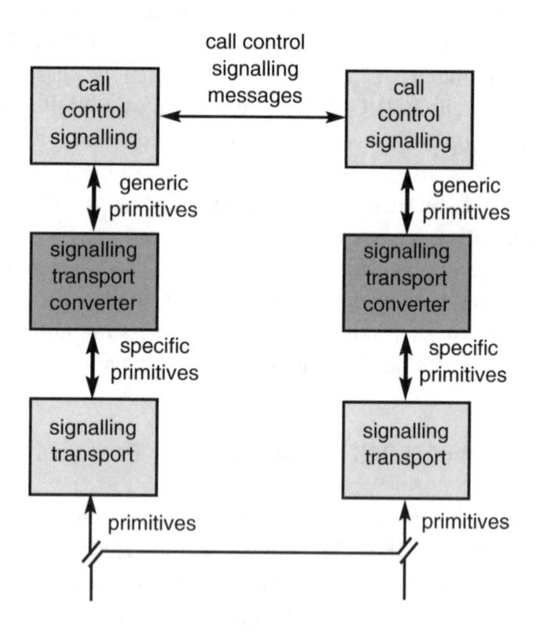

Fig 8.11 Signalling transport architecture.

A number of converters were documented (rather than specified), showing how the ISUP to MTP3 primitives were to be adapted to provide the same service over alternative transports.

Since all of the interfaces were internal, the specification was largely an informative documentation trick. The supported signalling transports are MTP3, MTP3B, SSCOP (including multi-link), SCTP and M3UA.

8.8 BICC Time-Scales

Bearer-independent call control was first brought into ITU-T as a concept for supporting narrowband services over ATM AAL1 in November 1998. The work was properly started in March 1999 and the first approval step completed in December 1999, with final approval for publication in June 2000. This is something of a record for the ITU, especially in the area of signalling (Study Group 11). The CS2 activity started in February 2000 and the first approval step was completed in December 2000, with final approval for publication achieved in July 2001. The pace of the work has surprised many people outside the ITU-T, and demonstrates a willingness to move into a new pattern of working that is more responsive to the needs of industry.

A complete listing of the various BICC documents and their status can be found in the Appendix.

8.9 Relationship Between BICC and ETSI TIPHON

BICC provides an architecture designed for providing narrowband services utilising an IP network, but there has also been other more general work on voice over IP (VoIP). In particular, ETSI TIPHON has produced a generalised architecture for the support of VoIP (see Chapter 16 for a detailed discussion of TIPHON). The following differences are summarised from the ETSI TIPHON and BICC documents:

- the CSF entity in BICC performs call control functions similar to the CC layer in the TIPHON architecture (H.225/SIP) and in addition the codec negotiation procedure that is specified as part of the BC layer in IP networks (H.245/SDP);
- the BCF entity performs functions similar to the BC layer (except codec negotiation) and part of the MC layer in the TIPHON architecture — this refers to the MC actions for transport signalling like address allocation, codec selection and QoS control (delay, bandwidth, etc);
- the MMSF entity performs the media mapping and switching actions as part of the MC layer.

The physical grouping of functions in a VoIP network differs from a BICC network mainly with regard to the positioning of the bearer control functionality, as shown in Fig 8.12.

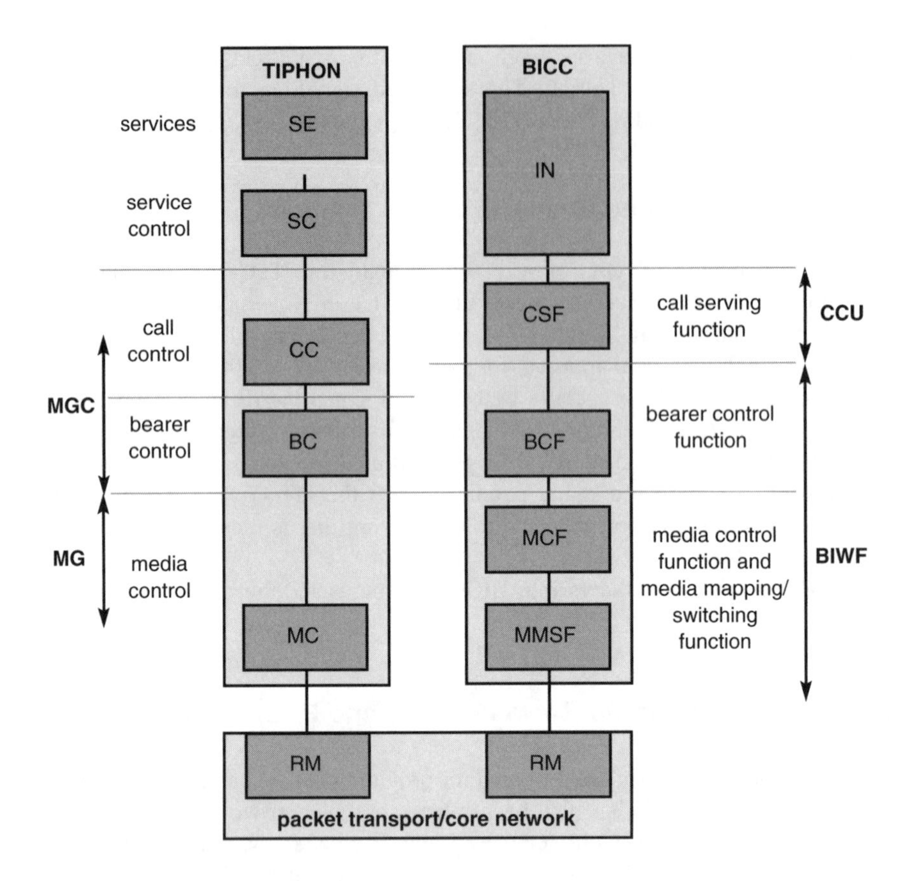

Fig 8.12 Mapping between the ETSI TIPHON architecture and the BICC architecture.

8.10 Summary

This chapter has given a comprehensive introduction to how the bearer-independent call control offers a complete network solution to provide fully all of the existing PSTN/ISDN services into networks utilising technologies other than circuit switched. While there has been considerable work on SIP to try to emulate PSTN/ISDN services, BICC offers an immediate solution to public network operators with a large customer base.

A further advantage of BICC could be that it will enable SIP to develop independently of the current circuit-switched services, and therefore more efficiently provide exciting new communications possibilities without being restricted by legacy technology and services.

Appendix

Bearer-Independent Call Control Documents

Table 8.A1 BICC CS1 protocol documents.

Document Number	Title	Date of Publication
Q.1901	Bearer-independent call control protocol	June 2000
Supplement to Q.1901	Implementors' Guide For Recommendation Q.1901	May 2001
Q.2150.0	Generic Signalling Transport Service	
Q.2150.1	Signalling Transport Converter on MTP3 and MTP3b	
Q.2150.2	Signalling Transport Converter on SSCOP and SSCOPMCE	

Table 8.A2 BICC CS1 requirements documents.

Document Number	Title	Date of Publication
TRQ 2140	Signalling requirements for the support of narrowband services via broadband transport technologies	December 1999
TRQ 2600	BICC signalling transport requirements, capability set 1	May 2001

Table 8.A3 BICC CS1 and CS2 mapping documents.

Document Number	Title	Date of Publication
TRQ 3000	Operation of the Bearer-Independent Call Control (BICC) Protocol with Digital Subscriber Signalling System No. 2 (DSS2)	December 1999
TRQ 3010	Operation of the Bearer-Independent Call Control (BICC) Protocol with AAL Type 2 Signalling Protocol (CS1)	December 1999
TRQ 3020	Operation of the Bearer-Independent Call Control (BICC) Protocol with Broadband Integrated Services Digital Network User Part (B-ISUP) Protocol for AAL Type 1 Adaptation	December 1999
AF-CS-VMOA-0146.000 (ATM Forum)	Operation of BICC with SIG4.0/PNNI1.0/AINI	

Table 8.A4 BICC CS2 protocol documents.

Document Number	Title	Date of Publication
Q.1902.1	Bearer-Independent Call Control Protocol (CS2) functional description	July 2001
Q.1902.2	Bearer-Independent Call Control Protocol (CS2) and Signalling System No. 7 — ISDN User Part, general functions of messages and parameters	July 2001
Q.1902.3	Bearer-Independent Call Control Protocol (CS2) and Signalling System No. 7 — ISDN User Part Formats and codes	July 2001
Q.1902.4	Bearer-Independent Call Control Protocol, basic call procedures	July 2001
Q.1902.5	Exceptions to the application transport mechanism in the context of Bearer-Independent Call Control	July 2001
Q.1902.6	Generic signalling procedures and support of the ISDN user part supplementary services with the Bearer-Independent Call Control Protocol	July 2001
Q.1912.1	Interworking between Signalling System No. 7 ISDN user part and the Bearer-Independent Call Control Protocol	July 2001
Q.1912.2	Interworking between selected signalling systems (PSTN access, DSS 1, C5, R1, R2, TUP) and the Bearer-Independent Call Control Protocol	July 2001
Q.1912.3	Interworking between H.323 and the Bearer-Independent Call Control Protocol	July 2001
Q.1912.4	Interworking between Digital Subscriber Signalling System No. 2 and the Bearer Independent Call Control Protocol	July 2001
Q.1922.2	Interaction between the Intelligent Network Application Protocol Capability Set 2 and the Bearer-Independent Call Control Protocol	July 2001
Q.1950	Bearer-Independent Call Bearer Control Protocol	July 2001
Q.1970	BICC IP Bearer Control Protocol	July 2001
Q.1990	BICC Bearer Control Tunnelling Protocol	July 2001
Q2150.3	Signalling Transport Converter on SCTP	

Table 8.A5 BICC CS2 requirements documents.

Document Number	Title	Date of Publication
TRQ 2141.0	Signalling Requirement for the Support of Narrowband Services via Broadband Transport Technologies, CS 2	December 2000
TRQ 2141.1	Signalling Requirement for the Support of Narrowband Services via Broadband Transport Technologies, CS 2 — Signalling Flows	December 2000
TRQ 2410	Signalling Requirements Capability Set 1, for Support of IP Bearer Control in BICC Networks	December 2000
TRQ 2500	Signalling Requirements for the Support of Call Bearer Control Interface	December 2000

Table 8.A6 BICC CS2 (only) mapping document.

Document Number	Title	Date of Publication
TRQ 3030	Operation of the Bearer-Independent Call Control (BICC) Protocol (CS2) with IP Bearer Control Protocol	December 2000

References

1 Griffiths, J. M.: '*ISDN Explained*', John Wiley & Sons Ltd, London (1990).

2 ITU-T Recommendation Q.2630.1: '*AAL Type 2 Signalling Protocol (Capability Set 1)*', (December 1999) — http://www.itu.int/publications/itu-t/

3 IETF: '*SDP: Session description protocol*', RFC 2327 (April 1998) — http://www.ietf.org/

4 IETF: '*SIP: Session initiation protocol*', RFC 2543 (March 1999) — http://www.ietf.org/

9

NUMBERING AND NAMING IN VoIP NETWORKS

Q G Collier

9.1 Introduction

As previously described in Chapter 1, the operation of VoIP networks, such as that shown in Fig 9.1, requires both signalling messages and media data to be sent between the constituent elements involved in the system. For a pure VoIP solution, both of these types of data will be transmitted in the form of the data payload of IP packets. These data packets contain the actual signalling information required to flow between the terminals and call servers to establish and control both calls and media streams. They also transport the actual media information directly between the media end-points. In either case, the data packets need to have defined points of origination and destination to enable them to be correctly routed across the data network. These identities are therefore an integral element of how VoIP works and are one manifestation of the subject of numbering, naming and addressing related issues that are the topic of this chapter.

In the case of a hybrid network environment that includes connectivity with the existing PSTN, however, the signalling and media streams will also need to flow through an appropriate gateway device and within the PSTN switching elements themselves. Among other things this creates two additional areas of complexity. Firstly, from the view of placing a call from an IP network into the PSTN, the call signalling must also ultimately provide information that enables a call to be successfully delivered to a telephone connected to the PSTN. Secondly, in the case where a call is originated in the PSTN, the PSTN must be able to route the call to an appropriate VoIP gateway that can onward route the call over the IP infrastructure to a VoIP end-point.

In either of these cases, the scheme used to identify and refer to the constituent elements and users of a network is an essential part of any VoIP solution. In particular, the identification scheme used must enable the network elements and users to be uniquely identified within the context of the overall system. This chapter considers the means used to identify and enable communication between these constituent elements. It also goes on to describe, and to provide definitions for, the alternative means of identification involved, and considers the functionality required for the routing of calls that traverse VoIP networks.

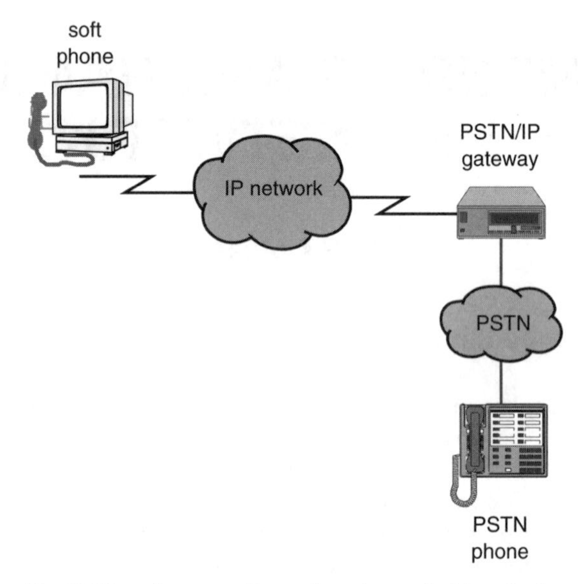

Fig 9.1 The VoIP environment from a 'naming and addressing' perspective.

9.2 Names and Addresses

As previously implied, entities within IP networks are identified by an IP address [1]. So in the case of VoIP networks, IP addresses will be required for customer terminals whether they are IP phones or PCs with VoIP clients, call servers or PSTN gateways. While IP addresses can easily be handled by machines, the way in which they are structured as either 32-bit (IPv4) or 128-bit (IPv6) binary numbers [2] is not easily amenable to human end users. In addition, for many kinds of public service, it is unreasonable to assume that all potential users will be highly computer literate. This lack of user friendliness is compounded by the fact that two versions of IP address may be in use — IPv4 and increasingly IPv6. Furthermore, due to the common use of IP addresses being dynamically allocated via DHCP, the IP address in use by a given end user will change between different log-on sessions. In such cases, an actual IP address is therefore of little practical use for permanently identifying end users directly since it changes too frequently; by analogy it is as if your postal address changed each time you went through your front door!

When considering such issues it is important to reflect on the fact that after more than a century of using the telephone, people are actually accustomed to the use of numeric identifiers and generally experience little difficulty with using them. However, what ideally is required by a human end user is an identifier that is user friendly and easily readable. It must also be time-invariant in that the identifier must not change over short periods of time since this would confuse potential callers as to which identifier should be used. Also it should be compatible with legacy networks and customer premises equipment (CPE) to ensure as wide a community of users as possible. Unfortunately, in practice, these requirements are not fully compatible. User friendliness implies the use of the full alphanumeric character set, while compatibility with legacy equipment implies the use of only the numeric characters, plus the characters * and #.

These requirements point to the need for two fundamentally different types of identifier to distinguish that used for the end user from that used for the network equipment. For this reason names and addresses are identified and recognised as separate and different entities. A useful definition of these two entities which recognises this distinction has been developed by the ETSI SPAN (Services and Protocols for Advanced Networks) group as follows:

- a **name** is a combination of alpha, numeric or symbols that is used to identify end-users;
- an **address** is a string or combination of digits and symbols which identifies the specific termination points of a connection or session and is used for routing.

In either case, it is important to recognise that both names and addresses need to be unique in the context of the system in which they are used. Otherwise calls cannot be reliably routed to the correct user or via the correct network element. It is important to note here that this can present an issue where two systems are interconnected if they have incompatible or overlapping name or address spaces. As a practical example, this has been found to be a common problem when enterprise users wish to connect two virtual private networks (VPNs) together — especially if the VPNs concerned have been developed independently.

Generally, however, names are more user friendly than addresses since they do not require the user to have a detailed knowledge of the structure of the network being used. This can also be an advantage when two public networks are connected together since the name rather than the address can be used as the identifier that points to the end user. As an example to illustrate this point, it is instructive to briefly examine how names and addresses have evolved as separate concepts as switching technology has developed. This examination also highlights a number of requirements that relate to the routing of VoIP calls.

Historically when electromechanical step-by-step switching systems, as typified by Strowger telephone switching systems, were first introduced, the digits dialled by the caller were used to directly step the selectors in the exchanges. The identity, or directory number, of the called customer as recognised by callers was directly determined by where the customer's line was terminated on the switching system hardware. Therefore, based on the above definitions, the name and address were identical.

Such a scheme made inter-exchange dialling possible. However, because the digits dialled by the caller directly actuated the selectors in the switching system, the caller in effect had to be aware of the physical structure of the network. Any change to the network structure as it evolved would therefore require a number change. In practice, experience has shown that number changes are very unpopular with customers due to the disruption it creates, such as requiring new stationery to be printed. Therefore this effectively imposes a practical limit on the complexity of the network that can be easily supported.

In the case of the UK, the introduction of 'Director working' in six major cities from the late 1920s meant that the network was able to examine and manipulate the digits dialled by the caller. The ability of the network to insert routing digits of which the customer was unaware was quite radical. This enabled a more complex network structure to be in place, which could absorb changes to the network necessitated by growth, changing traffic patterns and the like, transparently from the end customer — avoiding the need for unpopular number changes. Initially this only applied to traffic within the six cities involved. Long-distance traffic was still handled by manual operators. However, the ability for customers to dial their own long-distance calls was progressively introduced until by the mid-1970s, all long-distance calls could be dialled directly by customers in the UK.

A subsequent step was taken with the introduction of crossbar switching systems into the UK in the 1960s. In crossbar systems, once the call had reached the terminating exchange, a translation was made from the directory number to an equipment number which reflected the architecture of the switch and indicated where on the switch hardware the customer's line was terminated. This meant that the termination point of the customer's line on the switch hardware no longer directly determined the called directory number and one of the fundamental characteristics of step-by-step systems was removed. The differentiation between names and addresses that began with the introduction of routing digits was therefore further widened by this directory-number-to-equipment-number translation. This separation of directory numbers from equipment numbers has been carried forward from crossbar, through electronic systems such as TXE4 to modern digital switched-circuit systems such as System X and AXE10.

More recently, with the move of the telecommunications regimes in many countries away from monopoly and towards open markets and competition, it has been necessary to recognise that increasingly calls will not terminate in the network in which they originated. Arrangements have therefore been necessary for the interconnection of different operators' networks. However, in order to maintain network integrity, one network operator will generally prefer not to reveal the details of their network structure to another network operator. The interconnection agreements involved therefore tend to work on the basis of the transfer of a name, rather than an address, between networks. Since it is easy to see the potential for

VoIP networks to emerge as both another network type and also a replacement for the existing network, it is not unrealistic to consider this practice of using names rather than addresses being carried forward.

9.3 E.164 Numbers as Names

For high volume commercial deployment of VoIP in a public network environment, it is necessary to have access to and from all other public telephony customers since that is where the majority of existing users are served. In terms of naming, this means that the VoIP user must be able to make and receive calls using the same type of identification as is used on the PSTN. The requirements for the international public numbering scheme are set out in ITU-T Recommendation E.164 [3], and such numbers are therefore referred to as E.164 numbers.

Having established the clear distinction between names and addresses, it is worth clarifying that E.164 numbers are best thought of as names which are wholly composed of numeric characters. In the past it has been possible to argue that geographic numbers, described later in this chapter, had some of the characteristics of addresses. This is because the inspection of such numbers provided a degree of information about where in the network they were terminated. However, the introduction and increasingly widespread use of number portability means that it is no longer possible to know purely by inspection of the number where in the network it is terminated. So the address-like characteristics of E.164 numbers are therefore becoming increasingly diluted with the passage of time.

9.3.1 E.164 Numbers

Recommendation E.164 recognises three classes of international public telecommunications number (IPTN), covering IPTNs for geographic areas, global services and networks.

IPTNs for geographic areas constitute by far the greatest proportion of numbers currently in use, so these will be described in some detail. In contrast IPTNs for global services and networks, which represent specialised niche applications, are only described in outline. It is expected that VoIP networks may be called upon to support any of these three classes of IPTNs, and it would therefore be unwise for designers of VoIP systems to take any design decisions which would preclude the use of any given class of IPTN.

9.3.1.1 IPTNs for Geographic Areas

IPTNs for geographic areas comprise three significant groups of numbers that are structured as shown in Fig 9.2.

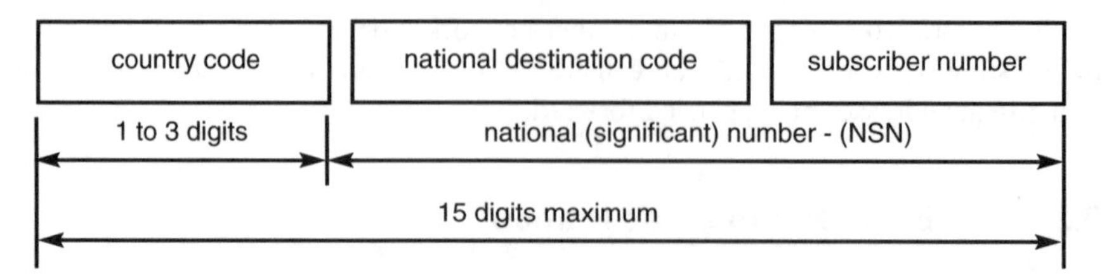

Fig 9.2 Structure of IPTNs for geographic areas.

The maximum number length is fifteen digits with the country code comprising from one to three digits, which are assigned by the ITU-T. For example, the ITU-T have assigned the country code for the UK to be 44. The national (significant) number (NSN) normally consists of two parts, the national destination code and the subscriber number. Responsibility for the NSN for a given country, as defined by the country code, is delegated by the ITU-T to national regulatory authorities. In the case of the UK, the national regulatory authority is Oftel. Numbers are then assigned by national regulatory authorities to public network operators in blocks, usually of 10 000 numbers. The national destination code (NDC) can represent:

- a geographic area, for example in the UK, the NDC 1473 represents Ipswich;
- a service, for example the NDC 800 is widely used globally for freephone services;
- a network, for example a particular mobile network.

9.3.1.2 IPTNs for Global Services

In contrast to IPTNs for geographic areas, IPTNs for global services are structured as two digit blocks, as shown in Fig 9.3.

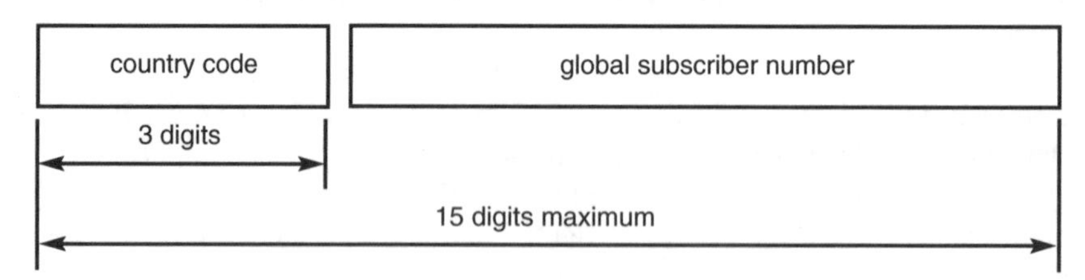

Fig 9.3 Structure of IPTNs for global services.

Country codes for global services, which are always three digits in length, are assigned by the ITU-T, and identify the specific service being delivered, e.g. Universal Freephone. Unlike IPTNS for geographic areas, responsibility for assignment of individual global subscriber numbers remains with the ITU-T rather than being delegated to a national regulatory authority.

9.3.1.3 IPTNs for Networks

IPTNs for networks are structured with three significant groups of digits as shown in Fig 9.4.

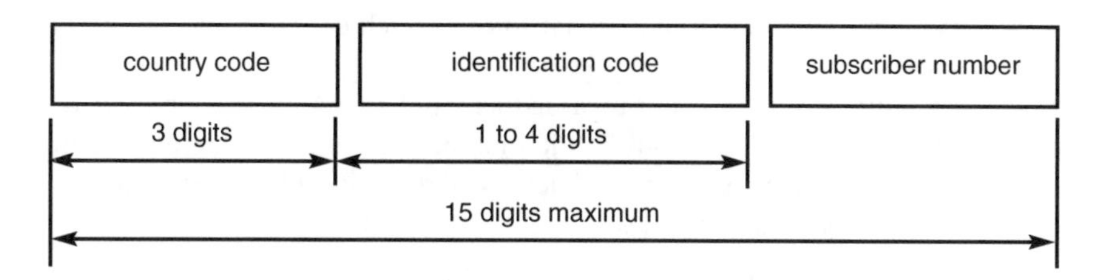

Fig 9.4 Structure of IPTNs for networks.

Country codes for networks, which are always three digits in length, are assigned by the ITU-T. The identification code identifies the specific network. It should be noted that private networks do not qualify as networks in this context, as they do not meet the criterion of providing 'public correspondence services'. Further, for a public network to qualify for a country code it must have a presence in at least two different countries. Responsibility for assignment of individual subscriber numbers is delegated to the organisation to whom the identification code is assigned. At present only country code, 882, is in use for this purpose.

9.3.2 E.164 in Practice

In all the above cases, where the full IPTN is dialled, it must be prefixed by the international prefix.

This is increasingly standardised as 00 in accordance with Recommendation E.164, but represented by the symbol '+' to cover those cases where 00 has not yet been adopted.

However, in the case of numbers for geographic areas, it is not necessary to always dial the full IPTN including the country code. For instance, it is also possible to dial a:

- local number, consisting solely of the subscriber number — provided the caller is in the same geographical area as defined by the same geographic NDC as the called number;
- national number, consisting of the NSN prefixed by a trunk prefix, for which the digit 0 is recommended by E.164.

So, for example, subject to the caller being in the correct geographical area, a call to the same number could be dialled in the following ways:

Local number	654321
National number	01494 654321
International number	00 44 1494 654321.

It should be noted that in some countries, local number dialling is not permitted, and all calls have to be dialled in national format. Such countries are said to have closed number plans.

Local and national numbers are regarded as valid E.164 numbers provided the number can be reached by dialling the number in full international format. Conversely, numbers that cannot be reached by dialling in full international format are not E.164 numbers. This includes:

- access codes (e.g. 999 or 112);
- numbers prefixed by number portability codes;
- numbers prefixed by carrier selection codes;
- numbers prefixed by carrier pre-selection codes;
- private number plans — these are number plans that are not part of the public numbering space running on private networks.

As VoIP technology becomes more widely used and terminal equipment capable of supporting the full alphanumeric character set becomes deployed in increasing quantity, there will be increasing opportunity to support alphanumeric names, probably of the 'user@host' format. The need to be able to coexist with legacy networks and CPE means that such terminals will need to retain E.164 numbers for the foreseeable future, but, subject to the capability of the network, will be able to support 'user@host' (e.g. quin.collier@bt.com) names in parallel for on-net VoIP applications. Such names are referred to as aliases. At some point in the future, a situation will be reached when the balance between new and legacy CPE is such that it is appropriate to consider moving to public names of the user@host format, possibly supported by E.164 aliases to allow interworking from residual legacy CPE. Formulation of such a naming scheme will need to consider not just the naming requirements in isolation, but also a wide range of related topics such as calling line identity (CLI), emergency call, lawful intercept, and portability.

9.4 IP Addresses — IPv4 and IPv6

Initial widespread deployment of the Internet was based on Internet Protocol v4 [4]. However, it soon became clear that the explosive growth of IP usage meant that the address space available within IPv4 was unlikely to be sufficient to meet potential demand. Accordingly an initiative was created in the IETF to define a new version of the Internet Protocol, with a vastly increased addressing range. The new version, known as IPv6 [5], has the ability to support multiple addresses for every person on the planet, as well as allowing individual devices to have IP addresses of their own. This latter point offers some interesting customer possibilities, such as your vacuum cleaner sending you a message to say that it needs emptying!

However, IPv6 networks are in the formative stages of deployment and, in the short term, a number of measures [2] have been introduced which allow IP addresses to be conserved. The main ones are the use of classless inter-domain routing (CIDR) which reduces the address consumption from a routing point of view and network address translation (NAT) which provides a point of decoupling between private and public IP addressing domains. NAT is a useful tool for address management since it permits an organisation with an intranet to allocate 'private' IP addresses to large numbers of individual users while limiting the number of 'public' IP addresses required. This concentration is analogous to the PSTN, where a large number of customer lines are concentrated on to a much smaller number of inter-exchange trunks based on the level of traffic carried.

NAT and similar devices such as firewalls are generically referred to as middleboxes and while their use violates the 'end-to-end' principle espoused by the Internet community, their use is widespread. In particular, it is believed that at the end of the 20th century, there was a marked slow-down in IPv4 address consumption while traffic growth continued. It is highly likely that this may be attributable in part to the deployment of NAT technology. From a VoIP perspective, however, the use of middleboxes is likely to be an inherent feature of carrier-scale deployment of VoIP systems, in which calls will typically traverse separate but interconnected IP networks. This is because their deployment enables a network operator to establish a closed high-integrity environment suitable for real-time traffic that would otherwise be difficult to establish and maintain. In particular, it is expected that individual networks may use private IP addresses internally while interconnecting them using middleboxes with public IP addresses acting as border elements between the networks. At present however, devising systems which can handle signalling and media streams in the presence of middleboxes presents a significant challenge to the designers of VoIP systems, as discussed in Chapter 5.

Another major challenge faced by VoIP designers is the fact that IPv6 is not backward compatible with IPv4, thus making migration to IPv6 less than straightforward. A number of alternative approaches to migration have been proposed, but it will be necessary for VoIP systems to be able to operate in any of the resultant environments.

9.5 Call Routing

Numbers and names play an integral part in routing VoIP calls. As an illustration, the following discussion concentrates on the case where E.164 numbers are used as names. This is highly likely to be the case in public network environments for many years to come for the reasons previously stated. However, while the discussion is predicated on E.164, many of the principles will also apply to cases where other naming schemes are used, such as 'user@host'.

Any routing in an E.164 environment is based on an examination of a digit string, which may be either the called party number or a routing number. The digit string will be contained in one or more signalling messages sent between the various network elements involved in a call, e.g. between a terminal and a call server or between one call server and another call server. As an aside, some signalling systems provide a separate information field for the routing number, whereas others provide a single field, the digits within which then being changed as the signalling messages pass across the network. However, in many, and probably most, cases calls are routed on the basis of the called party number alone, although a number of circumstances do exist where the called party number on its own does not contain sufficient information to enable the call to be routed as required. In such cases a routing number is used which will generally consist of the called party number prefixed by additional digits. Such circumstances include the following.

- Carrier selection

  This is where the caller chooses to route calls via a network other than that of their usual service provider. In this case the caller, on a call-by-call basis, prefixes the called party number with a numeric code indicating the network of choice. So for example a customer wishing to call 01473 304321 would enter the digits 136401473304321 via their keypad, where 1364 represents the chosen carrier.

- Carrier preselection

  This is similar to carrier selection, but the caller's decision to route calls via a carrier other than their usual service provider is an across-the-board one applying for all calls, and the prefix digits are inserted automatically by the network. Using the example in the case above, the caller would enter only 01473304321, and the prefix digits 1364 would be automatically added by the network. Routings within the network would be based upon the resultant digit string 136401473304321.

- Number portability

  Where a customer has chosen to change their service provider but to retain the same number, it is necessary to route incoming calls to the new service provider's network. One of a number of technical solutions, and the one adopted in the UK, involves prefixing the called party number with a prefix which allows the call to be routed to the appropriate point of interconnection between the networks.

9.5.1 Underlying Principles

Call routing is best explained by describing a typical call set-up. Before doing so in detail, however, it is important that a number of underlying principles are understood as described below.

9.5.1.1 Establishing an End-to-End Signalling Path

An end-to-end signalling path must be established before the media path can be set up. It is only once the characteristics of the media path have been determined, by the interchange of signalling information between the user terminals, that media path set-up is possible. However, as indicated in Chapter 1 there is no implicit need for signalling packets and media packets to follow the same path.

The purpose of establishing signalling paths is to allow call control entities to communicate with each other. The call control entities may be located in either user terminals or call servers. Since signalling messages may need to be acted on by call control entities located in call servers, server routed signalling, as described in Chapter 1, is used.

Given the current competitive telecommunications environment, a significant proportion of calls will terminate in networks other than the one in which they originated. Establishment of an end-to-end signalling path will therefore in many cases be a multi-stage process.

9.5.1.2 Routing Tables

The heart of the call routing function is a routing table based in a call server. This maps the digit string contained in a signalling message to the identity of the next call control entity which can deal with the call. In some instances it will be necessary to examine the entire digit string to achieve the mapping to the next call control entity, e.g. in the case of terminals logged on to the current call server, or for some implementations of number portability. In other instances, it may only be necessary to examine part way down the digit string. For example, the practice by which national regulatory authorities allocate numbers to network operators in blocks, usually of 10 000 numbers, means that it may only be necessary to partially analyse the digit string to identify the next network, and hence call control entity, to be used.

The mappings between digit strings and associated call control entities, as reflected in the routing table, may be transient or semi-permanent.

In the case of user terminals, an IP address will normally be allocated as part of the terminal registration process, and the mapping of a terminal E.164 number to an IP address in the call server will only last for the duration of the terminal registration session. However, for other network elements, the mappings may be of much longer duration, in the order of months or even years, and may only change when significant network reconfiguration takes place, e.g. when additional call servers are provided to increase network capacity. These issues must not be understated since, for large networks, they can represent substantial operational issues which need to be carefully planned and executed to maintain network integrity.

9.5.1.3 Call Identification

Signalling messages may be carried over either TCP or UDP, and well-known port numbers have been allocated for this purpose. For example, port 1720 is used for H.323 call signalling messages. Since in the case of a call server the same IP address and port number are used for signalling messages related to a large number of different calls, it is necessary to have a means of tagging a signalling message to identify the call to which it relates. This is typically done using a number referred to as the call reference value (CRV).

9.5.1.4 Media Establishment

Once a signalling path has been established between the two user terminals, end-to-end message transfer takes place — possibly mediated through the call server. For example, in the case of H.323 this is done using the H.245 protocol either directly between the end-points or via the gatekeeper. This covers such aspects as the exchange of terminal capabilities, flow control and opening and closing of media streams. One of the key items of information carried in the end-to-end message exchange is the identity of the RTP port numbers and associated RTCP port numbers to be used by the media stream(s) — these are determined by the end user's call control function. As stated above, it should be noted that there is no requirement for the signalling and media streams to follow the same paths. Furthermore, it is only in the special case of calls carried within a single network that a media stream will use the same IP address and port number throughout. Individual IP networks will in many cases use private IP addresses internally. Because it is probable that the same (private) IP addresses are used in other networks, it is essential that those addresses, and associated port numbers, are not passed across network boundaries if misroutings are to be avoided. This can be achieved by provision of middleboxes at boundaries between interconnected IP networks, as previously indicated.

9.5.2 Example Call Set-Up

An example of a typical call set-up is shown in Fig 9.5 and represents the case where VoIP is being used for a telephony-like service. Since the purpose of the example is to demonstrate the key routing principles described above, it does not represent the full complexity that could be met in a real network environment. In particular:

- only two networks are involved;
- there is no number portability required;
- only one call server is used per network;
- routing on the basis of E.164 number alone is possible.

In this example, the call set-up is based loosely on H.323 (see Chapter 10). However, the full message interchange has not been shown in order to focus on the numbering and naming aspects related to call routing.

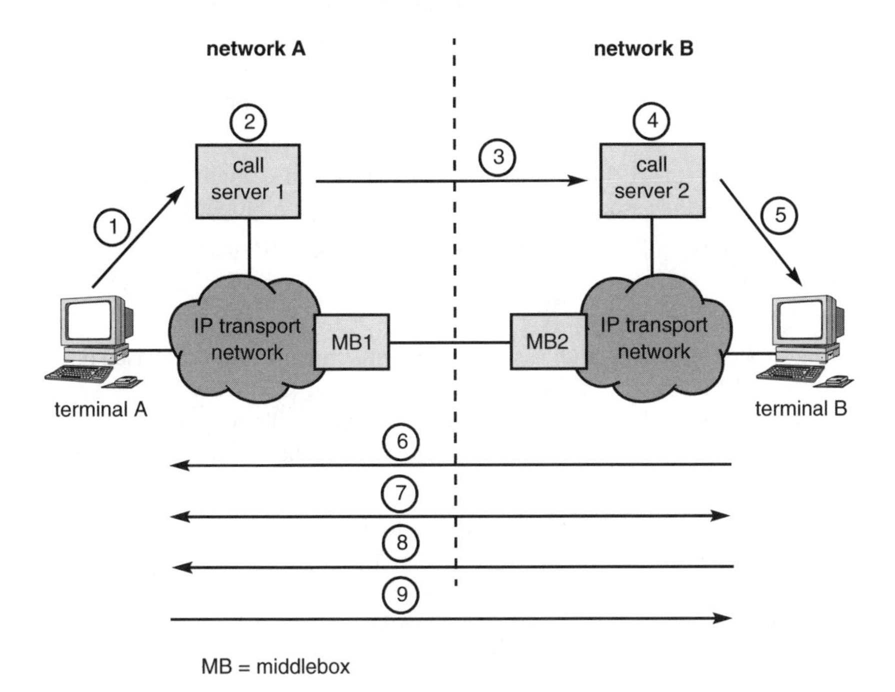

Fig 9.5 Illustrative call set-up in VoIP environment, showing information flows.

In the above example, it is assumed that terminal A has successfully completed the call server discovery/registration process, and is registered with call server 1 (CS1). Terminal B is similarly registered with CS2.

The core information flows shown in Fig 9.5 are illustrative of a possible scenario as follows.

1. Terminal A sends a signalling message to CS1, using the IP address of CS1 obtained at registration time and the well-known port number allocated for call signalling. A CRV allocated by terminal A is used to identify signalling messages used for this call. The signalling message contains the E.164 number of terminal B.

2. CS1 examines the E.164 number contained in the signalling message and determines from the routing table that the associated terminal is not associated with network A, and identifies that a further signalling message needs to be sent to CS2.

3. CS1 sends a signalling message containing the E.164 number of terminal B to CS2, using the IP address for CS2 and the well-known port number used for call signalling. Again, a CRV is used to is identify signalling messages used for this call. Note that the IP address used may be changed by the middleboxes, part of whose function is to prevent private IP addresses from network A being forwarded into network B.

4. CS2 examines the E.164 number contained in the signalling message and determines that the associated terminal is allocated to and logged to network B.

5. CS2 sends a signalling message to terminal B via CS1 using the IP address obtained from the routing table, using the well-known port number for call signalling and a CRV.

6. An end-to-end signalling path is set up between terminals A and B, based on their respective IP addresses and mediated via CS1 and CS2. However, as before, mapping of IP addresses at network boundaries will be performed by the middleboxes. This mapping may be achieved under the control of CS1 and CS2 respectively.

7. End to end message interchange between terminals A and B takes place, based on the IP addresses of the two terminals as mediated via CS1 and CS2. This will include negotiation of terminal capabilities. Once again, mapping of IP addresses at network boundaries will be performed by the middleboxes.

8. and 9. Once the necessary bearer capabilities have been established, media paths are set up. This is performed using a message interchange opening up a logical channel in each direction. These messages contain the dynamically allocated RTP (and associated RTCP) port numbers to be used for the media streams. RTP ports are even numbered, and the associated RTCP ports use the next higher odd number. Where a middlebox is not deployed between the two IP networks, the

RTP port addresses will be exchanged, via CS1 and CS2, to A and B respectively. Where a middlebox is deployed, CS1 may provide an RTP port address on MB1 to A and CS2 similarly may provide an RTP port address. CS1 and CS2 may then establish RTP streams between MB1 and MB2 to complete the end-to-end media path as three separate segments.

Although the above example represents a voice call, it should be noted that where multimedia calls take place, separate logical channels, with individual RTP/RTCP port numbers, are established for each media stream as explained in Chapter 1.

9.6 Future Directions

Over the last decade or two, a wide range of different electronic communication media have become available. In addition to the basic fixed voice line, consumers now have a choice of fax, mobile telephones — supporting short message service as well as voice calls — and e-mail to name but a few. Increased competition has also made it possible and attractive for customers to change service providers to get the most attractive and appropriate package of services they can.

The undoubted benefits of increased customer choice result in the need for customers to spend time managing their communications, if maximum advantage is to be obtained. For instance, every time a customer changes their contact details, for example by changing their e-mail provider, they will want to ensure that a wide range of potential communicants are notified of the change. Delivering such notification can involve significant effort and cost — and of course some of the communicants may have changed their contact details in the meantime! Increasingly intense life-styles mean that customers often wish to remain contactable at all times, and they may choose to optimise the chances of this happening by letting potential callers know the best means of reaching them at a particular time — although paradoxically, in some instances they may wish to be selective about who is able to contact them at particular times, such as when they are at home in the evenings. The customer may also wish to place constraints on the medium used for incoming communications. For instance a customer wishing to keep in touch while travelling on business might find it impossible to handle e-mail, but could quite easily cope with voice mail.

9.6.1 Universal Communications Identifier

Work currently under way in ETSI is considering the idea of a universal communications identifier (UCI), which would provide a flexible means of addressing the above and other challenges of a communications-intensive world. Although still at the concept stage, the likely key features are discussed below.

9.6.1.1 *Single, Unique Identifier for a Customer*

It is envisaged that a single and globally unique identifier is required. This is envisioned as consisting of a numeric part and a user friendly alphabetic part. The numeric part is likely to be based on an E.164 number. This has two advantages — firstly, the existing E.164 number allocation processes would ensure uniqueness, and, secondly, use of E.164 numbers would ensure that customers on legacy networks could still establish contact with UCI users.

9.6.1.2 *Network-Based Personal User Agent (PUA)*

It is envisaged that the information concerning the user would be handled through an electronic agent. This would contain a profile, updated by the UCI user, identifying a range of communications media by which they could be contacted. Use of a PUA-based profile would allow the following advantages to be realised:

- changes to contact information would need to be input once only by the UCI user, there would be no need to notify potential communicants of changes individually, and the potential communicants would be guaranteed access to the latest contact information;
- UCI users could indicate their preferred media for incoming communications, e.g. other things being equal, a UCI user might prefer to be called on a fixed line rather than a mobile because of potentially lower call charges and better speech quality;
- screening of calls based on caller identity would be possible, such as allowing calls from family members to be taken at home, but causing other calls to be steered to a voice mail facility;
- in the case of a profile updatable in near real time, the means of communication most likely to succeed at a given time could be chosen, e.g. when the UCI user has left the office to go to a meeting, it would be preferable to contact them on a mobile telephone rather than at the office number.

9.6.1.3 *Interaction between PUAs*

Where one person who is a UCI user, and hence has a PUA, wishes to communicate with another person who is also a UCI user, their two PUAs would exchange information with each other. By comparing the two parties' preferences for communications media, the PUAs may choose, and establish the optimum medium automatically without the need for manual selection by the originator. This would offer a high level of privacy to a UCI user who might, while content to be contacted by particular communicants, not wish their contact details to be explicitly available to them. Further, where no match was possible, inter-media transcoding could perhaps be provided in the network and invoked as necessary. For example, where the originator for some reason was only able to send an e-mail, but the recipient was not able to receive it, it might be possible to convert an originated e-mail into a fax which the recipient was equipped to receive.

Where only the recipient of a communication was a UCI user, their PUA would choose the appropriate way to contact the UCI user based on the capabilities of the calling person's terminal and services. Where a UCI user contacts a non-UCI user, the UCI user's PUA could only carry out basic communication management tasks such as making repeat attempts to establish contact. At an abstract level, Fig 9.6 indicates how such a scheme might work.

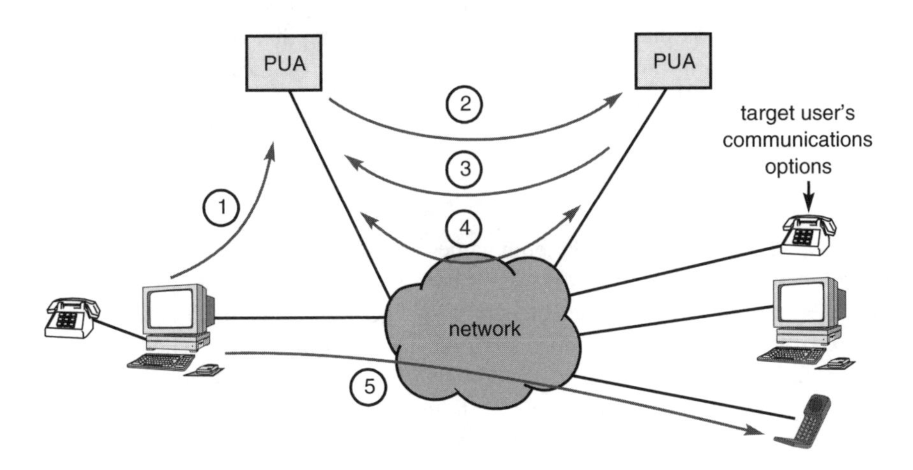

Fig 9.6 Possible operation of UCI.

9.6.2 Status of UCI

Work to date on the UCI has concentrated on the establishment of customer requirements. Further work is now taking place to understand the business and technical environment in which a UCI scheme could operate. For instance, if UCI deployment were widespread, significant benefits would be experienced by all users, since the benefits of interacting PUAs would be available on a large proportion of calls. However, in the early days when there will be few UCI users, such benefits would rarely be experienced and there would be little incentive for customers to become UCI users. Accordingly the issue of how a critical mass of UCI users would be built up in a reasonable timeframe is not certain. However, implementation of UCI solutions in a few large corporate networks would be one mechanism by which such a critical mass of UCI users could be rapidly achieved.

An IETF initiative, referred to as ENUM, is at the time of writing approaching the trial phase, and this will deliver some of the capabilities of the UCI. ENUM will use an E.164 number as a unique identifier, and the ENUM entries will be stored in a DNS-type database, where the alternative communications media will be recorded as NAPTR records. In this case, a client will convert the unique identifier into a URL format, and initiate a DNS query to the ENUM database. This will return a listing of the alternative communications media to the originating terminal, and these will be displayed by the client to the person originating the query, who can then, having selected a suitable communications medium, establish contact in the manner appropriate to the chosen medium. The work on ENUM has identified a number of issues relating to privacy and data protection. This has been compounded by the fact that, while ENUM and UCI are essentially global services, privacy and data protection laws that determine the commercial viability of such services vary greatly from country to country. Solutions to many of these issues have been proposed as part of the UCI work.

9.7 Summary

This chapter has considered why both names and addresses are necessary in a VoIP environment, and described the part that both play in the establishment of calls. It has highlighted the importance of names and addresses being unique within the context of the system in which they are used. The existing naming scheme based on E.164 numbers has been described and it has been explained why this will continue to be required for many VoIP applications. There are, however, a number of alternative options that may become available in the future and a method of handling some of these has been identified in outline.

References

1. Willis, P. J. (Ed): '*Carrier-scale IP networks: designing and operating Internet networks*', The Institution of Electrical Engineers, London (2001).

2. Roberts, P. A. and Challinor, S.: '*IP address management*', in Willis, P. J. (Ed): '*Carrier-scale IP networks: designing and operating Internet networks*', The Institution of Electrical Engineers, London, pp 217-233 (2001).

3. ITU-T Recommendation E.164: '*The international public telecommunication numbering plan*', (May 1997) — http://www.itu.int/publications/

4. IETF: '*Internet Protocol, DARPA Internet Program, Protocol Specifications*', RFC 791 (September 1981) — http://www.ietf.org/rfc/

5. IETF: '*Internet Protocol, Version 6 (IPv6) Specification*', RFC 2460 (December 1998) — http://www.ietf.org/rfc/

10

VoIP AND MULTIMEDIA WITH H.323

C D Wilmot and R P Swale

10.1 Introduction

Due to the ease with which applications can be developed, the Internet has created a revolution in the telecommunications and computing industries. Combined with the rapid growth in the use of personal computers and their inherent ability to conveniently access the Internet, interest has turned to enabling more dynamic applications and, in particular, those offering real-time conversational multimedia capabilities. However, to be able to widely deploy these applications in a useful way, it is necessary to have a common set of specifications for media transport, call signalling, and the like. These have to be defined as international standards, to allow equipment and services to be designed for a global market place, as discussed in Chapter 5. This chapter describes one such standard developed by the International Telecommunication Union (ITU) — H.323. It provides an overview of H.323, charts its development and highlights the areas that are currently under development.

As mentioned in Chapter 5, the H.323 standard has been developed alongside other H. series standards within Study Group 16 (SG16) of the ITU. The people who contribute to the development of the standard come from a wide range of companies, including network equipment suppliers, terminal equipment suppliers, communications stack and application developers, service providers and network operators.

Occasionally portrayed as a mammoth and complex protocol, H.323 is actually a comprehensive framework capable of constructing a wide range of conversational multimedia and communications solutions. It has been defined to enable voice, data and video to be sent over packet-based networks, initially starting with local area networks (LANs). As the standard brings different types of multimedia traffic together, through a common communications stack, it has enabled a range of new multimedia applications to be created, some examples of which are considered in

Chapter 13. In terms of deployed equipment and software, H.323 was the first complete standard adopted by the VoIP industry and is certainly one of the most widely used today.

While H.323 provides a framework for establishing and managing multimedia conferencing session and allows voice and video to be sent over the packet network, it does not as yet completely address the quality of service (QoS) issues considered in Chapters 2 and 3. Without special QoS mechanisms in the network, or careful network design practices, it is therefore only capable of delivering a best-effort service. Even so, substantial and complex applications can be constructed using this technology, as illustrated in Chapter 13.

10.1.1 Origins

Multimedia standards development in the ITU was originally with Study Group 15, who had developed H.320 — the conferencing standard for integrated services digital network (ISDN) networks. However, the development of H.323 was moved to Study Group 16 and began in May 1995 under the title of 'Visual telephone systems and equipment for local area networks, which provide a non-guaranteed quality of service'. It was originally intended as a standard only for audiovisual conferences over local area network (LAN), and H.323 version 1 was approved in June 1996. Interest in H.323 grew rapidly when the emerging VoIP industry recognised the need for an industry-accepted standard and identified H.323 as a candidate solution — pure voice over IP simply being an example of a multimedia conference with a single media type! The founding of the VoIP Forum, which subsequently became part of the International Multimedia Teleconferencing Consortium (IMTC), led to the wider adoption of H.323 version 1, with the standard becoming driven by the needs of the Internet telephony community. The scope for version 2 of H.323 was therefore widened and the title of the standard changed to 'Packet-based multimedia communications systems' to reflect its use in VoIP applications. This version was approved in January 1998. Subsequent versions 3 and 4 were approved in September 1999 and November 2000 respectively (see Fig 10.1).

10.1.2 Overview of H.323

As has been said, H.323 can appear to be a complex specification to navigate; for example Fig 10.2 shows an outline view of the current core H.323 protocol stack [1]. The body of the text and the associated annexes describe a set of functions and services required to implement a multimedia conferencing solution. This incorporates references to video and voice standards, including codecs and data collaboration standards. The H.323 standard also refers out to other lower level standards, some of which existed prior to the creation of the first version of H.323, such as the IETF RTP specification, and some of which have had to be developed in

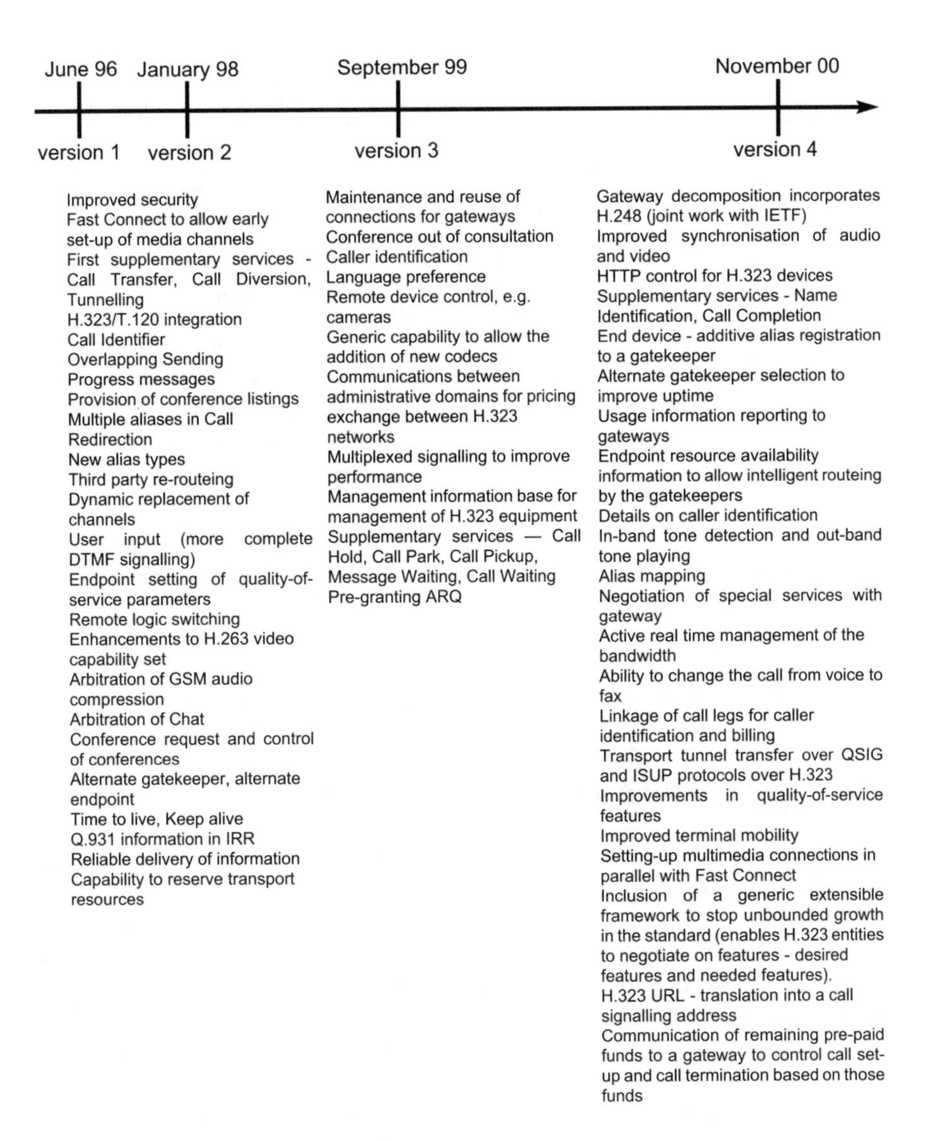

Fig 10.1 H.323 version timeline and functionality per version.

parallel with H.323, H.225.0 being an example of the latter. H.225.0 [2] is the recommendation for 'Call signalling protocols and media stream packetisation for packet-based multimedia communication systems' for H.323. It describes how various pieces of H.323 information are represented on the wire. The tables in the Appendix give a list of standards used in H.323 and details of the various annexes.

<table>
<tr><td colspan="3">media agents</td><td colspan="4">conference management and control</td></tr>
<tr><td>data applications</td><td>video applications</td><td>voice applications</td><td colspan="4">terminal control and management</td></tr>
<tr><td>T.124</td><td>H.261
H.263</td><td>G.711
G.723.1
G.728
G.729</td><td rowspan="2">RTCP</td><td rowspan="2">H.225
RAS
channel</td><td>H.225
call
signalling
channel</td><td>H.245
control
channel</td></tr>
<tr><td>T.125</td><td colspan="2">RTP</td><td colspan="2">X.224 class 0</td></tr>
<tr><td rowspan="3">T.123</td><td colspan="4">unreliable transport (UDP)</td><td colspan="2">reliable transport (TCP)</td></tr>
<tr><td colspan="6">network layer (IP)</td></tr>
<tr><td colspan="6">link layer (IEEE 802.3, 802.5, etc)</td></tr>
<tr><td></td><td colspan="6">physical layer (IEEE 802.3, 802.5, etc)</td></tr>
</table>

Fig 10.2 The H.323 core communications stack and relationship to other standards.

10.2 H.323 Version 1 — the Starting Point

Version 1 of H.323 was developed in a relatively short time, and driven by the commercial pressures to provide a conferencing-based communications solution for multimedia PCs with access to a local area network (LAN) connection path. However, while operation over IP-based networks was considered, it was not specifically designed for that. As previously described, since H.323 is actually a framework, it comprises a number of core specifications.

Version 1 of the standard identifies the main functional entities in an H.323 system. These include terminals, gatekeepers, various types of multipoint control units (MCUs) and gateways to other network types. The standard also references usage of H.323 on LANs using the IEEE802.3, IEEE802.5, IEEE802.10 and fibre distributed data interface (FDDI) protocols.

Gatekeepers are defined as entities to control access to the network for H.323 terminals, gateways and MCUs, by providing:

- address translation;
- admission control;
- bandwidth management;
- management of calls.

These functions are provided by the gatekeeper within its area of jurisdiction that is known as a gatekeeper zone. Due to their functionality, gatekeepers provide a convenient point for service integration as some or all of the end-point signalling may be routed through the gatekeeper, which may therefore also optionally provide:

- call control signalling;
- call authorisation;
- bandwidth management;
- call management;
- gatekeeper management;
- bandwidth reservation;
- directory services.

MCUs support conferences between three or more end-points by combining the received streams from end-points and sending the combination back out.

Gateways are defined as end-points on LANs providing real-time communications between H.323 terminals on the LAN and other terminals on a WAN, or to another H.323 gateway. Gateways provide:

- translation between different transmission standards;
- call set-up and clearing on both sides,
- possible translation between multimedia formats (audio, video, data);
- reflection of the characteristics of one end-point to another in a transparent manner.

A gateway is defined as also having the characteristics of a terminal or MCU. In version 1 the WAN terminals are those meeting H.310, H.320 (ISDN), H.321 (ATM), H.322, H.324 global switched telephone network (GSTN), H.324M (mobile) and V.70 standards.

The H.323 version 1 core specifications are listed in Table 10.3.

Table 10.3 H.323 version 1 compatibility.

H.323 ver 1	H.225.0 ver 1	h2250v1.asn	H.245 ver 2

10.2.1 H.323 Call Flows

Setting up and managing calls using H.323 occurs in five discrete phases, consisting of:

- Phase A — call set-up;
- Phase B — initial communication and capability exchange;
- Phase C — establishment of audiovisual communication;
- Phase D — call services;
- Phase E — call termination.

This sectionprovides an overview of the behaviour provided by H.323 in each phase. It is important to note, however, that there are a large number of different conditions that can be encountered in practice. The specification of the behaviour in these circumstances therefore becomes somewhat complex and further details are provided in the H.323 [1], H.245 [3] and H.225 [2] recommendations to which the reader is referred.

Call set-up takes place using the call control messages defined in H.225. The basic message set is self-explanatory and is outlined in one example, shown in Fig 10.3. The different types of call set-up that are supported relate to the possible interactions between the end-points and any associated gatekeeper. They include:

- direct party-to-party registration;
- both parties registered to the same gatekeeper, with the gatekeeper providing direct call signalling by the gatekeeper returning the end-point address of the called party;

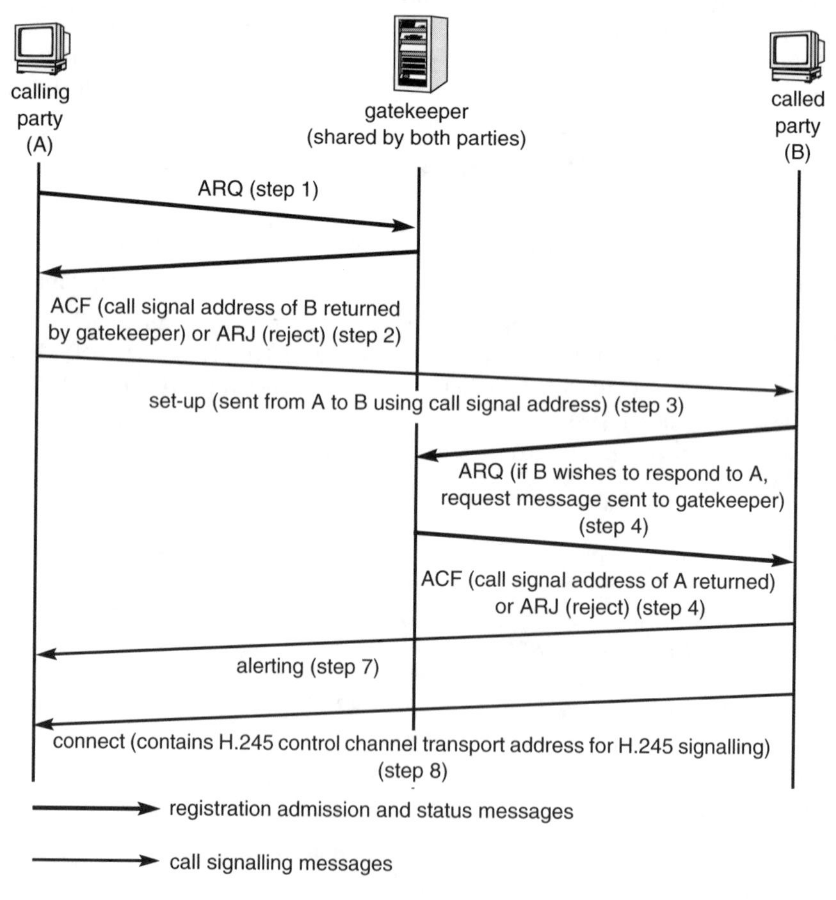

Fig 10.3 Call set-up messaging to a common gatekeeper.

- both parties registered to the same gatekeeper, with the gatekeeper providing routed call signalling by the gatekeeper returning its own address and then the gatekeeper setting up a call to the called party and, if the called party responds, then connecting the two parties together;
- only one party having a gatekeeper association;
- each party associated with different gatekeepers.

Initial communication and capability exchange is concerned with establishing H.245 control channels. To achieve this, the two parties exchange information about their capabilities (Fig 10.4) using capability set structures that are organised into definitions of simultaneous capabilities and presented as a capability table to the

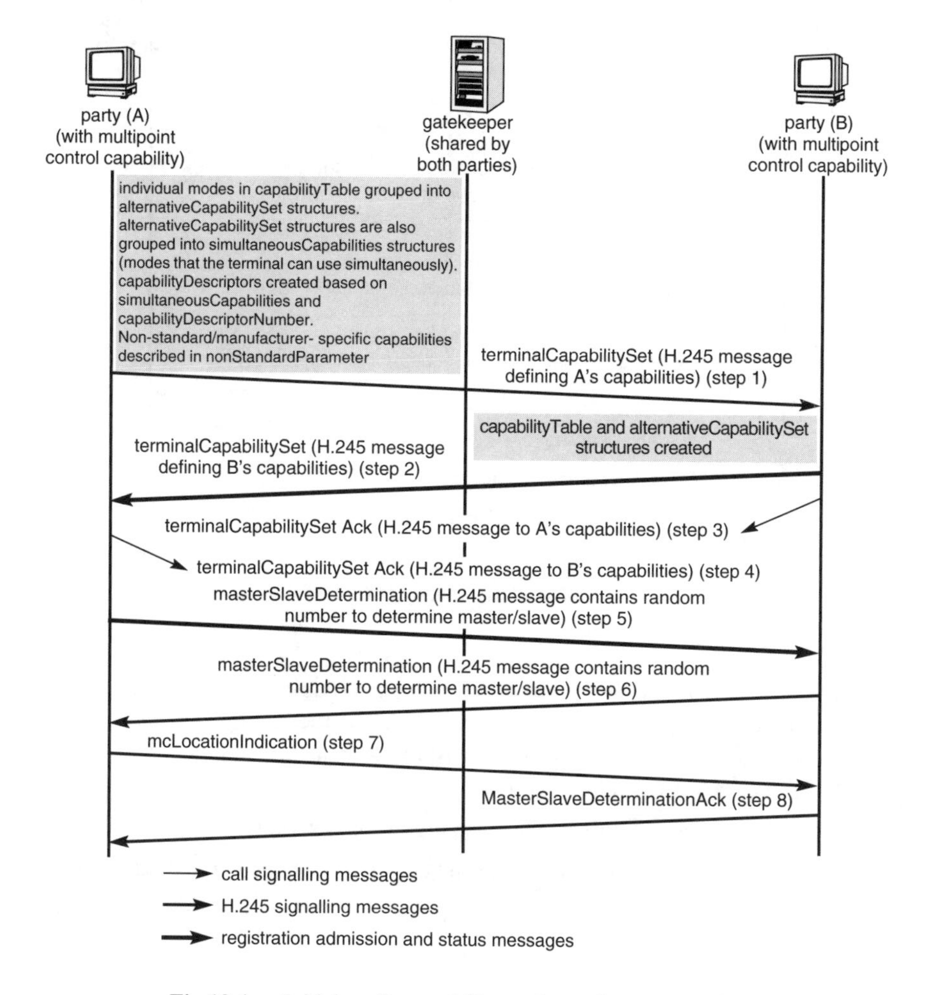

Fig 10.4 Initial media capability exchange between parties.

other party. Where both parties have multipoint control capability, the parties negotiate as to who shall be the master and who shall be the slave, by generating random numbers, sending them to each other and comparing them.

Establishment of audiovisual communication involves opening H.245 logical channels for the various information streams (see Fig 10.5). The audio and video streams, which are transmitted in the logical channels set-up in H.245, are transported over dynamic transport layer service access point (TSAP) identifiers using an unreliable protocol.

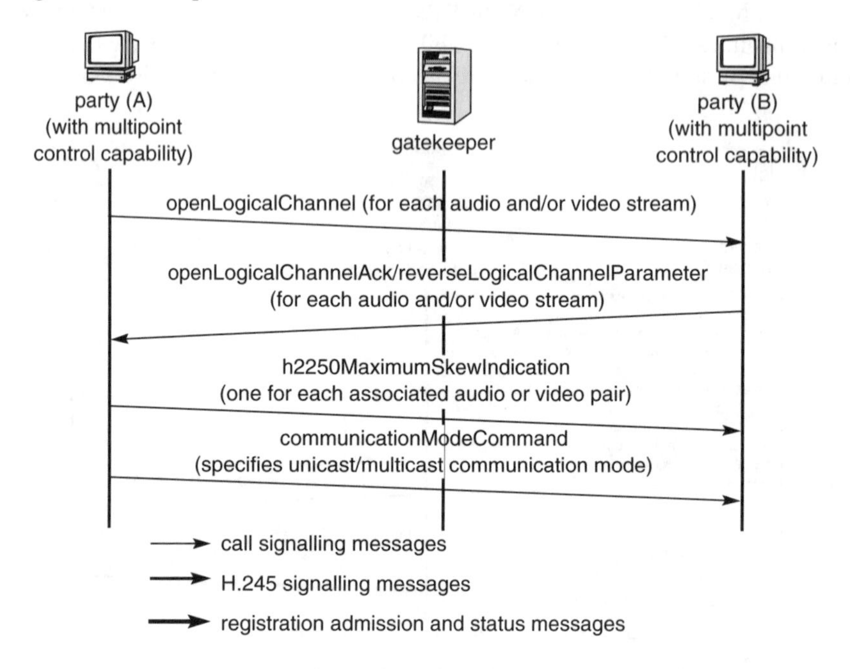

Fig. 10.5 Establishment of audiovisual communication.

Data communications are transported using a reliable protocol in the H.245 logical channels set-up for that data transmission. The messaging also determines if the logical channels are to be used for unicast or multicast communication.

Call services cover the following areas:

- bandwidth negotiation and changes (see Figs 10.6 and 10.7);
- party status — whether an end-point has been turned off, or entered into a failure mode;

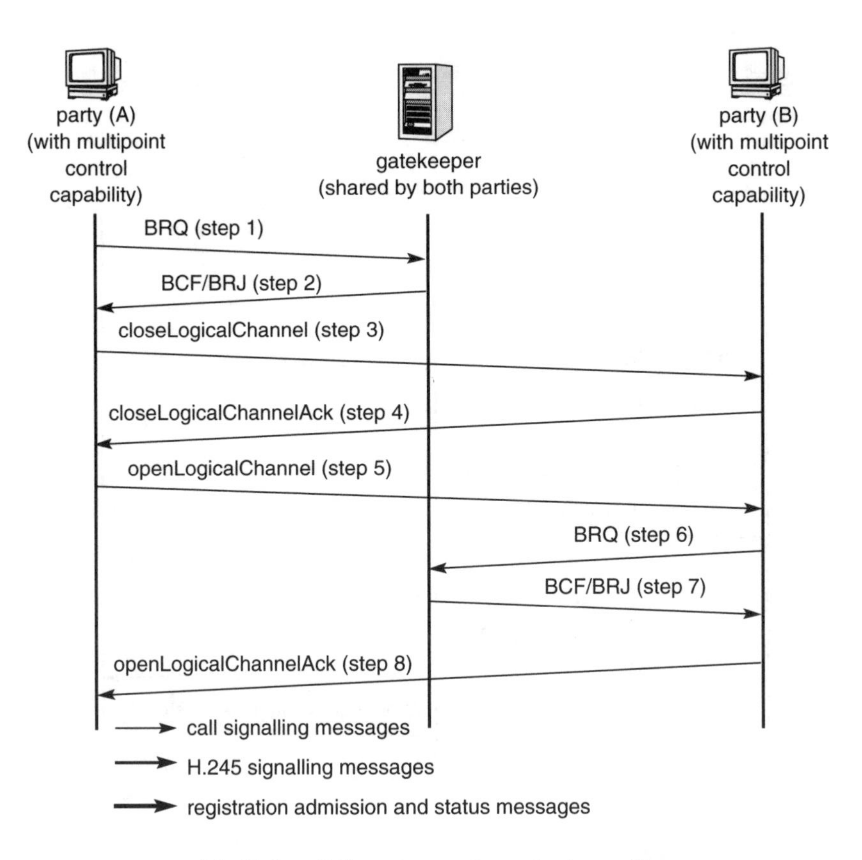

Fig 10.6 Call services exchange by transmitter.

- *ad hoc* conference expansion, i.e. from a point-to-point communication to a conference;
- cascading the multipoint capability to invite new people into a conference;

A basic call termination message sequence is given in Fig 10.8. Call termination allows either party to terminate a call in an orderly manner, allowing for the termination of any audio, video and data streams. Call termination covers the following cases of call clearing, being initiated by:

- either party with a gatekeeper involved;
- either party without a gatekeeper;
- a gatekeeper.

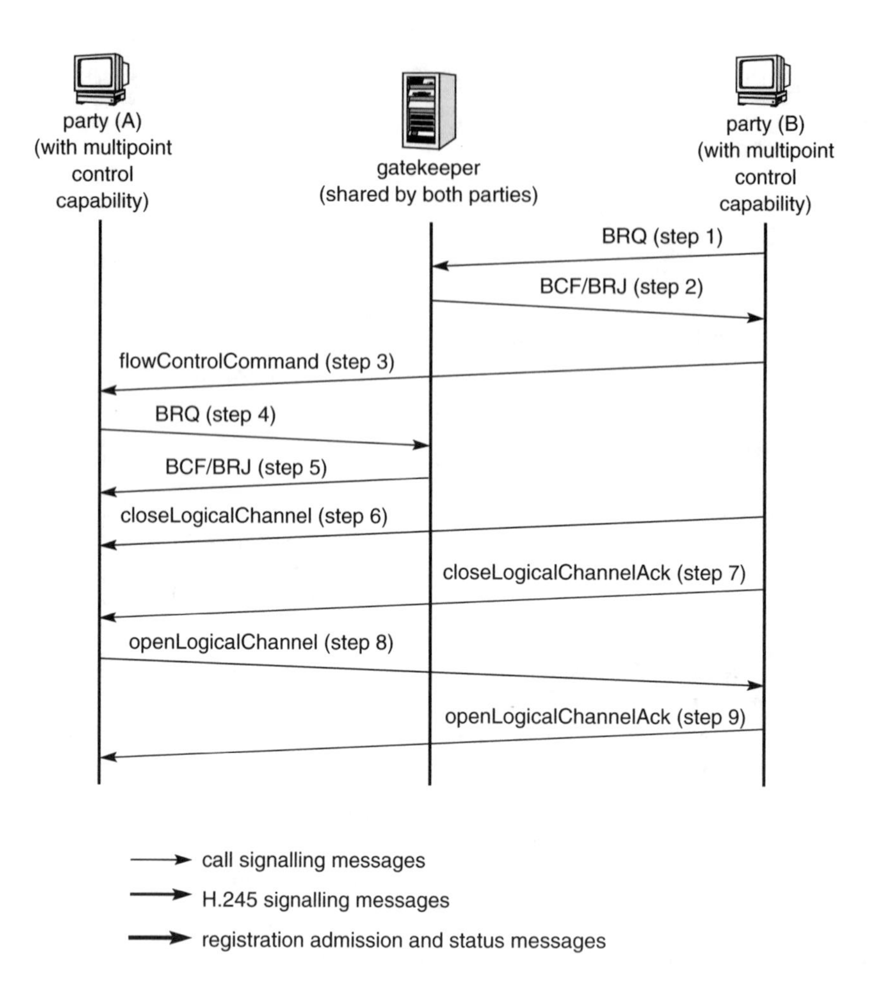

Fig 10.7 Call services exchange by receiver.

10.2.2 Error Handling

As with any communications technology, there are always cases where messages are lost, equipment fails or unforeseen events occur. An essential part of the H.323 specification therefore addresses these error conditions, allowing components of an H.323 system to respond gracefully and recover from the error situation. For this reason the detailed definition of the messages within H.245 and H.225.0 relies upon, and is declared, using ASN.1.

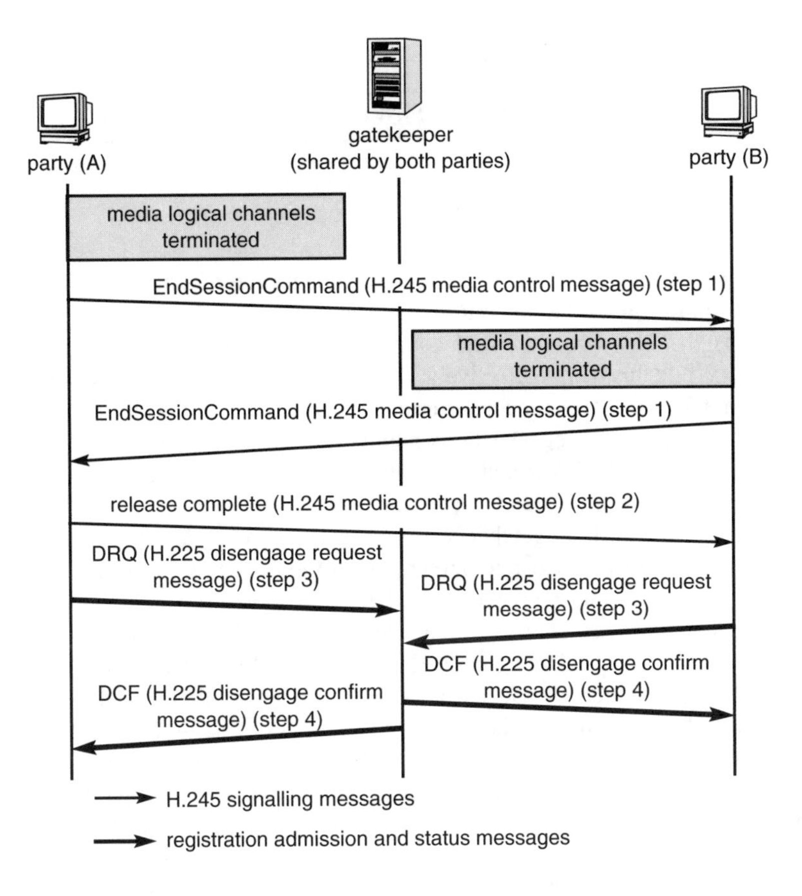

Fig 10.8 Call termination messaging to a common gatekeeper.

10.3 H.323 Version 2 — Speed and Scale

Once implementations of H.323 version 1 emerged, it rapidly became apparent that its use for large-scale VoIP applications would be highly restricted due to the serial signalling procedures and poor scalability of the server elements. In addition, version 1 only allowed the establishment and clearing of calls; it lacked many of the features taken for granted by a typical corporate PBX user. ITU-T SG16 therefore moved to address these shortcomings in developing version 2. The aim of the additions to the version 2 protocol were to increase the speed and scalability of operation, and the richness of the call handling and applications that could be supported. Table 10.4 lists the core specifications of version 2.

Table 10.4 H.323 Version 2 compatibility.

H.323 ver 2	H.225.0 ver 2	h2250v2.asn	H.245 ver 3

10.3.1 Speed and Scalability

Version 1 was constructed largely through the integration of a number of protocols under a common umbrella for multimedia conferencing. A consequence of this approach is that the signalling and control flows tend to occur in highly serial sequences. While this works effectively, it causes the call set-up and control process to be both laborious and time consuming, resulting in a significant number of message round-trip exchanges. This is mainly due to the call signalling (H.225.0) and bearer control (H.245) operating on different TCP connections that had to be opened and closed independently. This also affected the scalability of the H.323 network elements, such as gatekeepers, due to the number of physical ports that could be processed by the host computer. Two proposals emerged to address these concerns; the FastConnect procedure and the encapsulation of H.245 messages within call signalling messages — often referred to as H.245 tunnelling.

In the case of the FastConnect procedure, the normal H.245 capability exchange is abbreviated by including a special 'FastStart' element within the SETUP message sent by the calling end-point. The FastStart element consists of a number of H.245 'OpenLogicalChannel' structures that constitute a number of offers of media channels that the calling end-point may send or receive. The information included with each 'offer' enables the associated media channel to be activated immediately, if required. A message sequence demonstrating the fast connect procedure is shown in Fig 10.9. In the case illustrated, the SETUP message contains the FastStart elements and the called end-point elects to accept one of the capability proposals.

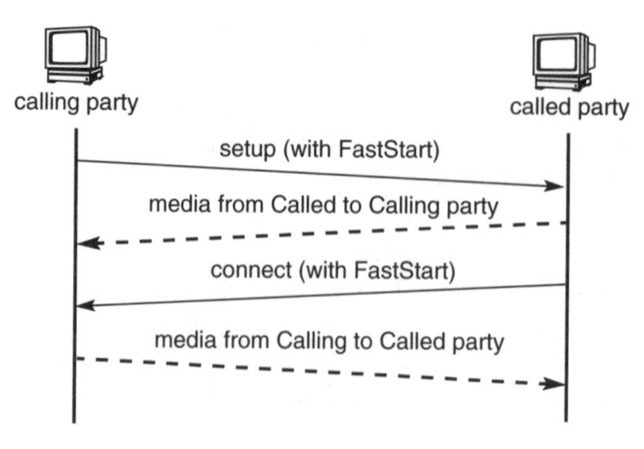

Fig 10.9 Example of set-up using fast connect.

The selected proposal is then included in one of the call signalling messages (CALL PROCEEDING, PROGRESS, ALERTING or CONNECT) returned to the calling party. In this case it is sent in the CONNECT message and the bi-directional media path is established. However, there is no mandatory requirement for H.323 end-points to support FastConnect. If the called end-point does not return a FastStart element in any of the call signalling messages up until the CONNECT message, then normal H.245 procedures are invoked. So the fast connect procedure can only be implemented in addition to normal H.245 procedures.

While the FastConnect method represents an effective solution for circumventing the highly serial process in H.323 version 1, it has two major deficiencies. Firstly, the FastConnect solution only really works in the case of simple call scenarios that do not require additional media streams or changes to existing ones; this is because these more complex multimedia scenarios invariably require the fuller features of H.245 to be subsequently used. Secondly, and possibly more importantly, this solution does not really address the problem of scalability since it still potentially leaves the need for an H.245 session to be subsequently established with its associated transport channels.

The encapsulation of H.245 messages within call signalling messages constitutes an alternative approach to the fast connect procedure described above. In this case, one or more H.245 messages are transported in any of the call signalling messages. This enables the call signalling channel to also be used for H.245 messages and so removes the need for separate transport sessions for that purpose. This has a number of benefits since the end-point, and more importantly any network element such as a gatekeeper, does not have to open as many IP sessions as would be otherwise required. This conserves the number of IP ports that need to be open on the computer hosting any H.323 server, such as a gatekeeper, increasing the number of calls that can be handled simultaneously.

It is also important to note that this method speeds up the overall signalling process for all types of media and call. It is therefore more generic than the fast connect procedure.

10.3.2 Changes to Call Handling

H.323 version 2 introduced two additional call handling features — the ability to establish a call without any media and the ability to put an existing media stream on hold. The first of these features enables an H.323 environment to potentially support a wide range of applications including the ability to pass call treatment information to a gatekeeper — since these are not otherwise 'callable'. The second feature is achieved by sending an empty capability set (CapSet 0) in an H.245 capability exchange. This is interpreted to mean that any receiving end-point should close any outstanding media sessions until a non-zero capability set is received. As a consequence an active media stream can be put 'on-hold' and the call manipulated by call logic in a gatekeeper. This feature effectively enables third party call handling from a gatekeeper.

10.3.3 H.450.x — Supplementary Services

As it was produced within a very short time-scale, H.323 version 1 only addressed basic call handling. There was no provision for the types of supplementary service feature, such as call diversion and transfer, typical of modern digital PBXs; only a crude blind transfer facility was available. For H.323 solutions to attract serious attention in the market, it was therefore necessary to enable similar call handling services to be provided in an H.323 environment.

When considering the development of supplementary services in an H.323 environment, it is important to differentiate the approach that can reasonably be taken from that taken in the digital PBX arena, the principal difference between the two being that the latter follows an exclusively centralised single point of control paradigm.

In contrast, the H.323 environment affords a much more open interface between the terminal and the network that enables the terminals involved in a call to co-ordinate service features directly between themselves. This corresponds to the inter-PBX private networking paradigm and it is not unsurprising to note that QSIG was chosen as the basis for the supplementary service framework for H.323, which is captured in the H.450.x framework. The H.450.x framework, shown in Fig 10.10 consists of a generic functional protocol (H.450.1) which provides access to the basic call handling capability with individual supplementary services being documented in subsequently numbered documents, as shown in Table10. 5.

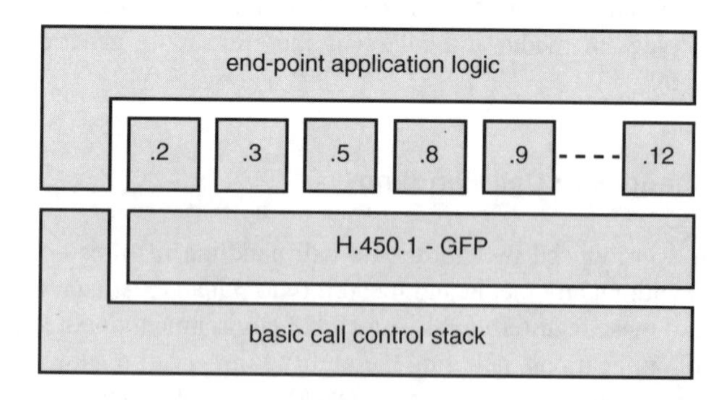

Fig 10.10 The relationship of supplementary service software to basic call logic and application code.

Table 10.5 List of supplementary services supported by H.323 version 2.

Recommendation	Supplementary service	Date of decision
H.450.1	Generic functional protocol — the underlying framework	01/1998
H.450.2	Call transfer	01/1998
H.450.3	Call diversion	01/1998

H.450.2 provides a call transfer mechanism with consultation capabilities, overcoming the deficiencies of the version 1 blind transfer solution. Similarly, the H.450.3 recommendation identifies a rich set of call diversion treatments, including unconditional call forward and call forward on busy, among others. As an example, Fig 10.11 shows the message flows for a typical call transfer with consultation. In this case there are three end-points, A, B and C respectively. A and B are in a call which is established in the usual manner. End-point B decides that end-point A should talk to end-point C and so invokes the call transfer operation. Since this is a transfer with consultation, once B has completed the transfer to C, it will exit from the call leaving A and C connected.

As shown in Fig 10.11, the transfer procedure in this case involves nine discrete steps comprising:

- A calls B using the normal call establishment message sequence (step 1);
- B determines that A should talk to C and creates a second call to C using normal call establishment procedures (step 2);

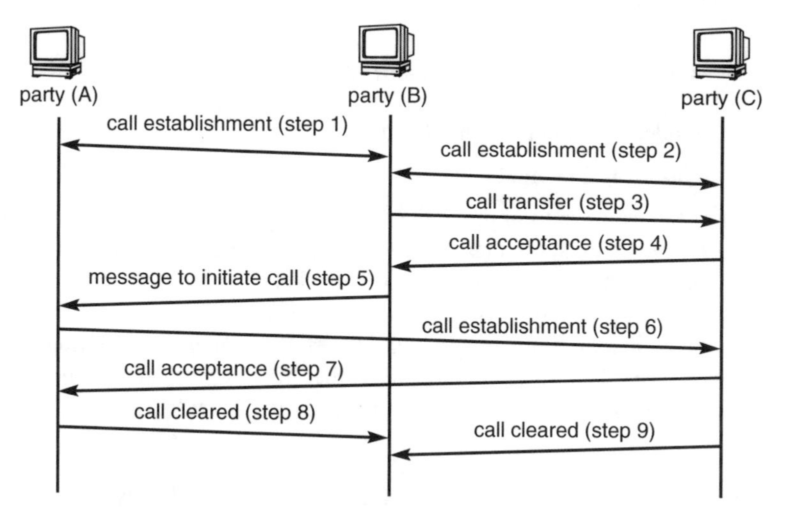

Fig 10.11 Call flows for consultation transfer using H.450.2.

- B decides to transfer the call, possibly after a conversation with C. B initiates this by informing C that it is about to transfer a call to it, and requests a token using a FACILITY message (step 3);
- C indicates that it is prepared to accept the call transfer by sending a FACILITY message containing the requested token (step 4);
- B indicates to A that it should make a call to end-point C containing the acquired token from C (step 5);
- A initiates a new call to end-point C with a SETUP message that contains the token previously obtained (step 6);
- C accepts the call by sending a CONNECT message — since C generated the token used in the SETUP message, it knows that this is the correct call to answer (step 7);
- when the transfer has been successfully completed, A and C can clear their individual calls with B and the transfer sequence is complete (steps 8 and 9).

This example has shown how the call transfer facility can work. However it is important to note that as there are a number of steps involved, there is always scope for errors to occur. To minimise the impact of any failure in the example given, H.450.2 specifies that end-point B can only leave the call once the transfer is complete. Also end-point C can lock out all other calls, since it generates the token used in the transfer. However, in spite of the call transfer procedure being specified as a peer-to-peer process, it is equally valid for the facility to be implemented using a gatekeeper and the empty capability set feature previously described.

10.3.4 Other Additions to Version 2

Version 2 included a number of other features.

- Improved use of H.245 and H.225

  As part of the exercise to improve the call set-up time, a method of using H.225 and H.245 over the user datagram protocol (UDP) was included. This removes the need for the initial TCP three-way handshake, and allows the application finer control of the retransmission timers in the event of lost packets. The latter is significant as in some implementations TCP starts with an initial retransmission time of 6 sec. The extra control can therefore significantly improve the responsiveness of the standard in the event of early packet loss.

- Security

  A new security standard, H.235, was added for use with H.323 systems.

- Pre-granted admission requests (ARQs)

  This allows a gatekeeper to give permission to an end-point to make calls as part of the registration process. If the gatekeeper gives the end-point such permission, it need not send admission requests prior to initiating or answering a call. This is of particular benefit in the gatekeeper routed model where the end-point will send the call signalling to the gatekeeper anyway, which will contain much of the same information that would have been validated in the admissions procedure.

- Support or overlapped sending of dialling information

  The latter is particularly useful in environments that do not have fixed-length telephone numbers (such as the UK), and when international dialling plans are used. It allows a gatekeeper to attempt to route the call based on partial address information in order to make a faster connection. For an MCU supporting multiple conferences, the calling end-point can be sent a list of the conferences using the Q.931 facility message. End-point address aliasing is allowed in order to transverse several gateways.

- Ordering of T.120 — H.323 calls

  Version 2 places restrictions on the order on which the H.323 call model is to be used compared with T.120 when dealing with conferencing. The T.120 call is treated as a logical channel for H.323. T.120 is optional when H.323 conferencing is established; it is at the discretion of each H.323 end-point.

- PROGRESS messages

  These were refined to show when a gateway is connecting with the SCN and to show the presence of in-band information. The call identifier was extended to give a unique identifier to show which call a packet is identifying when going through a gatekeeper.

- Seamless changing of codecs during a call

  To ensure that the media does not drop out, the seamless changing of codecs during a call is achieved by making certain that the new media stream is established first.

- Transmission of DTMF information

  The user input indication provides more information about the length of received tones, etc, for transmission of DTMF information out of band.

- Improvements to conferencing

  In H.245 remote logical switching was added to inform the end-point receiving conferencing information which media stream to present to the user, e.g. which video stream to present. More facilities for managing the mechanics of a conference call were added and H.263 video quality improved.

- Robustness

  Alternate gatekeeper and alternate end-point facilities were added to improve robustness.

- 'Keep alive' messaging

  A 'time to live' parameter can be specified in seconds, to prevent gatekeeper databases from filling with inactive end-points. Keep alive messaging was added so that an end-point can tell a gatekeeper that it wishes to keep registration active for a period.

- Administration

  For some direct end-point calls gatekeepers are informed of the call for administrative purposes and can request copies of the H.225.0 call messages they need.

- Reliability

  Reliable delivery of information responses, e.g. for billing, is ensured.

- Capacity and resource availability

  Gateway-to-gatekeeper communication on capacity and resource availability for the intelligent routing of calls is facilitated in order to optimise gateway usage. Gateways can indicate a type prefix and the data rate supported for the protocol.

- Call signalling

  The call signalling between the gatekeeper and end-points was modified to allow the gatekeeper to pre-grant admission requests to some end-points in order to reduce call set-up time.

- Quality of service

  Some QoS features were added to H.245 for the media streams. RSVP/QoS parameters are only conveyed between H.323 end-points, not down to the underlying IP network which could be QoS enabled using mechanisms like diffserv or MPLS.

10.4 H.323 Version 3 — Fax, Bug Fixes and Maintenance

Fewer additions were made to H.323 in version 3 of the standard compared with those included in the previous version 2 or the more recent version 4. In essence, version 3 was more of a maintenance release that addressed corrections or deficiencies in version 2 (see Table 10.6).

Table 10.6 H.323 version 3 compatibility.

H323 ver 3	H.225.0 ver 3	h2250v3.asn	H.245 ver 5

The main additions to H.323 version 3 include:

- support for simplified terminals, such as IP phones, known as simple end-point types (SETs);
- the ability for gateways to serve many calls from a particular gatekeeper, which ensures it is more efficient to maintain the same TCP connection for call signalling — for version 3 the call set-up efficiency between gatekeepers and gateways was also improved by providing a mechanism to allow gateways to re-use a call signal connection to support the same signal channel;
- the inclusion of caller identification and the ability for the caller to withhold their identity;
- an end-point can pass preferred language information to the receiving end-point enabling that end-point to perhaps forward the call to an appropriate voice response unit or to a customer service centre operator who speaks the required language;
- the ability to use remote device control through hooks to H.282 was added, so that it is possible to control a remote camera;
- the ASN.1 syntax specified in H.245 was modified to improve the interface to audio and video codecs — this allows new codecs to be developed without further altering the H.245 ASN.1 code;
- the inclusion of references to H.341 for SNMP management using a management information base (MIB);
- support for T.38 fax transmission.

10.4.1 Simple End-Point Types

Since H.323 has its origins in the support of rich multimedia conferencing, it can be rather difficult to implement on low complexity devices such as IP phones where memory capacity, processing power and media support can be constrained. To overcome this deficiency, version 3 introduced H.323 Annex F which addresses the needs of such so-called simple end-point types (SETs). Annex F provides a framework for specifying a profile of H.323 that is customised to the specific requirements of a given type of device.

The first device to be typified in this way was a simple audio-only device, representative of a basic telephone. A key feature of this particular profile is the constraint placed upon the messages used and the bit-wise specification of the message encoding that avoids the need to completely implement ASN.1 message handling in the device.

10.4.2 H.450.x — Supplementary Services

New supplementary services were added in version 3 to make implementations more competitive with the functionality available on PBXs. These are listed in Table 10.7.

Table 10.7 List of supplementary services supported by H.323 version 3.

Recommendation	Supplementary service	Date of decision
H.450.1	Generic functional protocol — the underlying framework	01/1998
H.450.2	Call transfer	01/1998
H.450.3	Call diversion	01/1998
H.450.4	Hold	05/1999
H.450.5	Park/pick-up	05/1999
H.450.6	Call waiting	05/1999
H.450.7	Message waiting	05/1999

10.5 H.323 Version 4 — Extensibility and Applications

Version 4 of H.323 represented a more significant release of the standard over version 3. The principal areas addressed in version 4 have been concerned with approaches to managing an increasingly complex base protocol, the support of inter-working H.323 with heterogeneous switched circuit network environments, and enabling a richer suite of applications. The core version 4 documents are identified in Table 10.8.

Table 10.8 H.323 version 4 compatibility.

H.323 ver 4	H.225.0 ver 4	h2250v4.asn	H.245 ver 7

As experience of deploying H.323 networks has been reflected by changes and additions to the base specification, it has matured and grown in size. However, there has been a slight tendency for these developments to swell the core protocol. In addition, the support of new applications continually drives the need for new features to be added. Ultimately this will become unsustainable and rather than allow this to happen, a general extensible framework (GEF) has been defined to enable new features to be added in a relatively lightweight manner. The GEF is described in further detail below.

The basic facilities offered by H.323 prior to version 4 largely addressed the needs of closely coupled multimedia conferences. However, a much richer suite of applications is possible in the converged network environment afforded by H.323. To enable richer and more complex applications to be created, version 4 includes support for an HTTP service control channel, to enable Web-based applications to be integrated, and provisions for stimulus signalling between end-points and feature servers. These aspects are also considered below.

10.5.1 Generic Extensibility Framework

As previously discussed, the increasing application scope of H.323 has resulted in continual additions to the specification. As this is ultimately defined using ASN.1 syntax, the addition of a new application results in a new version of the protocol being required. While ASN.1 allows variants of a protocol to be derived to meet the needs of a given application, this can be extremely complex to implement in practice. The GEF therefore addresses this problem by allowing new features to be defined without having to revise the core protocol functionality. It does this by providing a mechanism for both generically carrying data that is not an inherent part of the core specification and also negotiating any features supported.

Version 4 implements the opaque data transport element of the GEF by enabling most H.323 messages, including H.225.0 RAS and call signalling messages, to carry such data structures — known as 'generic data' — which are effectively like data containers. Each generic data structure is given a unique identifier and may in turn contain a number of individual application-specific data elements. Each of these data elements can be from one of a number of data types and is identified by a name declared as part of the standard or proprietary extension.

The data transport mechanism needs to be complemented by a feature negotiation capability since it is unreasonable to expect that all entities will either support the GEF approach or any given application-specific extensions. The mechanism defined as part of the GEF addresses three specific aspects in this area by allowing an entity to express the features it:

- supports;
- needs supporting elsewhere;
- desires supporting elsewhere.

Any entity that requires a given feature to be supported must check that the corresponding entity has acknowledged support of the feature before it can use it. This allows the GEF to be backwards compatible with non-GEF entities. Also any corresponding entity is able to reject a request where it is able to determine that it does not support a required feature.

A possible example of this would be where H.323 was being used to transport C.7 signalling and an intermediate node recognised that it was unable to support a given country variant at an egress point. The first specifications to use the GEF are H.460.2 (number portability and inter-working between H.323 and SCN networks) and the H.323 generic extensibility framework guideline document. H.460.2 addresses the issue of supporting service provider and location portability at the interface between an H.323 network and an interconnecting C.7-based PSTN. It uses the GEF to transport the relevant ISUP number-portability information elements across the H.323 environment.

10.5.2 Protocol Tunnelling

As more users are currently served by the PSTN than stand-alone VoIP networks, there has been a long-standing need to both interconnect and interwork H.323 with SCN protocols. Where H.323 is deployed in an extensive network using multiple gateways, there invariably emerges the scenario where a call originates on one part of the PSTN, is transported via a VoIP network and terminates on another part of the PSTN. Since the features available in such cases are determined by the information carried in the associated signalling messages, it is potentially desirable for information to flow end-to-end transparently through the H.323 network. In other words, it is desirable for H.323 to allow external signalling protocols to be 'tunnelled' through the H.323 network. As will be appreciated, the GEF would appear to be a natural basis for a tunnelling solution. In practice, however, a subset of the full GEF has been identified; this would appear to be a re-run of the fast connect versus H.245 tunnelling proposal debate, since it is another example of two solutions emerging for a similar problem when one solution would be fine.

10.5.3 HTTP Service Control Transport Channel

While it provides an effective basis for enabling real-time conversational applications, H.323 provides little support for enhancing the user interface. Version 4 has addressed this problem by specifying in Annex K a mechanism that enables URLs to be passed to H.323 end-points during the registration and call-signalling phases. This enables a suitably equipped end-point to access the associated Web pages which can lead to a wide range of interesting computer/ telephony integrated (CTI) applications. This mechanism also relates to Annex O, that describes the complementary use of Internet protocols with H.323 and in particular the use of H.323 URLs.

10.5.4 Stimulus Signalling between Terminals and Feature Servers

H.323 is essentially a functional signalling protocol; the messages that are exchanged and defined within the protocol relate to functions that are invoked. While this results in an effective solution for simple conversational applications, supporting more advanced features can result in the definition of an expanding range of specifications (e.g. H.450.x) and the need to continuously upgrade any deployed end-points. The alternative approach is to adopt what is known as the stimulus signalling paradigm. In this approach, an event-centric view is taken so that the messages that pass between the system entities relate to actions, such as a particular button being pressed on a terminal. The meaning of this event is then interpreted by a network feature server, such as may be implemented in a gatekeeper, which may depend upon the relationship between this event and previous ones. The benefit of this is that the functional behaviour of the end-point can be separated from the supplementary service enabling both to be developed independently.

10.5.5 H.450.x — Supplementary Services

H.323 version 4 also includes support for additional supplementary services. The full list of supplementary services is given in Table 10.9.

Table 10.9 List of supplementary services supported by H.323 version 4.

Recommendation	Supplementary service	Date of decision
H.450.1	Generic functional protocol — the underlying framework	01/1998
H.450.2	Call transfer	01/1998
H.450.3	Call diversion	01/1998
H.450.4	Hold	05/1999
H.450.5	Park/pick-up	05/1999
H.450.6	Call waiting	05/1999
H.450.7	Message waiting	05/1999
H.450.8	Name identification service	02/2000
H.450.9	Call completion	11/2000
H.450.10	Call offer	11/2000
H.450.11	Call intrusion	11/2000
H.450.12	Common information additional network feature	07/2001

10.5.6 Other Additions to Version 4

Other additions to version 4 are described below.

- Version 1 of H.248 (gateway control protocol) [4] was ratified with version 4 of H.323. This describes the decomposition of multimedia gateways into a media gateway controller (MGC) and media gateways (MG).
- Enhanced capabilities to synchronise video and speech by allowing video and audio to be multiplexed in a single RTP stream enable the end users to experience a more natural multimedia conversation.

- An additive registration feature was included, since network elements like gateways and MCUs which have to possess large numbers of alias addresses, need to register those with gatekeepers, but this is limited by the size of the UDP packets. A method was specified for devices with large numbers of aliases to be able to send an initial RRQ to a gatekeeper and then follow this with additional RRQs.
- Improvements were made to end-point notification for alternate gatekeepers. If an end-point supports an alternate gatekeeper then a gatekeeper may choose to redirect that end-point to a more lightly loaded gatekeeper to achieve load balancing. Gateways and conferencing end-points (MCUs), have to report call status information back to the gatekeepers. To achieve efficiency, this is done by sending an IRR containing information about a group of calls.
- End-points can now provide usage information to a gatekeeper at the start of a call, throughout the call, or at the end of the call. The usage information includes start and end times, cause of termination and other non-standard data. An end-point capacity mechanism was added primarily for gateways and conference units. The gatekeeper is notified of the capacity of the end-point so that, if, for example, a gateway is reaching its capacity, the capacity is signalled back to the gatekeeper and then the gatekeeper can possibly implement a strategy to re-route a call to another gateway. This potentially increases user satisfaction by increasing the success rate and it also increases call revenue.
- End-points can signal to a gatekeeper if they need the calling party (such as a gateway) to support a special service, i.e. fax. The gatekeeper can then choose an appropriate gateway which supports fax. One would expect in a network that there would be fewer gateways with fax and voice functionality than ones with voice, and for the former to be more expensive.
- Annex D (Real Time Fax over H.323) was improved to allow end-points to switch from voice to fax so that IP fax machines could behave in the same way as fax machines today. TCP was also specified for the transmission of fax information.
- The usage of caller identification was clarified.
- Procedures for indicating in-band tones and announcements were added. For calls routed through the SCN, tones and announcements are heard when the destination number is incorrect or engaged. Details about end-point bandwidth utilisation can be communicated from the end-point to gatekeepers using IRR. This can allow a gatekeeper better control of the bandwidth in its domain.
- The gatekeeper may signal to an end-point to play recorded announcements at various times. This mechanism facilitates two-stage dialling, for example, where the gatekeeper may request the gateway to prompt the user for additional information to gain access to another network.

- Functionality was added to allow gateways to support virtual private networks by alias mapping at the entry and exit points to SCNs.
- Limited improvements were made to QoS by introducing the use of RSVP in call set-up.
- Parallel starting of H.245 with FastStart was included to exchange capabilities and to quickly determine the features supported by end-points.
- Fields were introduced so that for calls transiting several network segments the network equipment can link the call legs together to come up with a unique call linkage record for use in generating overall call charges.
- DTMF digits can be sent and received in accordance with the IETF RFC 2833.
- Calls made out to the SCN or other external networks tend to be charged on a call-by-call basis, whereas, for calls made on-net, version 4 provided call credit capabilities to convey information about available funds and for a gateway to be connected to terminate calls where there are insufficient funds.

10.6 Core Design Issues

This section highlights some of the core design issues for VoIP systems based on H.323. It is important to note, however, that some of the issues raised here are common to any VoIP application.

10.6.1 Security

Since H.323 version 1 was constructed in a short space of time and represented the first substantial standards effort to support and enable VoIP, a number of detailed issues were either not completely addressed or omitted entirely. One of the main areas in which version 1 was deficient was in the area of security. In particular, since H.323 effectively provides a real-time communications facility over an inherently insecure network there are two main areas of concern. These are authentication and privacy. Primarily aimed at any H.245-based application, H.235 was created to address these concerns. In a similar way to H.323, H.235 is actually a framework for addressing security concerns rather than a simple mandatory protocol. Depending upon the deployment scenario, which includes any underlying commercial model, there are a number of options that can be deployed.

10.6.2 Inter-Gatekeeper Operation

As H.323 solutions started to be deployed, it rapidly became apparent that a single gatekeeper model would be unrealistic. The initial model assumed in H.323 version 1 therefore needed to be revised to allow multiple gatekeepers to exist within a single service provider's network. It also needed to take into account that multiple service providers were most likely to be involved in a call at some point. This issue can be addressed in a number of ways, including interconnecting gatekeepers in a hierarchical structure similar to telephone switches in the PSTN. An alternative is to use special nodes, termed border elements, between single service provider domains. Further details concerning inter-gatekeeper operation are given in Chapter 14.

10.6.3 Management

As with any communications network technology, H.323 systems need to be deployed, operated and managed in an economic manner if they are to be successful. H.323, as with any VoIP protocol, requires careful attention to be paid to these issues if a commercially successful result is to be achieved. Since the issues are common across a number of protocols, a summary of the main areas of concern are given in Chapter 11 and some practical examples are given in Chapter 13.

10.6.4 Application Support

H.323 was originally developed for simple, closely coupled, conversational multimedia applications. However, to realise the potential benefits of an IP-based communications infrastructure, it is necessary to provide mechanisms that enable developers to produce more sophisticated applications. Although a number of proprietary techniques for achieving this have existed since version 1, standards-based solutions are now emerging, including the use of HTML and the IETF's call processing language (CPL), outlined in Chapter 11, and application and service standards like F.700, defined by the ITU-T. A further consideration of building applications using H.323 is given in Chapter 13.

10.7 Future Developments

The ratification of version 4 of H.323 in November 2000 coincided with the completion of a four-year long study period. Version 4 of H.323 can be used to provide a robust VoIP multimedia telecommunications solution. The next study period for the H.323 study group (SG16) is from 2001 to 2004. Over this period, not only will there be a continuation of the present SG16 remit, but there will be a considerable enlargement in the number of work areas covered. The new work areas cover more on development of usage.

The ITU has identified, as priority issues for study, multimedia related to mobility (especially when applied in mobile networks), the Internet, and the convergence between conversational multimedia and interactive broadcasting. It includes the creation of a project — Mediacom 2004 — for the development and harmonisation of multimedia standards across other ITU-T and ITU-R study groups and across other related forums.

The scope of Mediacom 2004 will include the following applications:

- end-to-end multimedia systems and services over all network types, including the Internet, for videophone/videoconference, multipoint/multicast multimedia systems, multimedia on demand, electronic commerce, distance learning, tele-medicine, interactive TV services, Web-casting, MBone — including their distribution within the home environment, etc;
- end-to-end multimedia systems and services over wireless access systems, e.g. using radio frequency or infra-red (IMT-2000, Wireless Application Protocol Forum, Bluetooth, HomeRF, IrDA, etc) — in this environment, computer or consumer information appliance devices will be used;
- security system for/using multimedia systems (watermark in the video contents, individual authentication, etc);
- multimedia broadcasting systems that interactively handle audio and video;
- the extension of e-mail and the WWW for the transmission of multimedia 'documents'.

Multimedia architectural work will be progressed with roadmaps, to ensure that multimedia system architectures are consistent with communications networks as a whole. Multimedia applications and services will be standardised and based on F.700, which is a framework recommendation for the description of services. A map of multimedia services and applications will be created and the requirements captured for those applications and services within the remit of the study group. Service and network infrastructure interoperability will also be developed, taking into account mobility issues for multimedia systems. Quality of service and end-to-end performance work will take place and this will include APIs or interfaces to different QoS signalling methods. The security of the multimedia systems and services will be developed to meet future user needs to support eCommerce and eBusiness. Accessibility, the impact and benefits of multimedia transformation for age-related problems and disabilities are all to be examined. Detailed technical work will continue on the development of the standard, and specific study questions have been included to improve the quality of video and audio coding across a range of bit rates. Recently a new question has been added for advanced distributed speech recognition and speaker verification for IP-based multimedia systems.

The other standards bodies with which the project anticipates liaising are:

- ITU-T SGs and projects (e.g. GII, IP, SSG on IMT-2000 and beyond);
- ITU-R SGs, particularly ITU-R SG6 WP6M;
- ISO/IEC (MPEG-4, MPEG-7, MPEG-21);

- W3C;
- IETF;
- regional standardisation bodies (e.g. ETSI, T.1 (USA), TIA (USA), etc);
- other forums/consortia (e.g. 3GPP, 3GPP2, etc).

The ITU's view of of study areas and how they map to organisations/bodies is shown in Fig 10.12.

More details of the ITU-T SG16 Work Programme and Mediacom 2004 Project may be downloaded from the ITU Web site [5].

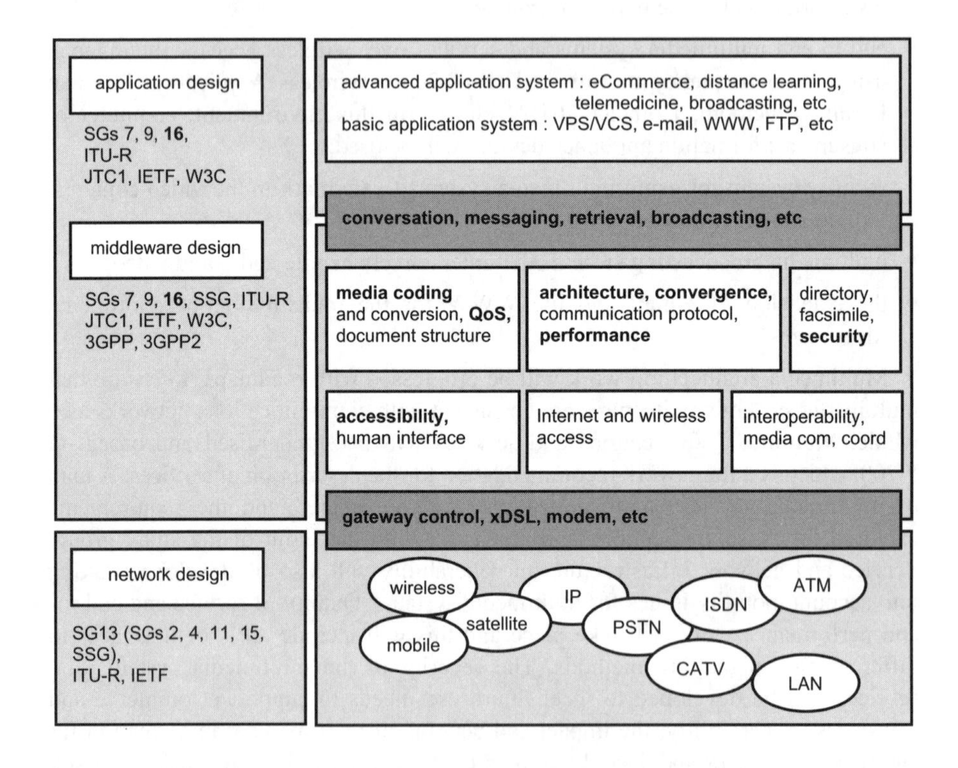

Fig 10.12 Study areas and developing organisations/bodies.

10.8 Summary

H.323 and its related standards support a rapidly growing and key area within the telecommunications market. The various standards bodies have already undertaken significant work to develop mature and robust standards. The development of H.323 will, if anything, accelerate as the new study period for ITU-T SG16 commences.

Designers implementing VoIP networks need to be aware of the capabilities of the different versions of their chosen protocol(s) — in this case H.323 — to be able to design optimal, robust networks on which functionality can increase through the introduction of new network equipment supporting new versions of the standards.

The benefits of H.323 are varied. At one level the standard enables the rationalisation of data and switched-circuit networks into a single packet-based network, carrying voice, video and data. Further cost benefits and other efficiencies will accrue as end users start to use more of the multimedia facilities, allowed by the H.323 standard, to freely communicate information and knowledge in the work and home environments. Multimedia applications may in the future be as significant as e-mail, spreadsheets and word processor applications are today. Multimedia applications may be widely used in eCommerce and in the fast and easy conveyance of corporate and personal thoughts and knowledge.

Appendix

Standards Used in H.323 and Details of the Various Annexes

Table 10.A1 List of standards used in H.323 (independent of version).

	Title of Standard	Organisation
FDDI	Fibre distributed data interface	ANSI
G.711	Recommendation G.711 (11/88) — pulse code modulation (PCM) of voice frequencies	ITU-T
G723.1	Recommendation G.723.1 (03/96) — Dual rate speech coder for multimedia communication transmitting at 5.3 and 6.3 kbit/s	ITU-T
G.728	Recommendation G.728 (09/92) — Coding of speech at 16 kbit/s using low-delay code excited linear prediction	ITU-T
G.729	Recommendation G.729 (03/96) — C source code and test vectors for implementation verification of the G.729 8 kbit/s CS-ACELP speech coder	ITU-T
H.225	Recommendation H.225.0 (11/00) — Call signalling protocols and media stream packetisation for packet-based multimedia communication systems	ITU-T
H.225	Recommendation H.225.0 (11/00) — Call signalling protocols and media stream packetisation for packet-based multimedia communication systems	ITU-T
H.282	Recommendation H.282 (05/99) — Remote device control protocol for multimedia applications	ITU-T
H.332	Recommendation H.332 (09/08) — H.323 extended for loosely coupled conferences	ITU-T
H.235	Recommendation H.235 (11/00) — Security and encryption for H-series (H.323 and other H.245-based) multimedia terminals	ITU-T
H.245	Recommendation H.245 (02/00) — Control protocol for multimedia communication	ITU-T
H.246	Recommendation H.246 (02/98) — Interworking of H-Series multimedia terminals with H-Series multimedia terminals and voice/voiceband terminals on GSTN and ISDN	ITU-T
H.248	Recommendation H.248 (06/00) — Gateway control protocol (also known as MeGaCo)	ITU-T/IETF
H.261	Recommendation H.261 (03/93) — Video codec for audiovisual services at p × 64 kbit/s	ITU-T
H.263	Recommendation H.263 (02/98) — Video coding for low bit rate communication	ITU-T
H.341	Recommendation H.341 (05/99) — Multimedia management information base	ITU-T
H.450	Recommendation H.450.1 (02/98) — Generic functional protocol for the support of supplementary services in H.323 (also in this series H.450.2 to H.450.12)	ITU-T
H.460.2	Number portability	ITU-T
IEEE802.3	Local and metropolitan area networks carrier sense multiple access with collision detection (CSMA/CD) access method and physical layer specifications	IEEE
IEEE802.5	Local and metropolitan area networks — specific requirements — token ring access method and physical layer specification	IEEE
IEEE802.10	Local and metropolitan area networks: interoperable LAN/MAN security	IEEE
IP	Internet protocol	IETF
RSVP	Resource reservation protocol	IETF
RTCP	Real time control protocol	IETF
RTP	Real time protocol: a transport protocol for real-time applications	IETF
RFC2833	RTP payload for DTMF digits, telephony tones and telephony signals	IETF
T.120	Recommendation T.120 (07/96) — Data protocols for multimedia conferencing	ITU-T
T.123	Recommendation T.123 (05/99) — Network-specific data protocol stacks for multimedia conferencing	ITU-T
T.124	Recommendation T.124 (02/98) — Generic conference control	ITU-T
T.125	Recommendation T.125 (02/98) — Multipoint communication service protocol specification	ITU-T
TCP	Transport control protocol	IETF
UDP	User datagram protocol	IETF
X.224	Recommendation X.224 (11/95) — Information technology — open systems interconnection — protocol for providing the connection-mode transport service	ITU-T

Table 10.A2 Annex breakdown within H.323 (independent of version).

Annex	Standard	Description
A	H.245 messages used by H.323 end-points	Describes rules to apply to the use of H.245 messages by H.323 end-points
B	Procedures for layered video codecs	Describes enhancements within the framework of the H.323 specification, to incorporate layered video codecs. Layered video coding is a technique that allows the video information to be transmitted in multiple data streams in order to achieve scalability. These may provide bandwidth scalability, temporal scalability, SNR scalability, and/or spatial scalability.
C	H.323 on ATM	An optional enhancement allowing H.323 end-points to establish QoS-based media streams on ATM networks using AAL5.
D	Real time fax over H.323	Facsimile and speech are typically sent using the PSTN with the same calling and addressing infrastructure. It is desirable to continue this approach in the context of H.323. Facsimile can be viewed as another kind of real-time traffic similar to a particular speech coder. Facsimile entering the packet world via a gateway from the PSTN should logically be treated in a similar way to speech if the customer expects a real-time, assured end-to-end transmission service.
E	Multiplexed call signalling over UDP	Describes a packetisation format and set of procedures (some optional) that can be used to implement UDP and TCP based protocols. Includes a signalling framework, wire protocol and use cases.
F	Audio simple end-point type	This introduces the notion of simple end-point types (SETs) that are designed to operate in a well-defined set scenario and with a restricted functionality range, reducing the overall system and hence the implementation complexity. It defines simple end-point types for audio communication (audio SETs) that operate using a well-defined subset of H.323 protocols suited for IP telephony applications, while retaining interoperability with regular H.323 devices.
G	Test simple end-point type	It adds text conversation to the H.323 packet multimedia environment and also defines a SET for text and audio conversation in packet networks.
H	User, terminal and service mobility	Functions to enable mobility in H.323 systems.
J	Secure simple end-point type	Describes the security for H.323 annex F simple end-point types. The specified security profile is based upon recommendation H.235 v2 and uses the featured baseline security profile of H.235 Annex D.

K	HTTP based service control transport channel	Describes an optional way of controlling supplementary services in an H.232 environment. By opening a separate connection conveying a service-independent control protocol, new services may be developed and deployed without updates to the H.323 end-points. The service control channel is intended for use in a wide range of services, some of which require the use of H.450 or proxy signalling for invocation/execution.
L	Stimulus control protocol to recommendation H.323 — packet-based multimedia communications systems	Describes stimulus signalling procedures between H.323 terminals and a feature server functional entity. This stimulus method allows the network service provider to implement new supplementary services for the terminals without changes in the terminal software, which results in easier maintenance. An example of such a terminal is a LAN-attached feature phone. A feature server may be collocated with the gatekeeper.
M.1	Tunnelling of signalling protocols (QSIG) in H.323	Gives guidance as to how the generic tunnelling mechanism can be used to tunnel QSIG over H.323 networks.
M.2	Tunnelling of signalling protocols (ISUP) in H.323	Gives guidance as to how the generic tunnelling mechanism described in H.323 can be used to tunnel ISUP over H.323 networks.
M.3	Tunnelling of DSS1 through H.323	Deals with the case where DSS1 subscribers are connected to an H.323 residential gateway. Obviously, in this case there is also a need for feature transparency when these subscribers access the PSTN via DSS1 media gateway.
N	End-to-end QoS and service priority control and signalling in H.323 systems	Concerned with end-to-end quality of service.
O	Use of complementary Internet protocols with H.323 systems	Addresses use of Internet protocols as network infrastructure.
Q	Far-end camera control and H.281/H.224	Provides far-end camera control protocol based on H.281/H.224. It also permits an H.323 end-point to run any H.224 application using the IP/UDP/RTP/H.224 protocol.
R	Robustness methods for H.323 entities	Specifies methods that can be used by H.323 entities to implement robustness or tolerance to a specified set of faults. Methods for recovery of call signalling (H.225) and call control signalling (H.245) channels are specified.

References

1 ITU-T Recommendation H.323: '*Series H: audiovisual and multimedia systems infrastructure of audiovisual services — systems and terminal equipment for audiovisual services for packet-based multimedia communications systems*', versions 2 to 4 (2000).

2 ITU-T Recommendation H.225: '*Series H: audiovisual and multimedia systems infrastructure of audiovisual services — call signalling protocols and media stream packetisation for packet-based multimedia communications systems*', versions 2 to 4 (2000).

3 ITU-T Recommendation H.245: '*Series H: line transmission of non-telephone signals — control protocol for multimedia communication*', versions 2 to 4 (2000).

4 ITU-T Recommendation H.248: '*Series H: audiovisual and multimedia systems infrastructure of audiovisual services — transmission multiplexing and synchronisation, Gateway Control Protocol*', (2000).

5 ITU-T SG16 Work Programme and Mediacom 2004 Project — http://www.itu.int/itudoc/itu-t/com16/

11

SIP — THE SESSION INITIATION PROTOCOL

D R Wisely and R P Swale

11.1 Introduction

There are two major revolutions going on in the world of telecommunications today — the widespread deployment of mobile communications services and the emergence of IP-based multimedia applications. On fixed networks, increasingly large amounts of the total traffic carried by operators, such as BT, is now data traffic. In addition, data traffic is now many times the current level of voice traffic — with Internet traffic, in particular, estimated to be doubling every 7 months. When examined more closely, virtually all this data traffic emanates from IP-based applications — even though it might be carried over ATM, SDH or private circuits in the public network. This growth in traffic is in part due to the World Wide Web and the deployment of corporate intranets, such as the BT Intranet, which are increasingly used for all forms of information dissemination and business applications. For example, the BT Intranet hosts many IP-based applications that, among other things, allow employees to order supplies, look up contact details for other employees, manage projects, view lectures, conduct training, watch BT business TV broadcasts, share documents and perform many other business-related functions. This revolution in IP-based data traffic has initially been fuelled by a number of factors including:

- the emergence of new charging strategies, such as flat-rate Internet access;
- ubiquitous and affordable computing due to the great reduction in PC costs;
- the availability of licence-free source code;
- a potentially rapid and pragmatic standardisation process for Internet technology;
- the growth in mass-use applications, including e-mail and WWW browsing.

When analysed in detail, virtually all IP-based applications follow the client/server paradigm, with Web browsing being a classic example. In this case, the client — a browser such as Netscape — requests files from a distant server that are rendered by the local client. Each request to the server contains the IP address of the client and the server then uses transport layer protocols, such as TCP, to deliver the requested file. A key feature of this process is that the server is unchanged after the file transfer. In other words it is stateless and has no memory of the transaction; request another WWW page and the whole transfer process starts again. Although e-mail is a store-and-forward application, it too exhibits very similar characteristics. This is because this kind of application also involves servers receiving messages from clients and clients requesting messages from servers. Even when you are logged on to your e-mail server you are usually polling it — continually asking if there are any more messages.

In contrast to client/server and store-and-forward applications, there are very few peer-to-peer applications running on IP networks. An example of a peer-to-peer application would be a simple real-time, two-way, voice application. For example, it is possible to talk to a friend in Australia for the cost of a local call if both parties install a suitable application [1]. However, there are severe limitations to this type of peer-to-peer application today. Firstly the parties must agree, probably by e-mail, what time they will try to connect. As discussed in Chapter 1, the codec and bit rate must also be agreed. In this example, this will again have to take place in advance of the time of the call. Finally, each other's IP addresses need to be known to enable users to initiate, negotiate and modify sessions such as this. As previously discussed these are usually allocated dynamically at log-on for dial-up connections.

As previously highlighted, mobility is the second big telecommunications revolution under way at present. Second generation systems, such as GSM, offering digital voice and text messaging, have shown phenomenal growth in user numbers. The emerging third generation systems, including UMTS, offers the possibility of true Internet capabilities with much higher data rates than those available from second generation solutions. Users will, however, continue to have a large number of possible terminals and contact points — for example, a fixed and a mobile telephone, a fax, a PC at home, and another one at work.

If the Internet is to move beyond information retrieval, such as WWW browsing, and non-real-time information transfer, such as e-mail and ftp, to encompass real-time voice and video, and if people want to be contacted on the most appropriate of their many devices, then a protocol for initiating, negotiating, modifying and tearing down multimedia sessions is essential. The session initiation protocol (SIP) [2, 3] has been developed by the IETF [4] for this very purpose. SIP is an IETF protocol and must therefore be considered from the point of view of an architecture comprising a number of layered protocols with each layer adding specific functionality. In particular, it is an application layer protocol that can make use of either TCP or UDP transport protocols. However, as SIP does not specify anything about resource reservation, or media transport, other protocols are required to deal with these aspects (Fig 11.1).

This chapter introduces the basic operation of SIP, considers its characteristics and describes how it could be used for rapid multimedia service creation. The SIP

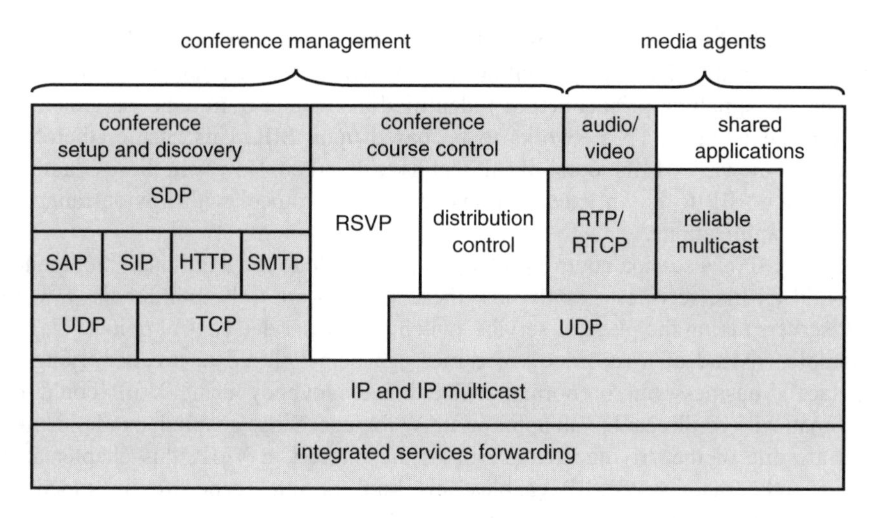

Fig 11.1 SIP in the IETF multimedia conferencing architecture.

architecture and basic messages, as well as the response codes are described in the context of example uses.

Since it involves a registration function, SIP offers users what is known as 'presence'. This is the functionality that allows a user to 'sense' the presence of someone they wish to call, react to their preferences and invoke scripts that depend on the time, location or media type of any associated call. SIP has therefore been designed with the intention of performing functions equivalent to those afforded by the intelligent network (IN) architecture, allowing facilities such as transfer on busy, call waiting and the like. However, SIP has the added benefit of also enabling multimedia. So the session — or call — can comprise not only voice but also complicated combinations of video clips, html pages and the like. It can also be used to set up sessions involving multiple users interacting with IP multicast and anycast network protocols.

In traditional TDM-based PSTN and mobile networks the media transport is inextricably bound up with the call control. In IETF-defined networks, the 'one function, one protocol' rule automatically means that media transport is handled by quality of service protocols, such as RSVP (resource reservation protocol), and application layer protocols, such as RTP (real-time protocol — see Chapter 1). However, the focus of this chapter is on SIP and, in particular, the call control aspects of SIP. The architecture, mobility and QoS features of IETF IP networks can be found elsewhere, such as Wisely [5].

In contrast the architecture of standards bodies like 3GPP, the UMTS standardisation body [6], consists of entities (base stations, home location registers, etc) and descriptions of the interfaces between these. An extended SIP protocol is being developed by 3GPP as one of the protocols for the mobile-to-network interface for the 'all-IP' UMTS Release 5. In addition, proposals, known as SIP-T after the initial draft submissions to the IETF, have also been made which address the inclusion of C7 signalling within SIP messages as a candidate for the inter-softswitch signalling protocol (ISSP) identified in Chapter 6. In doing so, this aims to allow existing PSTN networks to be based upon SIP. This chapter therefore considers the various 'flavours' of SIP that have emerged. However, these examples alone show SIP to be an extensible and powerful protocol with a wide range of potential applications.

Since SIP is a session control protocol it is possible to use additional intelligence to build IN-like services — only now these services are truly multimedia — with SIP servers taking the place of service switching and service control points [7]. For example, instead of a recorded voice message being played to a caller trying to contact a business out of normal office hours, anybody using VoIP could be automatically re-directed to an appropriate Web page. Since enabling examples like this are one of the driving factors behind the interest in VoIP, this chapter also reviews the main ways SIP enables this kind of intelligence to be added to applications.

Finally, since it is a relatively new technology, some difficult questions regarding the deployment of SIP still remain. These include the effects of firewalls, organising a SIP network deployment and other practical operational issues. The chapter therefore concludes by providing an overview of some of these aspects.

11.2 SIP Components and Call Flows

The session initiation protocol is based upon the request-response message exchange semantics of the hypertext transport protocol used by Web applications. It provides a means of negotiating contact between one or more entities, whether they are individuals or automatons. In this sense it constitutes a rendezvous protocol for real-time communications. On its outward face, SIP manifests itself as an application — the user agent. User agents comprise a user agent client (UAC) function to make requests and receive responses, and a user agent server (UAS) to receive requests and generate responses. The format of the SIP message comprises two elements — a header, consisting of SIP fields, and a body. Header elements provide parameters such as the identity of the caller, the identity of the receiver, a unique call ID, sequence number, subject, the hop traversed to deliver the message and so forth.

The body typically uses the session description protocol (SDP) to describe the session that is being negotiated. In fact a profile of SDP is embedded in the core SIP specification [2].

11.2.1 SIP Request Messages (Methods)

SIP works on the basis of a simple request/response message interchange analogous to the operation of HTTP. SIP messages are relatively few and entirely in plain text, requiring relatively little processing. In fact, the basic SIP Internet RFC (request for comment [2]) specifies six message request types that are known as SIP methods. At the time of writing a further two others, INFO and PRACK, have been proposed and seem likely to be adopted when SIP finally becomes an Internet standard.

11.2.1.1 INVITE

The INVITE method is the basic way of establishing multimedia sessions — it usually carries a message body which is often a proposed session description declared in SDP. Responses to INVITEs are always acknowledged using the ACK method described below and a media session is only considered established when the INVITE, 200:OK and ACK messages have been exchanged between user agents (UAs). Associated with each INVITE is a globally unique call-ID to identify the multimedia session set up by that INVITE — there might be several different sessions running between the two agents. A sequence number is also generated so that retransmitted INVITEs, due to lost messages, can be differentiated from fresh INVITEs used to modify the session.

11.2.1.2 ACK

The ACK method is used to acknowledge final responses to INVITE messages — basically all responses apart from the 1xx informational responses (see next section). ACKs mirror the original sequence number from the INVITEs, allowing them to be matched to specific INVITEs. ACK can only carry message bodies when the original INVITE was empty and is not used to modify sessions. For 200:OK responses the ACK is an end-to-end message — it will contain VIA headers from all the proxies in the path as described below and can carry a message body. ACKs to all other responses, such as 410 gone, will be regenerated at each hop and contain only a single VIA.

11.2.1.3 BYE

BYE is basically used to terminate a media session — which itself is established as soon as the messages INVITE, 200:OK and ACK have been exchanged between the user agents. BYE messages travel end to end and carry no bodies.

11.2.1.4 OPTIONS

The OPTIONs method is used to discover a user agent's response to an invitation without actually signalling the intention, such as 'ringing'. It can be used to discover the capabilities of user agents and SIP servers. An OPTION message is basically an INVITE message without the body and user agents respond to it as if it were an INVITE — a 200:OK response to an OPTION message can also contain information about, for example, the methods and languages that the agent supports.

11.2.1.5 CANCEL

The CANCEL method is used to cancel the session being negotiated by users' agents and to cancel forked INVITEs by proxies after receiving a successful response. CANCELs are hop-by-hop messages — meaning that a stateful server receiving a CANCEL message will respond to it, rather than relay it to the final user agent for action. Proxies receiving CANCEL messages send them to the same locations that they sent the appropriate INVITEs — recognised by the call-id and sequence numbers being identical in the two messages.

11.2.1.6 REGISTER

The REGISTER method is used by user agents to provide the SIP registration server with information about where they would like to be contacted. Registration is not essential and non-SIP mechanisms can be used to populate the location server, e.g. mobile IP care-of-address registration could be used. REGISTER messages can contain bodies — and we will see later that this might include scripts to be executed by a proxy server when INVITEs, or information about preferences, are obtained. There is no specified standard — bodies are simply transported by the REGISTER messages transparently.

11.2.1.7 INFO

INFO messages are used to send information between user agents during media sessions. INFO messages carry bodies, which might relate to signalling information, e.g. a trigger to switch on encryption. INFO messages do not modify sessions — re-issued INVITEs are used for that purpose.

11.2.1.8 PRACK

PRACK is basically an ACK for provisional messages, e.g. 1xx response such as 180 ringing (see below), where these are crucial to the operation of the service. Such an example might be: "My IP telephony application won't work if the 180 ringing response is lost". User agents are therefore set to re-transmit these until a PRACK is returned.

11.2.2 SIP Response Messages

In reply to a SIP request, the recipient generates an appropriate response. The response may be either end-to-end or hop-by-hop. SIP response messages follow a format similar to that used in HTTP since they are numbered within distinct number ranges depending upon the success or otherwise of the request. Responses sometimes carry message bodies and look very similar to request messages, as is shown in the example given in section 11.2.6. This section presents a number of examples from each range to illustrate the principles involved.

11.2.2.1 1xx: Informational

Informational responses are in the range of 100 through to 199 and indicate that the request has been received and the server, whether it is a proxy, redirect, User Agent or other kind, is continuing to process the request. For example, a 100 ***trying*** response might be sent back by a forking proxy server in response to an INVITE — since the search of several locations might take time and repeated INVITEs from the originator are not desirable. Similarly a 180 ***ringing*** response indicates the user agent has received the INVITE and that some from of alerting, such as flashing up the text of the subject field as well as the caller's name, is taking place. A simple ringing response is shown in Fig 11.2. Note that the 'To' and 'From' fields are not swapped around, as you might expect, in order to allow a simple 'copy and paste' by the user agent.

```
SIP/2.0 180 ringing
Via: SIP/2.0/UDP spain.tel:5060
To: Mel Bale <sip:mel@uk.net>
From: Clive Dellard <sip:clive@sipstreme.net>
Call-ID:10000001@sipstreme.net
C Seq: 1 INVITE
Content-Length: 0
```

Fig 11.2 180 ringing response to Clive's INVITE.

11.2.2.2 2xx: Success

There is only one 200 message at present. This is the 200:OK which means the action was successfully received, understood, and accepted. In fact the OK is not actually needed since the responses, like all SIP responses, are processed on the code number, so 200 ***great*** would work just as well. However, the inclusion of the OK as part of a standard nomenclature does assist human observers.

11.2.2.3 3xx: Redirection

3xx responses indicate further action needs to be taken in order to complete the request. An example of a redirection response is 301 ***moved permanently***. This is rather like a Web page displaying a 'These pages have now moved, please see our new URL' message. The 301 response contains a contact field with either a SIP URL or an IP address at which the user can be found. Analogously there is a 302 ***moved temporarily*** response, which could be returned by a redirect server in response to an INVITE containing an IP address to which the INVITE should be subsequently sent.

11.2.2.4 4xx: Client Error

A 4xx response indicates that the request contains bad syntax or cannot be fulfilled at this user agent server. Users familiar with the '404 page not found' in response to a WWW request will not be surprised to know that SIP has a 404 ***not found*** response also — indicating the SIP URL could not be found by the server. An example of where this would be used is a proxy server that finds no entry in the location database. Another of the 4xx codes includes the 407 ***proxy authorization required*** response that indicates that the proxy requires the user agent to authorise itself with the proxy before it will process the request. The response will inform the user agent of the type of authorisation required, which may be, for example, a digital certificate.

11.2.2.5 5xx: Server Error

5xx responses indicate that the server failed to fulfil an apparently valid request. 5xx messages include the 500 ***server internal error*** and 501 ***not implemented***. 501 messages might be sent in response to an unknown method and indicates that the requested method is not supported by the server. This may not be an uncommon event since SIP is readily extensible, as discussed in section 11.2.5.

11.2.2.6 6xx: Global Failure

6xx responses indicate that the request cannot be fulfilled at any server. 600 ***busy everywhere*** indicates that INVITEs to any location for this user will result in a busy response (i.e. 486 ***busy here***). Obviously this can only be sent from a server with knowledge about all of a user's possible locations. An example of this might be when users have indicated to their service provider that they do not want to be contacted at all at this time.

11.2.3 A Simple Example

To illustrate the above SIP messages and responses in action, a simple call flow between two user agents is shown in Fig 11.3. In this case, only setting-up and tearing down a basic call scenario is shown. The interchange commences with an INVITE method being invoked on the called party who accepts the call attempt by replying with a 200:OK response. This is subsequently acknowledged with an ACK and media then flows.

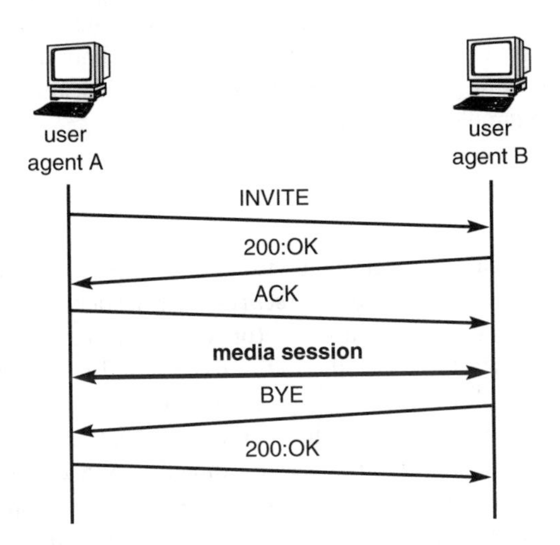

Fig 11.3 Simple message flows for media session establishment and tear down.

At the end of the call, one of the parties — in this case the called party — initiates the termination sequence by invoking a BYE method which in turn is acknowledged by a 200:OK response from the other party.

11.2.4 SIP Headers

SIP headers are the lines found in the messages, such as 'To', 'Contact', 'Via' and so on. There are four types of SIP header — request, response, general and entity. They are generally similar to HTTP headers and take the form of header: field. This section describes a few examples in each category.

11.2.4.1 General Headers

General headers include all the required headers in a SIP message, as well as others, and can be present in both requests and responses. They include headers such as: To, From, Call-id, Cseq, Contact, Subject and Content-length. The 'Via' header is particularly interesting as it is used to record the route taken by request and response messages.

A user agent always inserts its own address before the message is sent and subsequent servers then add their address on top of the growing 'Via' list. When replying to a request, a user agent copies the 'Via' list, in order, on to the response and sends the reply to the top address. The server receiving the message then checks that it is named as the top address in the 'Via' list, removes the 'Via' and forwards the message on to the next via address. In this way the response is made to travel the identical route through a chain of SIP servers that the request took — allowing all the servers in the path that might have state information about the request to process the information in the response. Some of the general headers are discussed further in the context of the example considered in section 11.2.6.

11.2.4.2 Request Headers

Request headers are used by user agents to modify or give additional information about the request. For example, there is ***record-route*** header that is used to force all subsequent requests to go through a proxy (or series of proxies). It has been shown that the via field is used to ensure the initial request and responses (e.g. INVITE, 200:OK and ACK) follow the same route through the proxies, but, after this exchange, both user agents know each other's IP address and can send messages, such as re-INVITEs to modify the session, directly. A user agent inserting the ***record-route*** header prevents this and ensures every request/response follows the proxy chain. This might be useful if those proxies are providing services that the user agent requires — for example the proxy might be noting when users are busy on VoIP calls and re-directing new VoIP calls to a voice mailbox when users are actively 'talking'. Only by seeing all the messages, however, could the proxy decide when it was OK to forward new INVITEs.

11.2.4.3 Response Headers

Response headers are added to responses by user agents or SIP servers to, typically, give more information about a response. Examples include a ***warning*** header that has a range of warning codes and a text message for display associated with it. For example, if Clive INVITEs Mel to a session with an obscure media type that Mel's machine does not support he might receive a response containing a header: ***Warning:305 mel@uk.net:5060 'incompatible media type'***.

11.2.4.4 Entity Headers

Entity headers relate to the message body and the resources requested. ***Content-encoding*** describes how the message body has been encoded, e.g. ***Content-Encoded***

:gzip. Another example of an entity header is ***Expires***. This is used to indicate how long the information contained within a response or request is valid. If this is included within an INVITE it signifies that the user agent must receive a final response, as opposed to a 1xx, before the time specified or the INVITE is automatically cancelled. When included with a REGISTRATION request, it indicates how long the user can be reached at that location.

11.2.5 SIP in Action

This section outlines some simple call flow diagrams showing the messages exchanged in a typical SIP session. As a starting point, consider the case of two users, perhaps connected via the Internet, who want to set up a session. To do this, SIP user agents are required on each of the participating machines. These are software processes that are capable of sending and receiving SIP messages using IP ports and of parsing SIP messages, that is analysing the SIP message headers and deciding what to do with the message contents. User agents actually comprise a user agent client (UAC) and a user agent server for sending and receiving SIP messages respectively. A simple SIP session set-up between two user agents is shown in Fig 11.4.

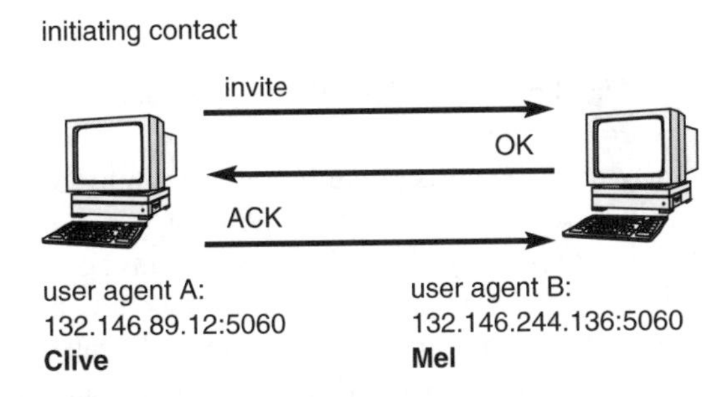

Fig 11.4 SIP signalling during a call set-up.

In this simple scenario, Clive's user agent is being used to initiate contact with Mel. Mel's IP address is known in advance — how this may be done is discussed later when the function of the various supporting SIP server elements are described. So Clive's user agent simply opens a socket and sends an INVITE message to the destination. The details of the actual messages are examined more closely in a later section. However, it should be noted that in this simple example both user agents are listening on port 5060, which is the default port for SIP. In this example, user agent B receives the invitation and now has to return a suitable RESPONSE message from the many defined in SIP. Responses come in different number ranges as we shall see, but 2xx means a successful operation. In this case the invitation is accepted by returning 200:OK.

In the above example, Clive might specify that he wished to invite Mel into a media session including audio. A possible encoding for this information in a SIP INVITE message is shown in Fig 11.5.

SIP header

```
INVITE sip:mel@uk.net SIP/2.0
Via: SIP/2.0.UDP spain.tel:5060
From: Clive Dellard <sip:clive@sipstreme.net>
To: Mel Bale <sip:mel@uk.net>
Call-ID: 10000001@sipstreme.net
CSeq: 1 INVITE
Subject: Urgent Call
Contact: Clive Dellard <sip:clive@sipstreme.net>
Content-Type: application/sdp
Content-Length: 160
```

SDP session description

```
v=0
o=clive 4534593492 3284729843 IN IP4 sipstreme.net
s=Session SDP
e=clive.dellard@bt.com
c=IN IP4 100.101.102.103
t=0 0
m=audio 9160 RTP/AVP 0
a=rtpmap:0 PCMU/8000
```

Fig 11.5 Typical SIP INVITE message.

It is reasonably easy to see what is going on by examining SIP messages when they are transmitted in clear text — which is the case in the example shown in Fig 11.5. The first line simply says that the message invokes the SIP INVITE method on Mel and is using SIP version 2. Each node that relays a SIP message adds its identity in a via field. In this case in the second line, the name of Clive's machine and the port number, the 'well-known' SIP port 5060, are added. The third and fourth lines identify that Clive (identified by his email-like SIP-URL — clive@sipstreme.net) is trying to contact Mel (mel@uk.net) — as seen in the 'from' and 'to' fields. Clive's SIP user agent also generates a unique call-ID which is shown in the fifth line. This is important as it enables the messages associated with this call to be uniquely identified and correctly correlated by both user agents. In doing so, this allows the status of the call to be subsequently modified by exchanging new INVITEs, CANCELs or other methods — even if Mel and Clive have several calls running at the same time. The sequence number, CSeq, is used to identify repeats of lost messages. A new INVITE, to modify the session for example, would increment this number. The subject field is simply information — something that might be displayed on Mel's screen, if it were able to display this message, to entice him to answer the call! The contact field is simply Clive's SIP URL. This leaves only a description and length of the message body — what SIP is carrying. The great flexibility of SIP is that it can carry any message body. In this case it is SDP although it could be GDP (games description protocol) or even RDP (recipe description protocol) — an imaginary protocol!

SDP is an important payload type for inclusion in SIP messages because it is used to set up audio and visual session. It is not necessary to provide a full description of SDP here — the example given below shows how it can be easily interpreted:

V = version number (0)

O = originating name (Clive)

S = subject (anything)

C = connection (Clive's IP address)

T = time

M = media (audio using RTP to create the IP packets)

A = attributes (coding used, in this case PCM μ-Law with a sampling rate of 8000 Hz)

SDP provides fields to specify the intended applications, codecs, and end-point addresses. If Mel can support Clive's suggestions, his user agent simply copies the SDP body back to A in an OK RESPONSE message, entering its own end-point addresses and port numbers for the medium. Session negotiation and set-up can therefore take a minimum of three SIP messages, which equates to just one-and-a-half network round trips. However, in the event that Mel's user agent cannot support one particular codec, but can offer another, it would amend this field in the SDP of the returned 200:OK response. If the change is acceptable to Clive, then the ACK follows as normal; otherwise Clive CANCELs the session, or renegotiates, sending another INVITE, with a new SDP, but the same call ID and a higher sequence number. Mel's user agent recognises the call ID and realises it is a re-negotiation from the higher sequence number and the process begins again. In the same way, in-session renegotiation is supported. For example, the existing video session is streaming and then Mel decides to add voice.

11.2.6 SIP Network Elements

As discussed in Chapters 1 and 9, due to the nature of the underlying IP network infrastructure, the terminal address of Mel's user agent may be dynamically allocated and forever changing due to the effects of DHCP. This section introduces the major SIP network elements — software, servers and so forth — associated with the use of SIP. In particular, in the example given previously, to overcome the limitation of Clive having to know Mel's user agent address directly, additional elements have been introduced into the SIP architecture. These currently include:

- universal resource locators (URLs);
- proxy servers;
- location servers;
- registration servers;
- redirect servers.

In a SIP network, every SIP user — including automatons — is given a SIP URL. SIP URLs resemble e-mail addresses, and are of the format:

sip:username@domainname.

Typically, the 'username' is the user's actual name, and the 'domainname' is the user's home domain. Within the user's home domain, there is a SIP registration server. The registration server listens for messages bearing the REGISTRATION method. When the user agent starts up, the first message it sends is a REGISTRATION, bearing:

- the URL of its user;
- its actual terminal address;
- port number;
- protocol, i.e. TCP or UDP;
- an optional timestamp for how long the registration is valid (default one hour);
- an optional preference for being contacted at this location.

The registration server authenticates the user, and adds the mapping between URL and network address to the location server's database. Figure 11.6 illustrates the process.

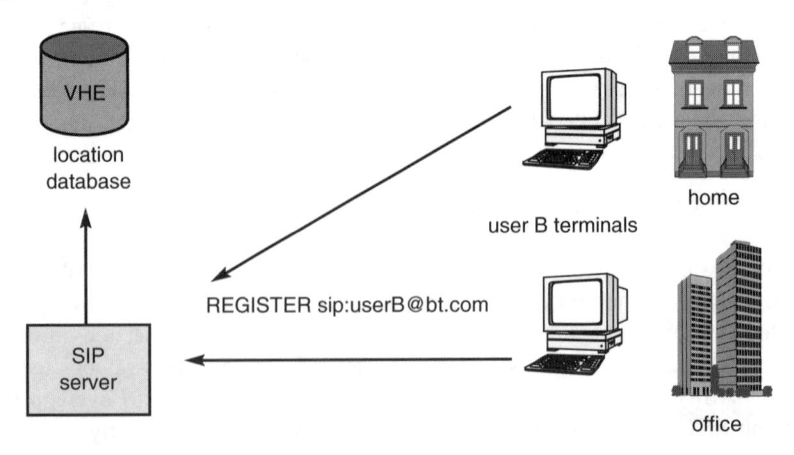

Fig 11.6 User B registers two terminals with the SIP proxy server.

SIP URLs make users contactable, irrespective of their current network address. User A simply needs to know the SIP URL of user B, which is constant, as opposed to its possibly ever-changing network address. Knowing a SIP URL is not sufficient to route a message to user B; to do so requires the service of a proxy or redirect server. Proxy servers, as their name suggests, act on behalf of user agents, routing SIP messages to correct destinations by invoking SIP URL to network address mapping by location servers. Figure 11.7 illustrates the revised message flows. Of course, if the IP address of user B is already known to user A or is discovered by another means (e.g. DNS), there is nothing to stop them exchanging SIP messages without the use of an intermediary server.

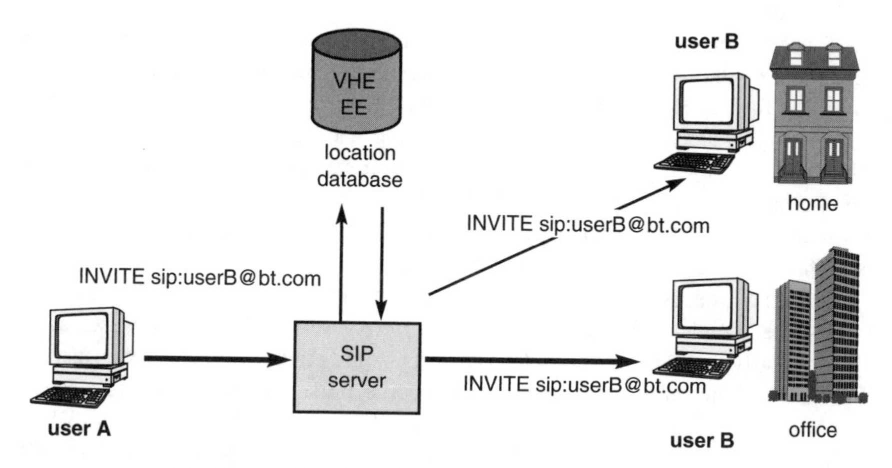

Fig 11.7 User A sending an INVITE message to user B.

User B currently is working from two terminals, each with a user agent that has registered its network address against B's SIP URL. Registrations are additive, although — as mentioned earlier — can be time-stamped for periods of validity, and can be prioritised according to preference in being contacted. When seeking to contact B, A sends an INVITE request to the proxy, specifying B's URL. The proxy determines that B currently has two terminal addresses and proxies a copy of the message to each — a process known as forking, inserting its own address into the path list (using the via field). B now sends it an OK response to the address at the top of the path list, which results in it being returned to the proxy. The proxy then returns the response to A's user agent, and remains in the path between A and B for the ensuing ACK. In terms of SIP architecture the proxy server makes use of a location server, which is basically a database. How the proxy interacts with the database is not specified by the IETF but might, typically, use the LDAP protocol. The location database might be populated by information from the registration server, but might also obtain information on users in other ways, such as from an authentication, access and accounting (AAA) server.

Coming back to the example of Clive and Mel, if Clive simply knew Mel's SIP URL — mel@uk.net — he would use a DNS look up on uk.net to find the IP address of the SIP proxy server relating to the uk.net domain. This is exactly analogous to looking up the DNS records for e-mail or WWW servers. Inside the uk.net domain there may be several servers that actually deal with the request — just as there are e-mail servers in a large ISP, say. One of these proxies will then query the location server for the domain and forward on the message to Mel.

A redirect server differs from a proxy server in that it does not forward messages but simply does a location look-up and returns one (or more) addresses for the user (B) and leaves it up to the original user agent, A in this case, to contact B at these addresses directly.

SIP proxy servers can be either stateful or stateless — a stateful proxy server keeps track of the requests and responses it receives and can use that knowledge in processing future requests. One example would be to retransmit a request for which no answer is received; another example being the server requiring user agent authentication. Forking is also only possible with stateful servers — only the first response is sent to the originator of the INVITE — the rest are cancelled. It will be shown later that IN-like services can be created based on stateful proxies. However, not all proxy servers are able, or required, to fork.

SIP, in fact, reuses the concept of client/servers and URL from HTTP — the SIP proxy server being analogous to the WWW proxy server. In fact, the user agent software really contains server and client parts — the client originates requests (like INVITEs) and the server part is waiting to receive, parse, and respond to requests.

The final element that might be found in an IETF-IP network, relating to SIP, is the SIP signalling gateway. This is an application layer gateway that translates SIP messages into those of other signalling protocols — such as H.323 or ISUP (integrated services user part) SS7 messages used in the PSTN. This is a complicated function since sometimes the functionality is a poor fit between the protocols and only a subset of SIP capabilities can be supported across the gateway.

11.2.7 Characteristics of SIP

SIP has a number of characteristics that make it particularly suitable for widespread use in very different applications, from voice over IP to 3G multimedia service creation. This section examines some of these characteristics.

11.2.7.1 Simplicity

SIP has been designed to be very lightweight, although as it develops this is becoming questionable; however, it can interoperate with just four headers and three request types. Also sessions can be set up in 1.5 round trip times. This minimal footprint means that SIP can run on very thin devices — such as pagers or baby alarms.

11.2.7.2 Generic Session Description

SIP separates the signalling of sessions from the description of the session. SDP is not mandatory and SIP could be used to initiate and control completely new types of session. For example, consider a virtual reality interactive game for network players. In order to play the game a complex description of the terminals, players' experience, preferences, and so on, might need to be exchanged. The VR game might use a new session, or game, description protocol while SIP could transport this transparently across the IP network.

11.2.7.3 Modularity and Extensibility

SIP is designed to be extensible allowing implementations with different features to be compatible. An example of this extensibility is the use of SIP for UMTS Release 2000; 3GPP is in the process of defining specific SIP mobile extensions for multimedia session control in UMTS.

11.2.7.4 Programmability

As will be described in section 11.3, the introduction of a SIP server offers the possibility of running scripts or code (e.g. Java servlets) that can alter, redirect or copy INVITE or other SIP messages. Not only can SIP servers be used to provide IN-like services traditionally seen on voice networks, but can also be extended to provide intelligent control of advanced multimedia services.

11.2.7.5 Integration with Other IP Component Technologies

The design of SIP is derived from already existing IP protocols such as HTTP (header field, authentication mechanism), SMTP (addressing server location), and border gateway protocol (BGP) (for stateless loop detection). SIP is designed to integrate with IP protocols such as the real-time streaming protocol (RTSP) — together these could be used to offer voice mail services or to invite a video server to play a movie during a multiparty conference.

11.2.7.6 Scalability and Robustness

SIP servers can be totally stateless, enabling scalability for solutions based on the technology. There are, however, reasons for having stateful proxies, such as for classic call control for example. SIP also supports multicast sessions, something that is very difficult for traditional circuit-based call servers that require a bridge to connect the parties.

11.3 Creating SIP-Based Applications

There are currently a number of ways already proposed for creating services such as the traditional IN call-waiting or transfer on busy, as well as to allow servers to make context-sensitive decisions on forwarding or modifying requests. An example of this would where a user set different priorities for being interrupted by different people — for example they may accept instant messages from anybody in the company, but voice-call invitations only from their boss and reject all INVITEs from double glazing companies! This section reviews some of the proposals for these service-creation mechanisms.

11.3.1 Call Processing Language

Call processing language (CPL) is interesting because it permits the user to construct their own services and upload them into their profile [8] running on a SIP server. CPL is considered safe as a scripting language because it does not permit loops — just conditional statements (see Fig 11.8). Server processing therefore cannot be tied up indefinitely either accidentally, by a malformed script, or maliciously by a virus. When a request comes into a location server, for example, the presence of one or more scripts is detected and executed accordingly, revealing just that subset of locations specified by the script.

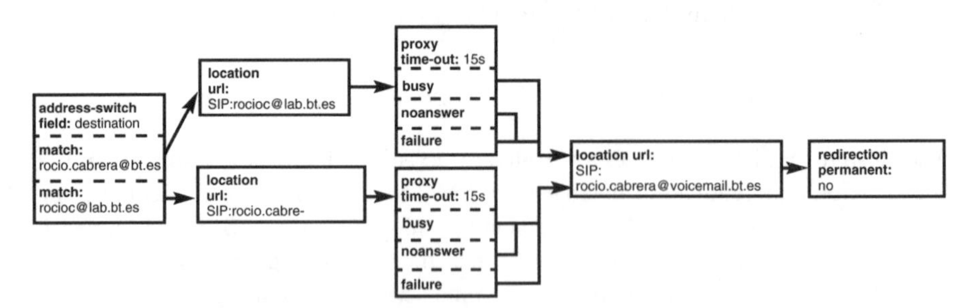

Fig 11.8 Example abstract representation of a CPL script.

In reality, the majority of users would probably not wish to write their own scripts from scratch. Rather they would most likely use some form of tool such as a number of options presented on a Web page. In this case details can be either selected or entered and scripts autogenerated behind the scenes. Autogenerated scripts can then be uploaded into the user's profile, potentially revolutionising service provisioning processes. Typical applications of CPL scripts could be call screening and call routing, where the user preferences and input are paramount.

11.3.2 CGI

The common gateway interface (CGI) [9] is a standard for providing an interface between external applications and network elements. CGI makes it possible for Web pages to trigger complicated programs on the hosting server. Typical functions provided by scripts include the ability to process forms or provide page counters. CGI is not a programming language as such, but an interface for applications to interact with network elements. CGI can be written in most common programming languages, such as C, C++ or PEARL, and can use SIP message headers as metavariables that are passed across the interface. CGI scripts can also generate requests and responses that are converted into SIP messages (Fig 11.9).

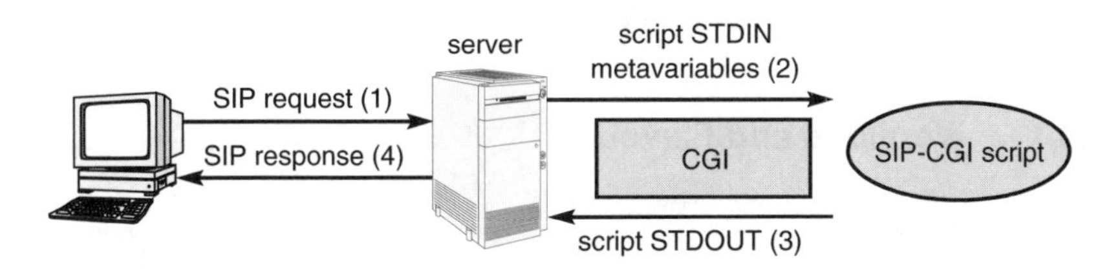

Fig 11.9 SIP CGI-based service creation.

11.3.3 SIP Servlets

SIP servlets are a Java extension application programming interface (API) to SIP [10]. Incoming messages received by a server can be associated with a particular servlet by a SIP servlet container. Each servlet is a process running on the server (Fig 11.10).

The server passes objects representing SIP messages to the servlet that then has access to all parts of the message and can modify or create new responses. A typical application for servlets is shown in Fig 11.10. Here phone user 1 sends an INVITE to user 2 via the SIP server. Phone 2 responds to the INVITE with a busy response and so the server passes this over the SIP API to the servlet container. The container examines the message and, based on pre-set rules, such as a service subscribed to by user 2, passes it to a servlet that executes on the message and results in a decision to re-direct to a voicemail server. The server modifies the busy response and passes it to the voicemail server that, subsequently, generates an accept response which is sent to user 1 to set up the call to the voicemail box.

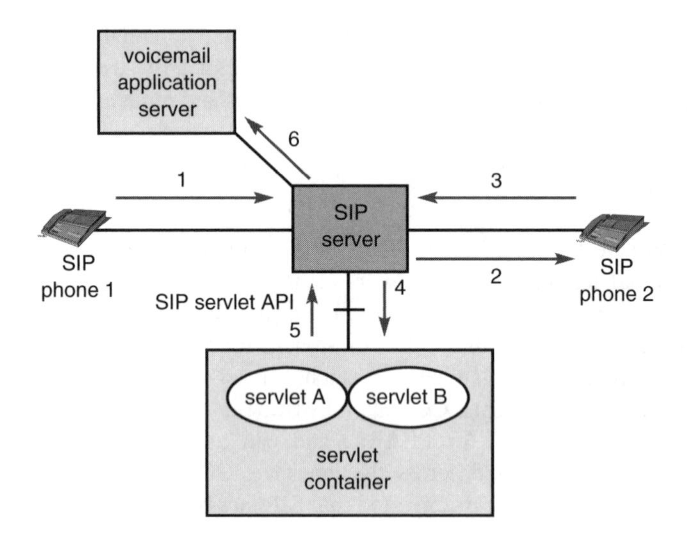

Fig 11.10 Basic SIP servlet operation.

11.4 Versions and Flavours

11.4.1 IETF

In many respects, the main version of SIP in common consideration is that specified by the IETF. Increasingly this is referred to as the 'core' SIP specification, because others build variations or derivative works from it. At the time of writing, the core SIP specification is documented in RFC 2543. A number of so-called 'RFC 2543bis' drafts have been produced as the core specification progresses along the IETF's standards track, as described in Chapter 5. The 'bis' drafts largely add further explanatory text and improve the readability of the RFC 2543 version.

11.4.2 SIP-T

Probably one of the greatest misunderstandings in the entire industry at the time of writing concerns SIP-T. In the earlier days of the VoIP industry when large-scale PSTN replacement solutions were first being considered, it became apparent that a signalling protocol was required to transport control messages between media gateway controllers over the IP network. As discussed in Chapter 6, an 'Inter-SoftSwitch Communications Protocol' or ISSP is required to deliver this functionality. One proposal that emerged to provide this functionality was based upon the use of SIP and comprised the encapsulation of SS7 messages within SIP messages. The original proposals were written as individual submissions to the IETF with the intention of becoming a 'Best Current Practice' document, as described in Chapter 5. During this early work the documentation became known as the 'SIP Best Current Practices for Telephony' — or SIP-T. However, to date no formal standards track document has been created with this name — in other words SIP-T does not actually exist in the sense of a specification that could be used contractually or is anything other than an individual Internet Draft submission. In fact, the situation has become much more complex since the functionality required to enable SIP to be used in this way has been distributed across at least three documents in the IETF SIP working group — chiefly the SIP INFO method, the MIME encapsulation of ISUP messages and their transport using SIP, and finally the specification of a set of telephony call flows using SIP.

11.4.3 PacketCable

The PacketCable initiative within CableLabs, has adopted SIP within its solution for providing telephony services over cable TV networks [11]. In this solution a derivative, or profile, of SIP is used as the basis of the signalling protocol between call management servers in the PacketCable environment. The extensions proposed, which are known as 'distributed call signalling', enable and define how multimedia and basic telephony services can be offered over a cable TV network infrastructure. The extensions identified so far have also largely been submitted as Internet Draft submissions to the IETF. However, the work presents both an application of SIP in a real network context and also another possible divergence point from the core SIP protocol.

11.4.4 3GPP

UMTS Release 5 allows all traffic to be carried as IP packets and, in addition, it introduces an Internet and multimedia (IM) domain (Fig 11.11) for users to access multimedia services [6]. A version of SIP has been developed by the UMTS standards body (3GPP) to provide the call control aspects of the IM domain. In earlier releases of UMTS, only transparent IP connectivity to an ISP was possible, rather than the operator's network being able to offer IP-based services. In this case, the call server control function (CSCF) acts as a SIP proxy server. It locates users, translating SIP URLs to IP addresses, proxies INVITE messages and maintains information state on the session.The CSCF keeps information to allow other multimedia streams to be added to existing sessions, both for charging and because it controls the multimedia resource function (MRF) which is essentially a bridge that allows multiparty sessions in a network that does not support multicast. The 3GPP flavour of SIP provides new components to allow users to register and make

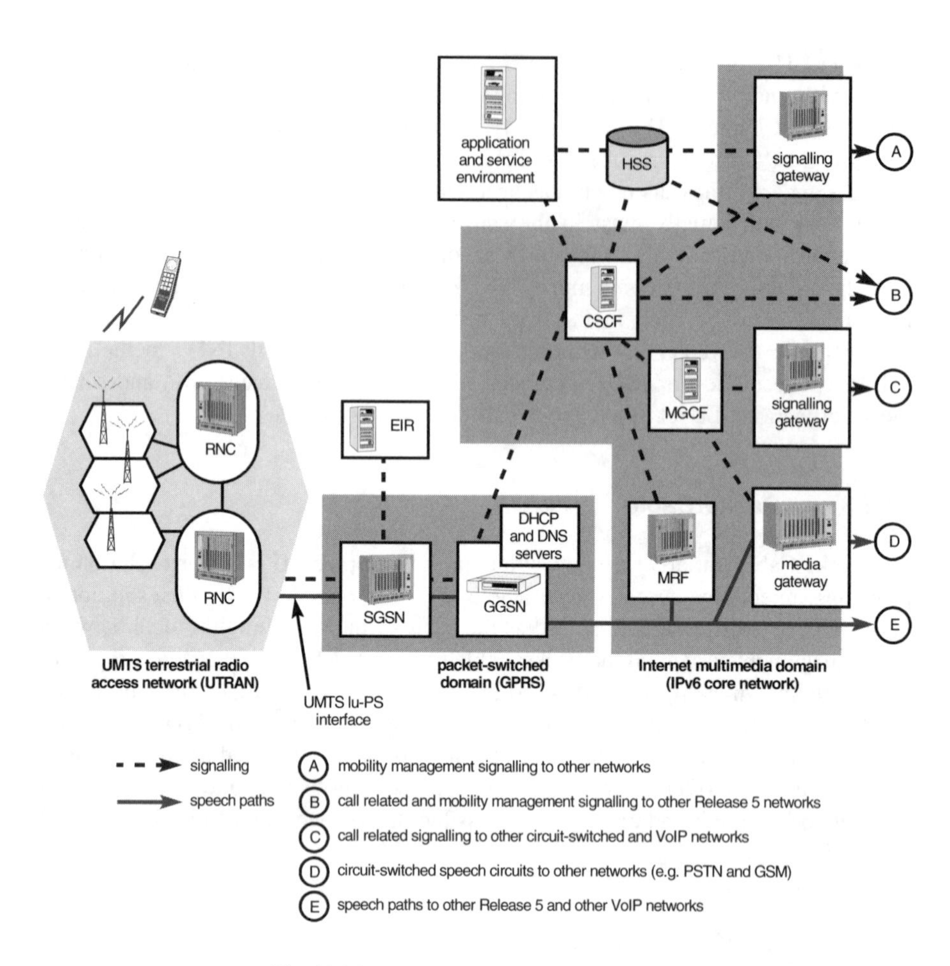

Fig 11.11 UMTS Release 5 architecture.

multimedia calls using the IM domain functions, while being billed from their mobile prepay or post-pay account.

11.5 Core Design Issues

11.5.1 Emergency Service Support

Although the issue is important for any public VoIP implementation, the support of emergency calling services has become a particular discussion point within the SIP development community. In essence there are a large number of features that are currently provided within the scope of emergency call services. These include establishing a processing priority for the call within the network, identifying the correct location or geographic area from which the call originates and ultimately providing control over the media path. Many of these aspects remain difficult or impossible to reliably achieve with VoIP technology — and SIP is no exception to this rule. In particular, this is likely to be highly problematic for pure peer-to-peer modes of operation. The planning and deployment of large-scale SIP networks offering public services therefore needs to pay careful attention to these issues.

11.5.2 Routing Calls

As SIP has been produced using the IETF's philosophy of one protocol for one purpose, it lacks a number of features that are required to build large-scale VoIP networks. One of the main areas that SIP does not address is the handling of routing decisions across a SIP network. As a consequence, while SIP messages allow the interaction with proxy servers and the like which enable routing decisions to be taken, SIP itself does not address how such decisions should come to be made. Fortunately the IETF have addressed this issue through the development of a 'telephony routing over IP' protocol known as TRIP. TRIP has a number of similarities with BGP [12] in that it enables a hierarchical set of routing information to be constructed and exposed to adjacent domains. At present TRIP is in the process of being developed and only experimental implementations are available at the time of writing.

11.5.3 Firewalls

Another issue that is problematic for session-oriented protocols such as SIP is firewall traversal. Most firewalls do not currently pass SIP messages. Future developments, such as allowing pin holes to be created, will solve this problem. Tunnelling SIP packets, possibly over a secure tunnel, or encapsulating them in HTTP packets are other possible solutions. Finally, network address translation (NAT), used, for example, to allow private address space to interface to the outside world by sharing a much smaller number of globally routable addresses, will potentially also cause difficulties for SIP services. Any private addresses embedded within the SIP messages, that is they are not in the headers that are dealt with by the NAT, will not be translated and therefore cause errors if used for routing in the rest of the Internet. Although one of the prime reasons for employing NAT — shortage of address space — will disappear as IPv6 is introduced, it is not unlikely that NAT solutions will be required for some considerable time to come. In particular, there are potential advantages in using NAT and firewalls since this combination may enable message routing to be forced through proxy servers as a method of policing network usage.

11.5.4 Management and Operational Issues

Irrespective of the particular VoIP technology in question, the creation and operation of a network and any associated services presents a number of challenges that need to be addressed for the system to function correctly. These issues fit broadly under the headings of:

- fault management;
- configuration management;
- accounting management;
- performance management;
- security management.

In the area of fault management, a wide range of issues need to be considered such as the availability of SNMP MIB functionality on the SIP network elements and user agents. This is because, without the ability to verify or report error conditions appropriately, expensive manual processes will be required. At the time of writing, the development of appropriate support for SIP in this area remains as 'work in progress'.

From the point of view of deploying SIP-based end-points and servers, configuring the equipment to interact correctly — such as locating the correct registration server or proxy — is a requirement that must be automated as far as possible to avoid expensive manual processes. As with the case of fault management this remains a largely incomplete area with respect to full industry-accepted, and so interoperable, solutions.

Since business models for VoIP services are still being established, requirements for addressing accounting issues and handling usage-related information remains a very much open issue.

Performance management is an essential element of any large-scale network deployment. Without being able to correctly operate, plan or build a network to an appropriate dimension, there can be an over-reliance upon crude over-provisioning to achieve an intended level of service offering. So if capital costs are to be under appropriate control, performance management issues need to be carefully addressed. At the time of writing, many of these issues for SIP-based networks are either within the exclusive domain of proprietary solutions or insufficient attention has been paid to them due to the relatively small scale of current deployments.

As will be appreciated from the discussion in Chapter 1, potentially any VoIP solution is completely open across the associated IP infrastructure. It is therefore of great importance to ensure that any users and their equipment are appropriately authenticated and authorised to access services provided by a SIP network. Many of the issues in this area are common across all VoIP-based technologies, although each must be addressed appropriately and individually.

At the time of writing, there remains a substantial amount of work to be done to ensure SIP-based networks can address these issues in a realistic manner that also allows the full opportunities offered by SIP to be realised.

11.5.5 QoS for Media Transport

As SIP is essentially just a session set-up protocol, it has no explicit mechanism for addressing how an end-to-end quality of service level of the type identified in Chapter 2 can be established. For example, if someone wanted to have a VoIP call to another party, they might send an INVITE message, negotiate a particular codec and receive a RINGING message. However, in the event that the underlying resources, such as network bandwidth or terminal resources, were not available, then the call would fail to deliver the expected or desired results. While overprovisioning of resources is an expedient step towards mitigating such problems, fundamentally this approach is unlikely to scale with the demands of increasingly complex multimedia applications. Standards groups such as TIPHON, 3GPP and PacketCable are actively exploring a range of solutions to this problem. In the meantime, this presents a challenge to delivering revenue-generating, cost-effective, SIP-based services and is a potentially significant weakness.

11.6 Summary

Current Internet applications, such as e-mail and Web-browsing, are client/server based and do not require session set-up or control. They work very effectively on best-effort IP networks. In the future, it is envisaged that both fixed and mobile services will be provided by a QoS-enabled all-IP core network. The next generation of peer-to-peer style applications, however, require advanced functionality in terms of:

- user location by name;
- media negotiation;
- session renegotiation on handover, and the like.

This functionality is analogous to call control and intelligent network-based call treatment in the PSTN world, except that it is many times richer and potentially more complex. SIP is an excellent candidate for any peer-to-peer applications capable of communication over IP because it is lightweight, flexible, scalable, extensible and programmable. This chapter has shown how the concept of a SIP proxy server can be used to provide personal mobility and allow complicated 'intelligent' services to be delivered. This has included familiar PSTN services, such as call forwarding, as well as new mobile multimedia services. SIP offers a great opportunity for both network and service providers, such as ISPs, to create carefully tailored services, control the process of session initiation to enable higher quality and more reliable services, and enable the use of chargeable network facilities, such as archivers, media codecs and the like.

References

1 University of Michigan — http://china.si.umich.edu/telecom/internet-telephony.html

2 Handley, M.: '*Session initiation protocol*', IETF RFC 2543 (March 1999).

3 Columbia University — http://www.cs.columbia.edu/~hgs/sip/

4 IETF — http://www.ietf.org/

5 Wisely, D. R., Eardley, P. L. and Burness, L.: '*IP for 3G*', John Wiley and Son (to be published in 2002).

6 3GPP — http://www.3gpp.org/

7 Dufour, I. G.: '*Network Intelligence*', Chapman and Hall, London (1997).

8 IETF: '*CPL: a Language for user control of Internet Telephony services*', draft-ietf-iptel-cpl/04.txt (November 2000).

9 IETF: '*Common Gateway Interface for SIP*', RFC 3050 (January 2001).

10 IETF: '*SIP Servlets*', draft-kristensen-sip-servlet-00.txt (September 1999).

11 PacketCable: '*CMS to CMS Signalling Specification*', PKT-SP-CMSS-I01-001128, Cable Television Laboratories Inc — http://www.packetcable.com

12 Willis, P. J. (Ed): '*Carrier-scale IP networks: designing and operating Internet networks*', The Institution of Electrical Engineers, London (2001).

12

SIP AND H.323 INTERWORKING

A Stephens

12.1 Introduction

Chapter 10 described how H.323 started life in the mid-1990s when ITU-T SG 16 started work on an umbrella recommendation for offering multimedia services over packet-based networks. The intention for H.323 was to create a framework that would allow multimedia services to be offered over LAN-based networks where quality of service was not an issue. This was based on the older H.320 recommendation that addressed the provision of multimedia services over ISDN-based switched-circuit networks. In keeping with this decision, ITU-T SG-16 adopted the call model, information elements and features of Q.931 and incorporated them into H.225.0 from the outset. They had also been working on H.324, a recommendation for offering multimedia services over dial-up connections. Since H.324 and H.323 require media and bearer control facilities, a common protocol, known as H.245, was developed to address the needs of both applications.

As the concept of offering real-time conversational services over packet networks entered mainstream thinking, the scope of H.323 [1] grew enormously to encapsulate many different applications. As a consequence, H.323 started to become more complex, containing a large set of annexes describing new protocol features and procedures, as described further in Chapter 10. To mitigate this growth in complexity, a number of profiles of H.323 have been created by industry groups such as the IMTC in addition to ETSI TIPHON and the ITU-T SG16 itself in the form of H.323 Annex F.

Around the same time as the ITU-T were developing H.323, the IETF's MMUSIC Working Group began creating an architecture to add conversational real-time capabilities to the existing Internet protocol suite. As discussed in Chapter 1, an integral component of the architecture developed by MMUSIC included a session set-up protocol known as the session initiation protocol (SIP) [2] that, as described in Chapter 11, was designed around the end-to-end principles of the Internet.

As will be apparent from the discussion in Chapter 1, both H.323 and SIP have to address a similar set of problems concerning methods of addressing, locating end-points, call establishment and media negotiation. Consequently there is an overlap in the functionality provided by these two technologies. While SIP is a lightweight, newly emerging technology, it is still relatively immature when compared with H.323. In particular, there are many features within H.323 that are not currently provided by SIP, such as a more complete bearer control mechanism. The number of H.323 deployments also exceeds those of SIP at the time of writing. This presents a problem for the industry in that it is desirable to be able to carry forward investments made in H.323 into an environment where SIP will be increasingly deployed. There is also the potential for both SIP and H.323 to co-exist, since in many respects they provide complementary functionality. This has led to some interesting proprietary developments by equipment manufacturers where H.323 functionality has been retrofitted into SIP solutions to provide a form of *ad hoc* interworking. However, with the potential for SIP and H.323 interworking to be a long-term part of the VoIP landscape, it is necessary to have agreed approaches to handling these issues. This chapter provides a consideration of the design options that surround SIP and H.323 interworking.

12.1.1 SIP Examined

SIP has passed through the early stages of development and is evolving into a more complete protocol that is able to support many different applications, in addition to VoIP. As with the majority of Internet protocols, and in contrast to the binary encoding method used to define H.323, SIP is a text-encoded protocol. As a minimum, any interworking between the two protocols must therefore overcome this encoding difference. As described in Chapter 11, SIP borrows many elements of HTTP, utilising the same transaction request/response model with similarly structured headers and response codes. It also adopts a modified form of the URI addressing scheme used within SMTP-based e-mail.

Fundamentally, SIP's functionality is enshrined within the headers and message bodies of its methods and responses, therefore becoming the focal point of any consideration of interworking. The procedures for establishing sessions are also relatively simple with information exchanges being carried out mostly by the overloaded INVITE, 200 OK, and ACK transactions. Additional procedures used to extend the existing capabilities of SIP are handled by introducing new SIP methods, new headers, the definition of new MIME types, and the ability to carry additional information within the message body of method requests/responses.

On the basis of its perceived simplicity and ease of implementation, SIP has been heralded as the successor to H.323 by service providers, network operators and equipment manufacturers. However, since its inception, many extensions have been proposed for inclusion into the base SIP specification in order to meet the increasingly complex needs of service providers who attempt to create large-scale network systems. Unfortunately, if not managed properly, any proposed extensions could potentially negate the simplicity and ease of implementing SIP solutions, and, in doing so, defeat the primary motivation for the interest in SIP in the first place.

12.1.2 H.323 Examined

The H.323 umbrella recommendation comprises a set of standards that describe the functionality delivered by a number of entities, and the protocol messages and procedures that these entities use to communicate. H.323 has evolved through a number of versions since 1995 and, as described in Chapter 10, the H.323 protocol set is extended by the inclusion of annexes that describe both core and peripheral protocol features. These extensions either add to the base ASN.1 syntax or define procedures and profiles for protocol feature usage. In keeping with the usual ITU-T practice for protocol development, extensions are generally added to H.323 in a manner that promotes backward compatibility with earlier versions of H.323. Earlier implementations are therefore not 'broken' when they communicate with more recent ones. As a consequence, the explosion of requirements for supporting VoIP has led to the expansion of the ASN.1 syntax since this is the core repository for every feature or application extension that may be required for an H.323 call. This fact contributed to what were perceived as major weaknesses of early versions of H.323, namely modularity and extensibility. However, these issues have largely been addressed by the inclusion of a framework into version 4 of H.323 that allows application feature extensions and feature negotiation to be handled in a way that does not require endless additions to the base ASN.1 syntax. As discussed in Chapter 10, this framework is known as the general extensible framework (GEF) and has the designation H.460 within the ITU-T. It is anticipated that almost all future extensions to H.225.0 will be defined in terms of H.460 packages and procedures. A similar mechanism, known as 'generic capabilities', has existed within H.245 since version 2 of H.323.

12.1.3 Future Directions for SIP and H.323

H.323 currently has a much richer bearer control capability than SIP, which, when coupled with a complete conference control capability, makes H.323 the ideal choice for LAN-based audiovisual conferencing services where issues such as bandwidth and device memory are not a concern. However, as indicated in Fig 12.1, even though they have had very similar overall goals in mind, SIP and H.323 have come from different perspectives. For example, H.323 has been shaped primarily by the telecommunications industry and has a voice-centric heritage. Conversely SIP started out life as a protocol for establishing simple sessions on the Internet and builds on the success of related protocols such as SMTP and HTTP by adopting a similar approach.

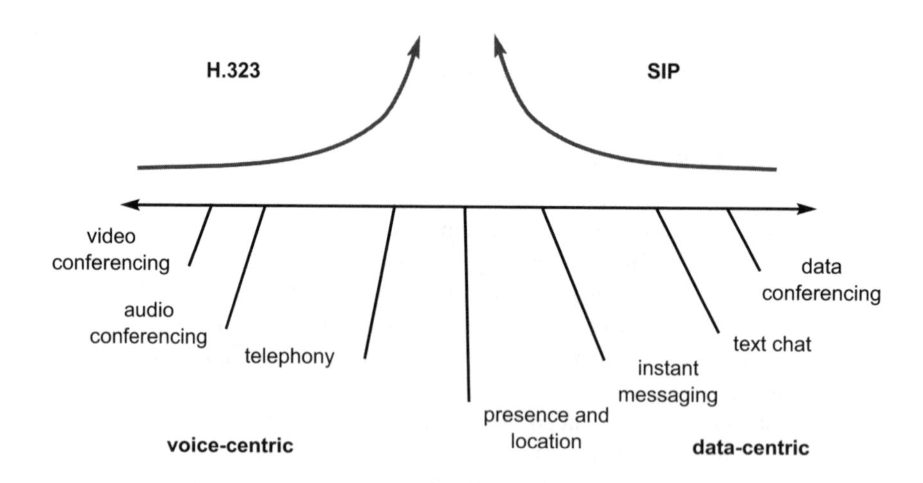

Fig 12.1 Perspectives on the developments of H.323 and SIP.

Since they address many similar issues, it is highly likely that future versions of both H.323 and SIP will continue to track the developments made by each other. As a consequence both protocols may naturally start to converge since they offer increasingly similar services and capabilities, albeit in slightly different ways. Due to the existing installed base of H.323 and the continued development of both protocols, it is also highly likely that both SIP and H.323 are going to be around for some time to come. This means that equipment manufacturers and network operators must find ways of allowing these two different worlds to co-exist through protocol-agnostic service solutions.

12.2 Protocol Interworking between SIP and H.323

From a practical point of view, SIP and H.323 are both typically deployed in the form of a set of terminals, back-end services and other logical entities under the control of a single administration. This is known as an administrative domain. Similarly, a signalling domain can be identified as comprising a set of entities that communicate using the same session and media control protocols. In this case these will be either SIP or H.323. Among other things, signalling domains comprise the lower layers of call servers as well as terminals and gateways. An administrative domain may contain one or more signalling domains. In turn a single administrative entity, such as a service provider, may control multiple administrative domains.

Entities providing supporting services including location, security and billing functions are considered to exist within a back-end services (BES) layer. Administrative domains may share end-point and service information directly between BES layers via inter-domain protocols, such as those described further in Chapter 14.

As an interworking device may bridge two administrative domains, a simple view of solving protocol interworking between SIP and H.323 would be to consider a device with two interfaces; one that communicates using H.323 and associated protocols and another that supports SIP. However, as discussed in greater depth in section 12.3.1, in order to provide interworking between the two protocols the message formats, semantics and error handling need to be correctly considered. Since there are a number of different ways of deploying SIP and H.323 networks, and many different ways for these entities and domains to interact, this suggests there are a number of possible scenarios where protocol interworking needs to be considered. Network design and topology therefore affect the interaction between entities at the call/session control and back-end service layers, since the interworking function may have manifestations at both the call/session-signalling layer and the BES layer. There are therefore two parts of an interworking function that must be addressed; an interworking function that handles interworking at the call and media control layer, known as a protocol interworking function (PIWF), and an interworking function that handles protocol interworking at the BES layer, known as a BES interworking function (BESIWF).

As discussed in Chapter 9, the precise identity of all the elements which will be involved in any given call are unlikely to be known when the call is originated. For example, it may not be known whether a call will traverse a number of administrative domains or signalling domains, or which end-point will eventually terminate a call. However, irrespective of the particular scenario concerned, each potential chain of devices used to route a call for interworking can be described in terms of an interworking architecture and the various call-flow scenarios that it supports.

In practice, interworking architectures fall into two broad categories — intra-domain interworking and inter-domain interworking. In either case the main issues that need to be considered include:

- how the registration or provisioning of end-points is handled;
- how call routing is addressed;
- how applications and features are made available.

The design decisions taken to address these issues therefore give rise to a number of options, some of which are described in the remainder of this chapter.

12.2.1 Intra-Domain Interworking

Intra-domain interworking occurs within a single administrative domain. Although it will always be easier to maintain and support a network that utilises a single call control protocol, intra-domain interworking remains a likely scenario for any large network due to the effects of mergers, acquisitions, multi-service provision and multi-vendor procurement.

The principal objective of interworking in these scenarios is to enable an administrative domain to offer the same services and applications to customers irrespective of the underlying protocols their equipment uses. If the ultimate goal is to migrate from one protocol to another, then a network operator may deploy a new product offering in the form of a new administrative domain. In such a case, users may then be migrated to the new domain with the old domain eventually being taken out of service. However, it is not unreasonable that customers, and hence network operators, will wish to continue using a particular protocol because it offers features that others do not. This will certainly be true while SIP matures and until it is able to offer the same rich feature set as H.323.

12.2.1.1 Service Supported on Neither Signalling Domain

The simplest intra-domain interworking scenario is shown in Fig 12.2, where H.323 clients and SIP UAs are configured to use an interworking function (PIWF) directly. In this case there is no gatekeeper or proxy server so the H.323 and SIP end-points must have *a priori* knowledge of the interworking function. As will be apparent from Chapter 9, each end-point will need to determine from the destination name whether a particular session crosses the domain boundary and route it accordingly to an appropriate interworking function. This clearly needs the end-points to be aware of both the existence of other domains and the location of relevant PIWFs through the naming and addressing policy applied.

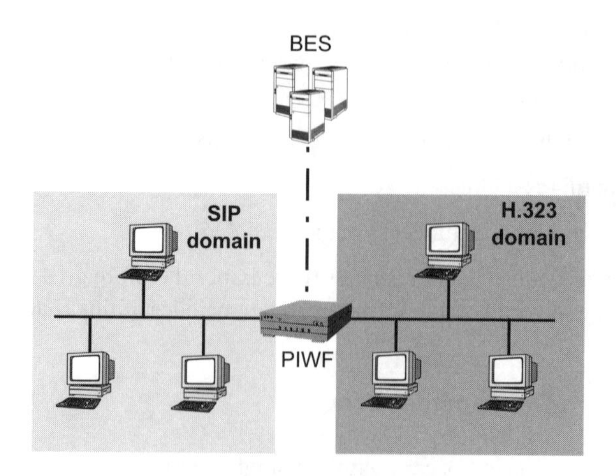

Fig 12.2 Simple interworking architecture.

A variation of this approach is to collocate the interworking function with a gatekeeper or SIP proxy. Through the gatekeeper or proxy functions, the PIWF can communicate with the BES layer and determine the transport address of the destination end-point in the next or terminating domain. This kind of architecture has primarily been used with prototype devices that have been created to demonstrate the principles of protocol interworking in a controlled laboratory environment. This has the advantage of negating many of the routing issues that need to be considered for more complex scenarios. However, this approach is unlikely to ultimately yield a useful solution for mass deployment as it is not readily scalable, because it relies on interpreting the naming and addressing policy at the end-points.

12.2.1.2 Service Supported on One Signalling Domain

An alternative scenario is shown in Fig 12.3. This represents the case where the IWF acts as a gateway for H.323 devices to communicate with a SIP proxy. Here the IWF handles the conversion between the native registration and location messages, and, for fully call-server-routed calls, the call signalling messages. This scenario can be mirrored where SIP CPE is managed as part of an H.323 domain for example.

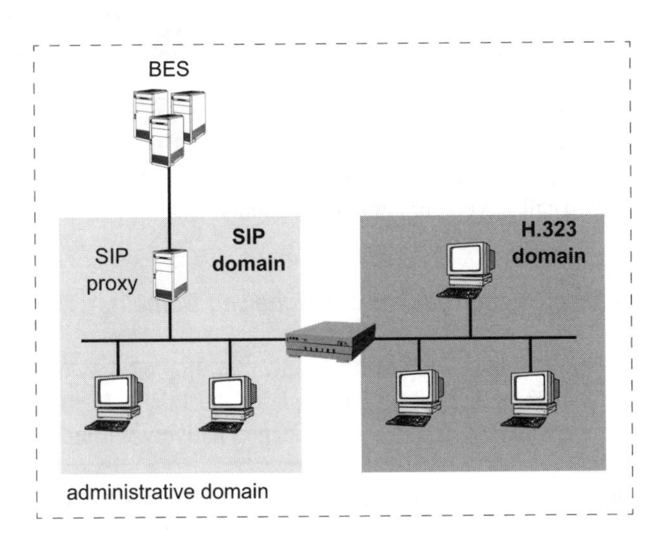

Fig 12.3 H.323 CPE managed by a SIP domain through a PIWF.

In the specific example shown in Fig 12.3, once interworking has been performed locally, it is assumed that the SIP network will then onward route the call to user agents in an external administrative domain. In this case the IWF will appear as the originating device as far as the SIP network is concerned.

In the reverse direction, incoming calls to the H.323 domain from an external SIP signalling domain may be handled in one of a number of ways depending upon whether or not the BES layer is aware of the IWF. In the case where the local BES layer is not 'protocol aware', a default IWF can be provisioned. For example, a SIP proxy can be configured to send calls to the IWF when the resolution of a local routing fails. Conversely, if the BES layer is able to discriminate, through the naming and addressing policy, between SIP and H.323 names, then the BES layer may select an appropriate IWF to handle the interworking process and onward routing of the call within the H.323 domain. In this case, the BES layer may be provisioned to know about one or more IWFs. Alternatively each IWF may register its own address with the local SIP registrar indicating its ability to receive calls for each H.323 domain. Generally, allowing the IWF to register with such a call server simplifies the signalling required. This is because the IWF appears as a native device to the call server and does not require specific detailed knowledge of protocol interworking within the BES layer. However, this does mean that specific IWF instances are tied to specific domains.

A disadvantage of this interworking scenario is that calls between end-points on the side of the IWF without the service infrastructure may end up being tromboned through the IWF, consuming resources unnecessarily. The IWF also needs to provide gatekeeper functions in the form of registration and end-point location services. One possible optimisation to assist in this case may be to provide a cache of transport addresses within the PIWF in order to minimise communication with the SIP proxy and speed up call set-up times. Large-scale networks comprised of multiple H.323 domains will also require multiple PIWFs, one for each domain. As a consequence, the scalability of the solution may be impaired. However, managing H.323 end-points as part of a SIP domain may become more likely if SIP becomes more prevalent and SIP-based call servers (proxies and registrars) become more widely deployed.

12.2.1.3 Service Supported on Both Signalling Domains

A variation of the previous scenario is shown in Fig 12.4, where BES layer components and call servers exist on each side of the PIWF. In this arrangement, end-points and end-point registrations are managed natively within their respective signalling domains.

For calls within the same signalling domain, the SIP proxy or H.323 gatekeeper provides translation services by communicating with the location services supplied by their respective BES layer. A call from one signalling domain to another is directed to an appropriate PIWF by the native call server. The PIWF is then able to use native location services in order to obtain the transport address of the terminating end-point before onward routing the call directly to the termination. Alternatively, all calls arriving at the PIWF may be onward routed to the call server in the terminating domain, which is probably the more likely scenario.

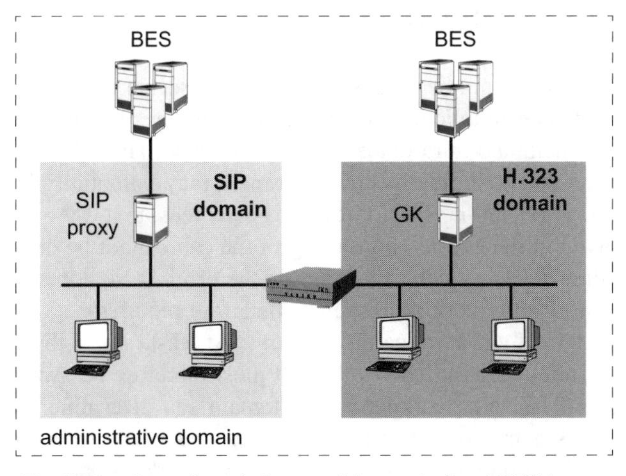

Fig 12.4 Intra-domain interworking — isolated BES layers.

A variation of this scenario is shown in Fig 12.5, which depicts the case where the BES layers have been combined into a single entity. Since the two signalling domains and the PIWF are controlled by the same administration it may be possible to establish a link directly from the PIWF to the BES location service. The benefit of communicating directly with the BES location service is that the need to communicate with a multiplicity of call servers in the terminating domain may be negated. This reduces the number of location service hits that are required. The number of signalling hops are therefore reduced, leading to faster call set-up times and fewer resources being needed to route the call. This does require that the services provided in the BES layer can be made completely protocol agnostic.

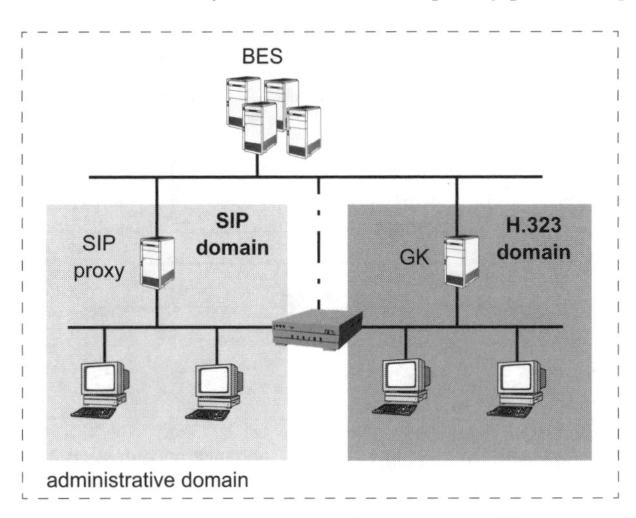

Fig 12.5 Intra-domain interworking — unified BES layers.

12.2.2 Inter-Domain Interworking

In contrast to intra-domain scenarios, inter-domain scenarios are concerned with interworking **between** individual administrative domains. The essential difference between inter-domain and intra-domain interworking scenarios is that inter-domain scenarios assume that each administrative domain may potentially be owned by a different network operator. This virtually guarantees that the arrangement of elements involved in the end-to-end routing of the call cannot be determined at the point that a call arrives at a call server within the originating domain. Call routing and the selection of interworking resources therefore requires negotiation between the originating, transit and terminating domains. At least one of the administrative domains participating in an inter-domain call must therefore be aware of the interworking function. The interworking-aware domain can determine these situations when it originates a call or when a call arrives and needs to be terminated. In situations where both domains have a protocol interworking capability, the design of the interconnection between the networks will have to identify how the selection of an interworking function is made. This will also have to address how a PIWF is used and where it is located. This may be determined via inter-domain negotiation between BES layers or via bilateral agreements between network operators.

The first arrangement, shown in Fig 12.6, has no communication link between BES layers. This implies that for protocol interworking to occur there must be a pre-arranged bilateral agreement between the two network operators. In this case, the operators need to agree on which interworking devices are to be involved in handling traffic between the two domains and how these interworking devices communicate with the associated call servers.

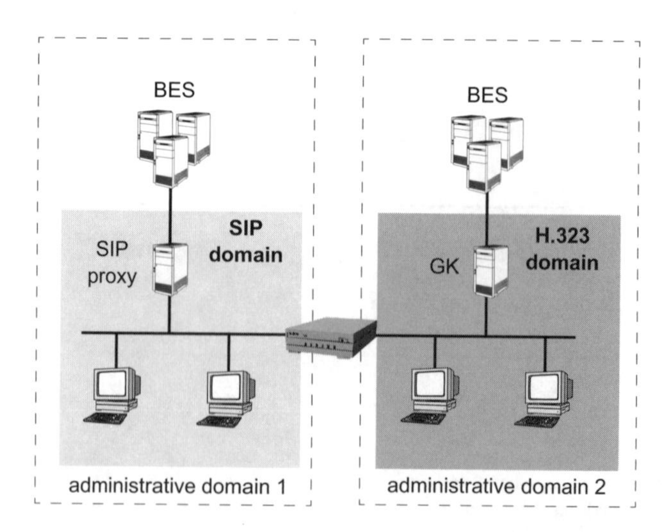

Fig 12.6 Inter-domain interworking — isolated BES layers.

There are a number of possibilities for realising this solution, including the use of native location services to obtain the transport address of the terminating end-point and then directly routing the call.

The second scenario, shown in Fig 12.7, shows a communication link between the two BES layers. This provides a more flexible arrangement than the previous one since it allows resources to be negotiated on a call-by-call basis. As long as there is a trusted third party that can authenticate an originating domain to a terminating domain, then an *ad hoc* arrangement could be set up between two domains that had never previously communicated. This is an example of the clearinghouse model that is described further in Chapter 14.

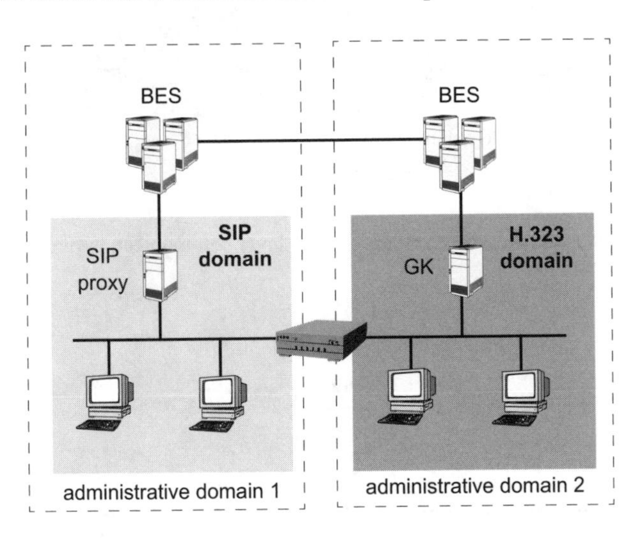

Fig 12.7 Inter-domain interworking — linked BES layers.

Inter-domain interworking may also require the addition of an IWF at the BES layer as shown in Fig 12.8. This allows negotiation between networks that use different protocols at the BES layer, such as H.225.0 Annex G and TRIP, which are discussed further in Chapter 14.

12.2.3 Interworking Using Clearinghouse Services

Fuller details of the clearinghouse model are given in Chapter 14. However, Fig 12.9 shows how that model is applied to the design of inter-domain interworking solutions. In Fig 12.9 a network operator, or service provider, provides a clearinghouse service to handle calls between domains. In this case it may be possible for interworking to be handled transparently as the originating and terminating domains may not be aware that interworking is taking place. However, calls may only be able to proceed with reduced capabilities or features.

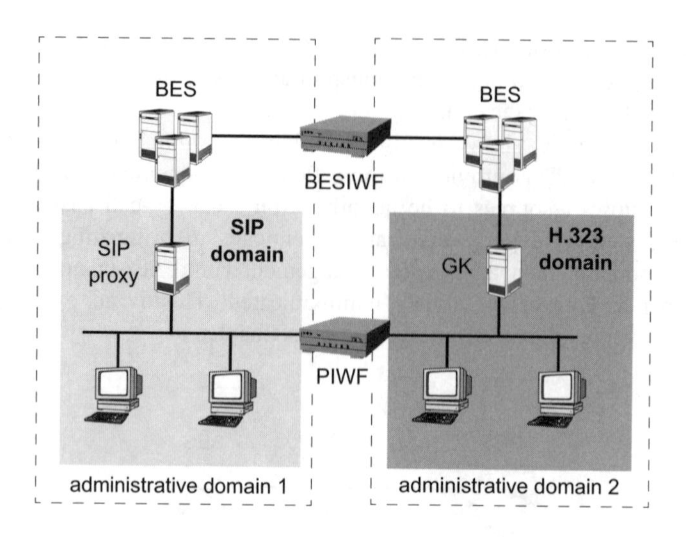

Fig 12.8 Inter-domain interworking — BES layer communication through BESIWF.

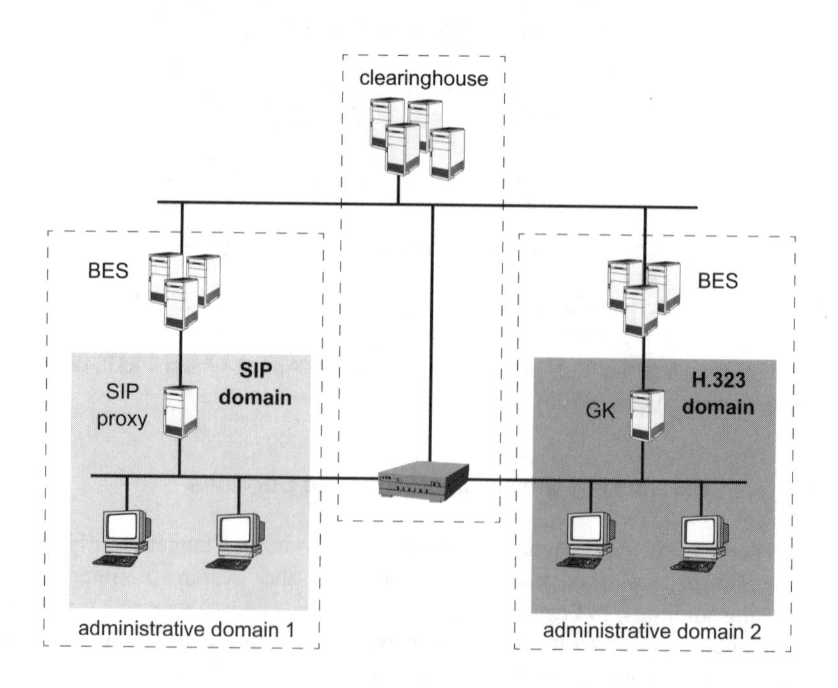

Fig 12.9 Inter-domain interworking — using clearinghouse services.

12.3 Routing Calls via an Interworking Function

Irrespective of the specific scenario concerned, a call must first be routed to an interworking device before it can undergo any form of interworking. The information required for routing inter-domain call segments may be provided by one of the BES layers involved in routing the call. This process may occur in conjunction with a third party, such as a clearinghouse, that provides address-translation and service-negotiation features. It may also provide the interworking devices that make protocol interworking transparent to both the originating and terminating domains.

When an end-point wishes to establish a call, the location of the recipient end-point, whether the call is to proceed within the same administrative domain, between administrative domains, or between signalling domains, must be determined. In the case where a call goes between signalling domains, the location service within the BES layer of the originating domain must locate, or negotiate with the terminating domain to find, an appropriate device (PIWF) for converting between the messages and procedures of each signalling domain. The fundamental requirements for the BES layer, to support inter-signalling domain calls, is therefore to provide the:

- transport address of an appropriate PIWF;
- onward routing of the call after interworking has been performed.

This information may not have to be resolved in a single operation, as the latter can be determined on the egress side of the interworking function. In practice, there are a number of ways this information can be obtained by the PIWF. For example, when a call arrives at the PIWF the transport address may already be available within the SIP INVITE or H.323 SETUP messages. However, this will largely depend on the functionality provided by the SIP proxy or H.323 gatekeeper that originates the call and the ability of these entities to populate the required information in the available structures within the INVITE or SETUP messages.

An alternative is to send the information within an opaque data structure carried within the INVITE or SETUP message. For H.323, this would involve using the non-standard data parameter within H.323 v2 and v3 PDUs, or using H.460.x feature negotiation/transport within H.323 v4. Similarly for SIP, a set of proprietary headers or multipart MIME could be used to convey this information. In either case, using non-standard data would still require modifications to any deployed call servers in order to implement the procedures required to negotiate, access and use the information provided within these parameters.

However, if the transport address of the terminating end-point or next-hop device is not carried in the INVITE or SETUP message sent to the PIWF, the PIWF will need to acquire this information from the terminating domain. This may be solved by pre-configuring any interworking devices to always communicate with a specific call server, whether gatekeeper or SIP proxy, in the terminating domain. In which case the transport address of the terminating end-point or next-hop device can be obtained when the call arrives at this device.

Alternatively, the PIWF may use native location services to obtain the required transport address from the call server before calling the terminating end-point or next hop device directly.

When using native location services in the terminating domain, this assumes that the originating domain is able to determine from the naming policy used that cross-domain signalling is required. This ensures that the call is initially routed on to the PIWF. Unfortunately this requires that the BES layer in the originating domain be aware that interworking is taking place since it will need to handle both SIP and H.323 aliases and act on them appropriately.

It may be possible for the interworking device to communicate directly with the BES layer, either in the originating or terminating domains, depending on which one owns the PIWF, instead of signalling the call server in the terminating domain. This has the advantage of minimising the number of signalling hops, reducing the resources required to route the call, and improving call set-up time performance. This also means that the PIWF would not have to be tied to a given call server, allowing the PIWF to be used for handling multiple calls between multiple domains.

Each call-flow scenario will depend on the direction of the call, for example whether the flow is from H.323 to SIP, or SIP to H.323, and when session information becomes available. Even though H.323 and SIP are both attempting to solve the same problem space, there are subtle differences that make interworking awkward. For example, one of the main issues to contend with is the mismatch that occurs concerning the point in time when information describing the type and structure of the media flows is exchanged by each protocol. For example, in order to construct a session description to send to the SIP UA, the PIWF requires access to information from the H.323 end-point describing media types and modes offered along with their associated RTP/RTCP [3, 4] ports. This information may arrive at different points in the H.323 call-establishment process depending on which H.245 procedural mode of operation is used.

Within SIP, the session description protocol (SDP) [5] describes media and media ports to be used to receive the RTP streams. As there is only a single port described within the media description, this information must describe not only the RTP port, but also the RTCP channel receive port. In SDP the relationship between the media ports (RTP) and media control ports (RTCP) is fixed. An associated RTCP port is chosen as the RTP port number plus one, i.e. RTP/RTCP ports are paired together.

Within H.323 the RTP and RTCP information is conveyed in the OLC/OLC ACK message exchange. If the H.323 receive RTP and RTCP ports do not have the same relationship as defined within SIP, media will not be established successfully. A media switch could be used to provide interworking of these ports by providing an appropriate mapping of RTP/RTCP ports from one protocol to the other.

An example call flow, from an H.323 end-point to a SIP UA using the nominal H.245 procedures, is shown in Fig 12.10a. At the point where the initial SETUP from the H.323 end-point (EP) is received at the PIWF there is not enough information for the PIWF to construct a session description. All the PIWF can do at this stage is send an INVITE to the SIP UA without a session description. After the initial message exchanges the SIP UA sends a 200 OK final response to the PIWF to accept the call and in response the PIWF sends a CONNECT to the H.323 end-point to complete the first phase of the call establishment process. The session description, received in the 200 OK final response, contains the SIP user agent's session description. The PIWF converts the session description to a terminal capability set (TCS) and sends it to the H.323 end-point. When the PIWF receives the H.323 end-point's terminal capability set there is enough information for the

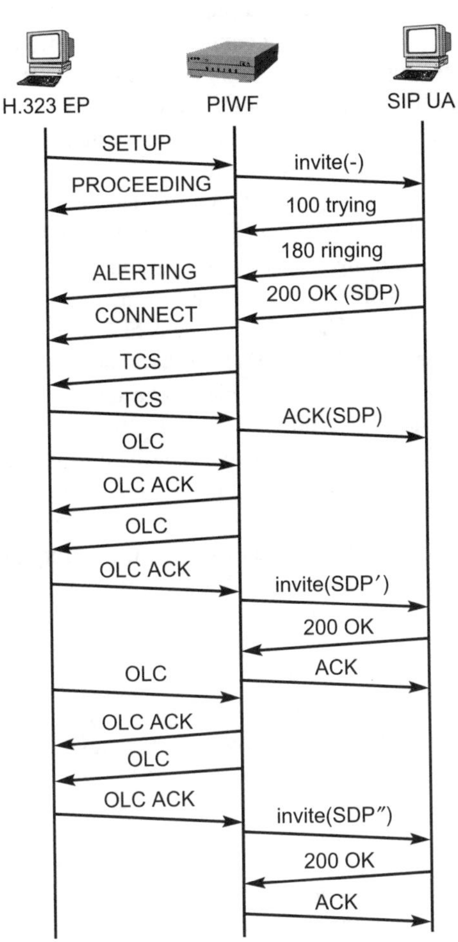

Fig 12.10a Example H.323-to-SIP call flow using nominal H.245 procedure.

PIWF to form a session description based on the H.323 end-point's receive capabilities. However, at this stage the media port information is not known and the session description sent to the SIP UA will need to contain arbitrary port values. As the media is still to be determined on the H.323 side of the PIWF, any media description lines formed in the session description, to be sent to the SIP UA, must be paused. This is carried out by setting each media line connection 'c=' attribute to 0.0.0.0.

As logical channels are opened by the H.323 end-point to the PIWF, a modified session description (SDP′ and SDP″), containing the updated media port and address information, can be sent to the SIP UA. As media channels are not explicitly opened by a SIP UA, this forces the PIWF to make media choices on behalf of the SIP UA. This requires appropriate media channels to be opened from the PIWF towards the H.323 end-point. In the example shown in Fig 12.10a the PIWF uses the principle of symmetry. Any logical channels opened towards the PIWF by the H.323 end-point are also opened from the PIWF back to the H.323 end-point. When there is asymmetry in the H.323 end-point's receive and transmit capabilities, the next most appropriate codec is selected. If a media switch is present, the PIWF may provide media transcoding if this is required.

Figure 12.10b shows the corresponding call flow initiated from a SIP UA to the H.323 end-point through a PIWF using the nominal H.245 procedures. In this example an INVITE, containing a session description, is received from the SIP UA at the PIWF. The PIWF is able to generate a terminal capability set, based on the receive capabilities of the SIP UA, which is sent to the H.323 end-point when the CONNECT message arrives at the PIWF. When the H.323 end-point sends its terminal capability set to the PIWF, the PIWF can complete the first phase of the call establishment process by sending the 200 OK final response. The session description sent in the 200 OK final response is derived from the receive capabilities within the terminal capability set received from the H.323 end-point. Again, at this point the media has not been selected on the H.323 side of the call and the session description sent to the SIP UA will need to contain arbitrary port values and have each media connection address set to 0.0.0.0. Note that at this stage the procedures for handling the establishment of media channels is identical to the case shown in Fig 12.10a.

Also, note, in Figs 12.10a and 12.10b, that the message flows labelled TCS represent the TCS/TCS ACK exchange between the PIWF and the H.323 end-point, and that the MSD/MSD ACK message flows are not shown.

12.3.1 Protocol Tunnelling

A number of techniques, such as the GEF for H.323 described in Chapter 10, have been defined to allow both H.323 and SIP to act as transport mechanisms for other protocols. Generally these have been centred upon scenarios where either a SIP or

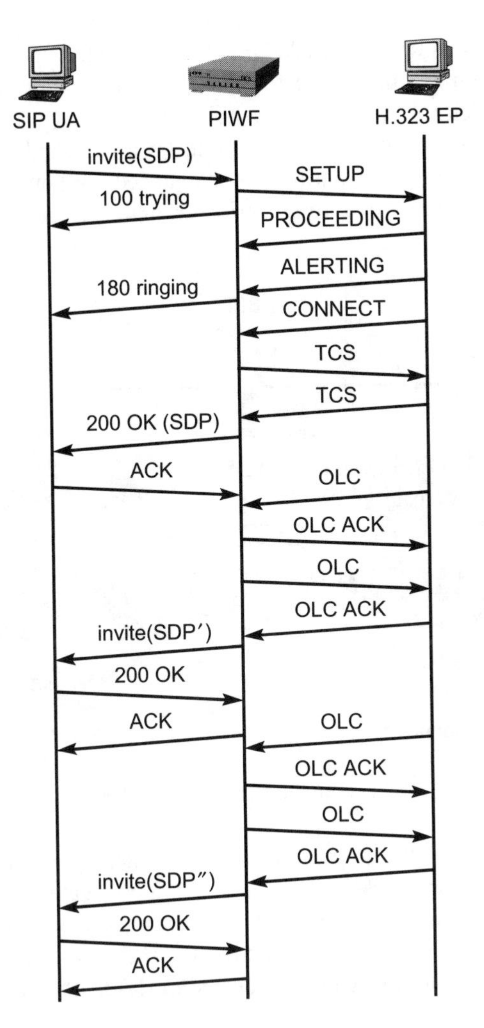

Fig 12.10b Example SIP-to-H.323 call flow using nominal H.245 procedure.

an H.323 network is being used as a VoIP overlay network to the PSTN. These 'tunnelling' mechanisms present a number of problems where VoIP networks need to interconnect since the embedded signalling is not necessarily understood by either the SIP or H.323 network being interconnected. As tunnelling provides a means of conveying information between two domains, via an intermediate transit domain without feature loss, the handling of specific protocol tunnels is therefore of major importance when considering the design of any large-scale VoIP network solution, otherwise complex problems may be created between the originating and terminating entities involved in a call. For example, incorrect C7 signalling information may be transported in the case where the VoIP network forms a trunk level overlay network.

In order to establish a tunnel, a PIWF involved in a call must therefore determine that it is communicating with another PIWF supporting the appropriate capabilities to handle the tunnelled information. To enable this to occur, procedures need to be defined to describe the tunnel and how it is to be established and torn down. As indicated in Fig 12.11, this negotiation may be carried out between the BES layers before the session proceeds or through prior arrangement as part of the network interconnect agreement.

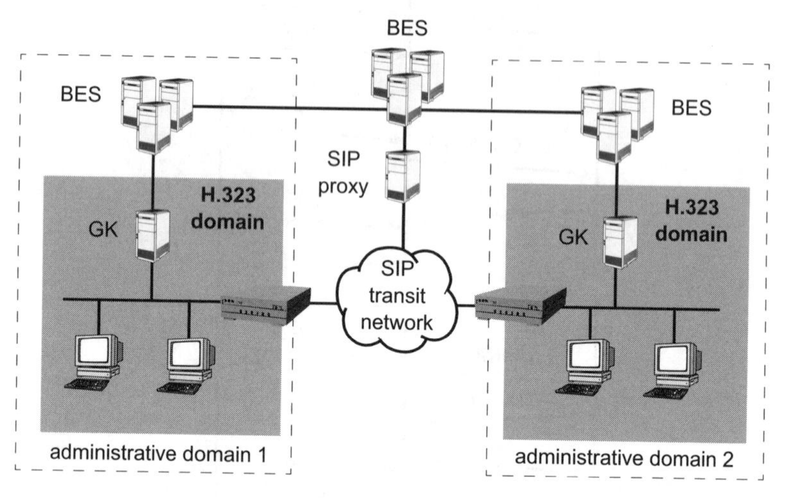

Fig 12.11 H.323 tunnelling through a SIP transit network.

In the example shown, two H.323 domains are interconnected via a SIP-based transit network. Where the H.323 domains also support gateways interconnecting with the PSTN, it is not unreasonable to envisage a scenario where H.323 administrative domain 1 transports embedded C7 messages via the H.323 tunnel over the SIP transit network to a gateway located in H.323 administrative domain 2.

12.3.2 Interaction at the BES Layer

At the point of interconnection between two administrative domains, there are a number of options that may be considered concerning how the interconnection is achieved. Some of these are discussed further in Chapter 14. However, in the context of attempting to interwork SIP and H.323 protocols, however, it is possible to address some of the service interworking issues at the BES layer previously described. Where concrete admission and access controls are required at the interconnection, a logical element known in H.323 terminology as a border element (BE) is used. The border element may be considered to be part of the BES layer and may be collocated with any other entity such as a gateway or gatekeeper.

Through the use of border elements, the BES layer can determine the transport address of the call server or an end-point within the terminating domain. The protocols used for inter-domain communication may also advertise and negotiate protocol interworking as a network service, and allow authentication, billing, and the determination of which physical devices will be involved during a cross-domain call.

12.4 The Protocol Interworking Function

The protocol interworking function itself (Fig 12.12) comprises a signalling engine and an optional media switch. The signalling engine handles the translation of protocol messages received at reference points A and B, and maintains call and transaction states between SIP and H.323 end-points.

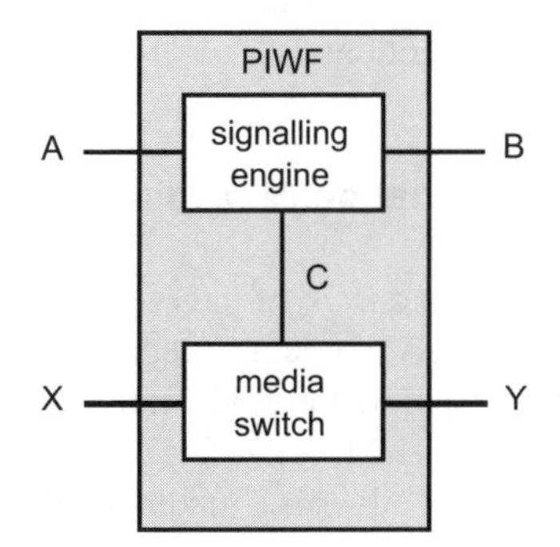

Fig 12.12 PIWF internal architecture.

The media switch provides media event notifications that are communicated to the signalling engine via reference point C. Media arriving at reference points X and Y may be transcoded, if required, by the media switch.

12.4.1 Feature Loss and Parameter Defaulting

As discussed in Chapter 16 (section 16.3.1), when protocol sets are interworking, the information exchanged during call establishment may become lost or degraded by the interworking process leading to a reduced ability to negotiate and establish sessions. For example, some features may have no parallel when mapping from one protocol set to another. In these instances these features are lost, leading to a reduced level of capability during the session. For basic call handling this is not generally an issue but this may be more problematic when implementing more complex services. If extra features are required beyond basic call handling that cannot be interworked, the call will have to be failed or at best dropped back to a basic call.

Some message parameters may not have the granularity required to sufficiently describe mappings from other protocols. This may lead to a reduced ability to negotiate features during the session and have unforeseen side effects related to differences in protocol implementation between vendor products. H.323 is much richer in describing call attributes than SIP. Consequently, in mapping parameters from H.323 to SIP there is a greater opportunity for feature loss. The impact of this feature loss on the call depends on whether the features described are mandatory or optional. Typically it would be expected that many H.225.0 or H.245 parameters would need to be set to default values when mapping from SIP to H.323. Standard profiles may be required to assist in this parameter mapping process. However, there will invariably be occasions where there is no appropriate placeholder for a SIP header within an H.323 message parameter. The interworking function will need to determine if a call may proceed under these circumstances and what corrective action, if any, may be taken. The PIWF will need to determine if an alternate means of conveyance, such as an H.460.x package, can be used, and, if not, whether the call may proceed under these conditions.

12.4.2 End-to-End Negotiation of Media

It is a significant advantage to the design of interworking functions that both H.323 and SIP systems use RTP channels to convey media between end-points. The main challenge is therefore to address how the RTP streams are controlled and how the payload is selected. In the case of H.323, H.245 is used to negotiate and select media modes, while SIP uses SDP to advertise media modes port information. Both protocols therefore convey similar information but there are subtle differences that need to be considered. For example, H.245 provides a more complete bearer control function than is available with SDP and is more explicit in the way media information is conveyed between end-points.

Typically the H.245 procedures comprise request/response exchanges between end-points. These H.245 messages may be handled by the PIWF in a number of ways when interworking to SIP/SDP, including:

- mapping into INVITE, 200 OK, ACK transactions with a modified session description and any relevant SIP headers (e.g. OLC/OLC ACK);
- always rejecting (e.g. mode requests);
- always accepting (e.g. round-trip delay).

Due to the evolution of H.323 aimed at improving call set-up time, there are four modes in which H.245 can be used (these modes are explained further in the ITU Recommendation [6]):

- nominal H.245 (or slow start);
- early H.245;

- FastConnect (or FastStart);
- tunnelling (or encapsulation).

SDP is simpler than H.245 and, in conjunction with SIP, it attempts to convey the same information as H.245 in a single round trip. In a SIP system, all SDP message bodies are conveyed via SIP messages. However, SDP is not as flexible as H.245 in being able to describe sets of codecs or indicating how they are to be used in a session. An SDP media description may contain multiple media modes for a given channel and the SIP UA is free to select any valid combination and later switch between them without explicitly signalling this to the remote UA. This differs from H.245, which explicitly signals which modes are going to be used.

Each time a session description needs to be updated this requires INVITE, 200 OK and ACK transactions to occur. For complex services, this may require a greater number of message exchanges than needed with H.245 to implement the same service.

In the SIP/SDP to H.245 direction, the session description and any related SIP headers must be analysed whenever an INVITE is received which introduces an additional processing overhead. Fortunately the latest features incorporated into SIP use new methods to support this negotiation, instead of further overloading the INVITE method. This will to some extent help to negate this processing overhead.

12.4.3 Mapping Session Descriptions and Terminal Capability Sets

Since H.323 and SIP use different methods for describing the media flows involved in a call, any PIWF must be able to translate between them. In essence this means that the PIWF must be able to map between flows described using SDP session descriptions and H.245 capability set declarations while also providing appropriate control of these flows. While H.245 and SDP address much the same problem of describing a particular media flow, they achieve this goal in subtly different ways. For example, as shown in Fig 12.13, SDP enables a number of port pairs to be specified along with the declaration of the type and format of the media that can be received on them.

```
...
m=audio 9160 RTP/AVP 0 3 15
a=rtpmap:0 PCMU/8000
a=rtpmap:3 GSM/8000
m=video 9170 RTP/AVP 31
a=rtpmap:31 H261/90000
```

Fig 12.13 Media description and attribute lines within a session description.

H.245 adopts a similar approach to SDP by identifying the individual media types and formats that can be supported, as illustrated in the example given in Table 12.1.

Table 12.1 Terminal capability set.

Capability number	Direction	Capability
1	Rx	G.711-ULAW64
2	Rx	GSM
3	Rx	H.261

However, H.245 also determines the combinations of the different types of media, as shown in Table 12.2. These parameters therefore form the basis of a negotiation between the entities involved in a particular session by enabling them to exchange information concerning the various combinations of media they can support concurrently. This is a particularly important feature for open multimedia solutions since the number of media formats is likely to be large while the demands each can place on the underlying terminal can vary greatly. As a consequence a given terminal may be able to support a number of media formats but not all at the same time, as illustrated in Table 12.2.

Table 12.2 Simultaneous capabilities.

Simultaneous capability set	Alternate simultaneous capabilities
1	{1, 3}, {2, 3}

Interworking the media descriptions between SIP and H.323 systems therefore becomes a matter of mapping between the SDP media format declarations and the H.245 capability declarations. However, it is important that the simultaneous capability declarations are also handled correctly — especially if the PIWF handles the media streams at the point of interworking.

12.4.4 Handling FastConnect

For the FastConnect procedures, the offered transmit and receive media modes and associated RTP/RTCP port information are available within the FastConnect element of the SETUP message allowing a session description to be sent in the subsequent INVITE message from the PIWF to the SIP end-point.

One of the main issues with SIP is that it does not have the equivalent logical channel procedures of H.323 and this causes problems with FastConnect as well as the nominal H.245 logical channel procedures. A SIP UA selects media to transmit from the receive capabilities contained within the received session description. Even for FastConnect there is no way for the PIWF to determine which receive options, derived from a session description, should be included within the FastConnect element to be signalled to the H.323 end-point. This would be required to allow the H.323 end-point to enable reception of the SIP UA's transmitted media. It is possible to infer which media a SIP UA will transmit but with varying levels of confidence, depending on how much is known about it. Determining which H.323 end-point receive options to select is an issue irrespective of whether it is a SIP UA or H.323 end-point that is initiating the call.

The PIWF can handle the absence of media description information from the SIP UA in a number of ways:

- if only a single media description is present in the session description, this is converted to the equivalent receive and transmit option OLC structures in the FastConnect element sent to the H.323 end-point;
- if several media descriptions are present within the session description, these can be converted to their equivalent receive and transmit option OLC structures in the FastConnect element sent to the H.323 end-point;
- if a media switch is used, the PIWF waits for the first media mode to be transmitted from the SIP UA before sending a response to the initial SETUP — the response will contain a single receive option corresponding to the media mode transmitted from the SIP UA, while subsequent media established from the SIP UA will be signalled to the H.323 end-point using the nominal H.245 logical channel procedures.

The second of these cases will require the H.323 end-point to expect to receive media on all its receive ports. However, this is not efficient and will waste local resources, but will allow the H.323 end-point to handle SIP UA mode switches as it will be listening on each media port. Unfortunately, how the H.323 end-point decides which media to present if multiple media are present would be implementation dependent. This will complicate multivendor interoperability in this case. Irrespective of the methods described, different H.323 implementations may handle these situations in different ways with the possibility that these methods may not work for all H.323 end-points. If none of the media descriptions can be converted to an equivalent set of receive-and-transmit OLC structures, the call may need to be failed.

12.4.5 Impact of SIP Mode Selection and Switching

There are no explicit media channel procedures provoked by SIP and all media that a SIP UA is prepared to receive are advertised in the session description. Without any constraints, the SIP UA may transmit any media described in the session description. This presents the PIWF with the problem that it will not know when media is being transmitted from the SIP end-point and therefore it will not be possible for the PIWF to carry out the logical channel procedures required on the H.323 side of the call to establish media end to end.

As the SIP UA does not explicitly signal the opening of media then certain assumptions must be made when using the logical channel procedures to allow media to flow from the SIP UA to the H.323 end-point. One way of handling this problem is to allow the H.323 end-point to indirectly constrain the media transmitted from the SIP UA. Initially the media lines within the session description, which are derived from the H.323 end-point's terminal capability set, are temporarily disabled. This stops any media from being transmitted from the SIP UA. This is carried out by setting the connection 'c=' line associated with each media description to 0.0.0.0, as shown in Fig 12.14.

```
...
m=audio <port 1> RTP/AVP 0 15
a=rtpmap:0 PCMU/8000
a=rtpmap:15 G728/8000
c=IN IP4 0.0.0.0
m=video <port 3> RTP/AVP 31
a=rtpmap:31 H261/90000
c=IN IP4 0.0.0.0
```

Fig 12.14 Disabling media ports by setting the connection line to 0.0.0.0.

As the H.323 end-point opens logical channels towards the PIWF, using the OLC/OLC ACK exchange, the PIWF enables the media description lines and updates the port information in the session description corresponding to the opened media on the H.323 side of the call. The modified session description, shown in Fig 12.15, is then sent to the SIP UA.

```
...
m=audio <port 1a> RTP/AVP 0
a=rtpmap:0 PCMU/8000
c=IN IP4 a.b.c.d
m=video <port 3> RTP/AVP 31
a=rtpmap:31 H261/90000
c=IN IP4 0.0.0.0
```

Fig 12.15 Enabling media in the session description.

Notice in Fig 12.14 that, in order to stop media mode switching, other media described within the same media line are disabled, e.g. payload type 15 is removed from the media line. This procedure is carried out for any media the H.323 end-point opens which is supported by the SIP UA.

As the H.323 end-point opens media towards the SIP UA, the PIWF enables reception of the same media mode within the session description, for transmission from the SIP UA to the H.323 end-point. A modified session description is sent to the SIP UA in a re-INVITE. By limiting the receive capabilities described for each channel in the session description sent to the SIP UA, the media modes that may be transmitted towards the H.323 end-point are restricted to those that have been opened by the H.323 end-point.

Despite constraining media in this way, it is still up to the SIP UA to select and transmit media chosen from the media descriptions that are enabled within the session description. How a SIP UA will behave, in choosing enabled media to transmit, will be implementation dependent. This is likely to create problems for multi-vendor interoperability.

The above describes a media selection mechanism based on symmetry. However, other schemes may be used. The decision as to which media modes should be enabled by the PIWF may be based on a number of criteria, including:

- network policy (e.g. mix of call types within a network);
- bandwidth available for the call;
- required media mix;
- available media mix;
- prioritisation for each media component.

The PIWF may also communicate directly with the BES layer to help in the determination of media types to be used during a call.

12.4.6 Using a Media Switch

It will be desirable to allow the SIP UA to have more influence in the decision of which media type to use during a session. In order to allow this, a media switch capable of signalling media events to the control layer of the PIWF is required. Upon detection of transmitted media, the switch can notify the control layer of the PIWF, which in turn can carry out the logical channel procedures to open the media channel from the SIP UA to the H.323 end-point, as shown in Fig 12.16.

SIP media mode switching can also be easily handled using this mechanism (see Fig 12.17). When mode switching occurs this can be detected by the media switch and signalled to the control functions of the PIWF. This notification can be carried out using a media gateway control protocol or using a proprietary interface. Once these media events are signalled to the control function of the PIWF, they can be mapped into the appropriate H.245 logical channel procedures, opening and closing media channels on the H.323 side of the session as appropriate.

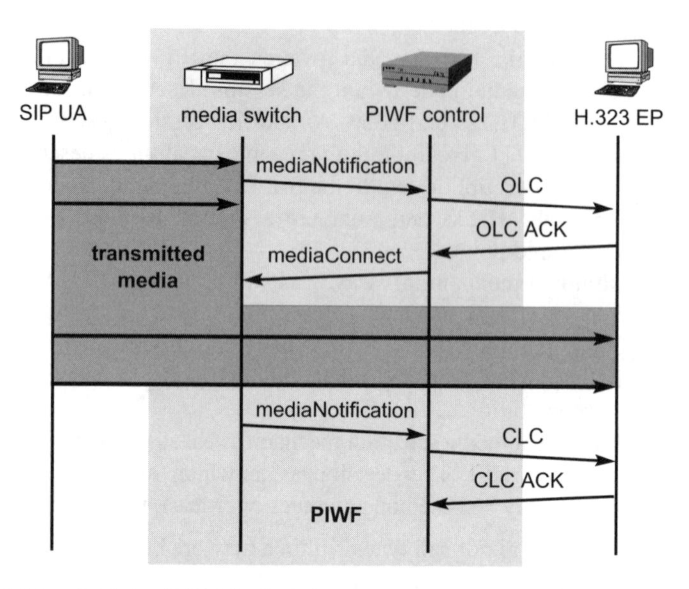

Fig 12.16 Initiating H.323 logical channel procedures from RTP media events.

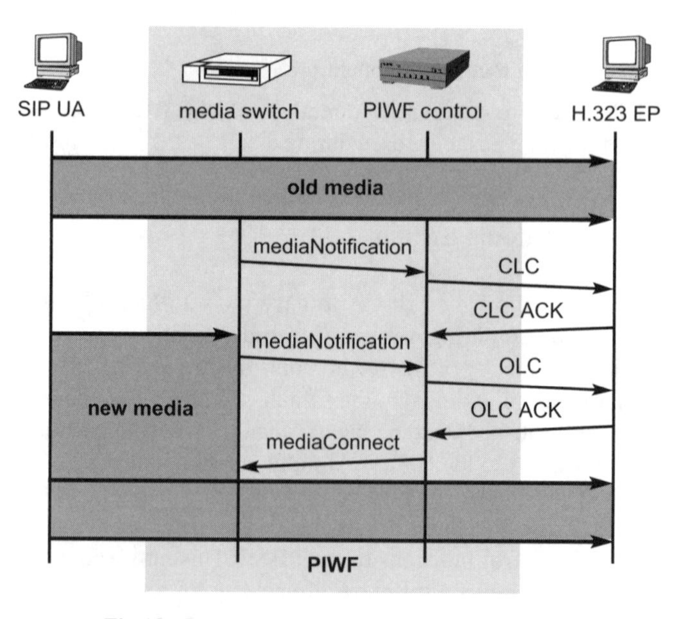

Fig 12.17 Handling SIP UA mode switching.

The media switch may also be required when interworking RTP/RTCP port parameters as H.245 allows greater flexibility over the actual port values that are chosen compared to SDP.

Ideally, media switches should be avoided as their use in an interworking solution will increase the processing requirements and hence cost of interworking devices. A media switch will also typically result in sub-optimal routing and a loss of voice quality as the media packets have to go first to the interworking function and then on to their destination. However, the practicalities of ensuring network integrity, for large-scale networks offering sustainable QoS, point to the use of such approaches. In general though, where media switches are required, transcoding should be avoided as performing codec operations adds further delay and may introduce significant extra distortion.

However, if such an arrangement is required, it can be made to scale by associating multiple media switch blocks with each signalling engine (Fig 12.18). This architecture is similar to that defined for the decomposed GW architecture where a media gateway controller (MGC) communicates with a number of media gateways (MGs), as discussed in Chapter 7.

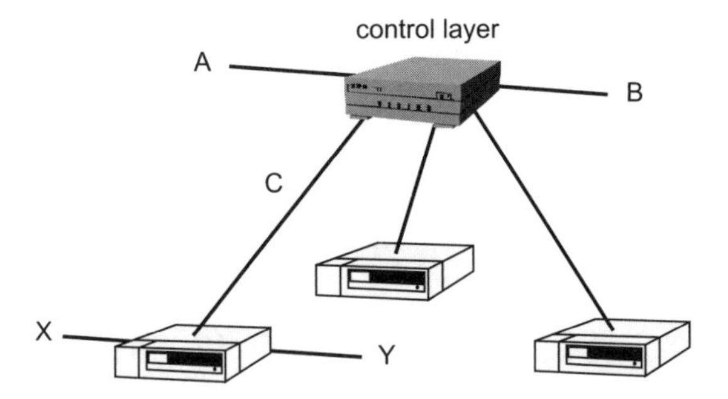

Fig 12.18 Scaling the PIWF by using multiple media switch blocks.

12.5 Summary

The existence of multiple equipment manufacturers, service providers and standards-forming groups fundamentally ensures that the early vision of a single unifying global communications protocol for the VoIP industry remains an elusive goal. The requirement to be able to exchange traffic between networks, based upon different technologies, is therefore largely a necessity. However, as has been shown, interworking is not a simple problem and there are a large number of complex decisions that need to be taken when designing any solution which requires it. In particular, the network architecture into which an interworking solution is to be deployed needs careful consideration. As a consequence, interworking solutions are not easily engineered and deployed.

References

1 ITU-T Recommendation H.323, v4: '*Packet based multimedia communication systems*', (November 2000).

2 Handley, M. et al: '*SIP: session initiation protocol*', IETF RFC 2543 (March 1999).

3 Schulzrinne, H., Casner, S., Frederick, R. and Jacobson, V.: '*RTP: a transport protocol for real-time applications*', IETF RFC 1889 (January 1996).

4 Schulzrinne, H.: '*RTP profile for audio and video conferences with minimal control*', IETF RFC 1890 (January 1996).

5 Handley, M. and Jacobson, V.: '*SDP: session description protocol*', IETF RFC 2327 (April 1998).

6 ITU-T Recommendation H.245, v5: '*Control protocol for multimedia communication*', (June 1999).

13

BUILDING AND LAUNCHING VoIP APPLICATIONS

S J T Condie, G Hare, T Baden and D R Wisely

13.1 Introduction

Voice is an integral element of many of the emerging breed of conversational IP applications found on the Internet. This chapter examines the opportunities to provide these types of application at scale, and considers some of the difficulties in translating what may be interesting toys on the lab bench into genuinely useful public-scale communication services. It characterises early VoIP network service solutions, charting their evolution towards contemporary services and considers some of the types of application that may be possible in the future.

The long-term promise of IP is to provide a coherent infrastructure on which to offer a plethora of multimedia services — including voice. Voice over IP, and more specifically voice over the Internet, as described in Chapter 1, has been seen as a major threat to the revenues of existing voice service providers by enabling business opportunities for new entrants to the voice transport market by exploiting regulatory arbitrage between voice and data services. This has been coupled with the perception of many end users that telephone calls will be 'free' over the Internet. However, the definition of 'free' needs to be carefully examined and has largely been exposed as a fallacy in the market. In the UK, Internet service provider (ISP) fees are usually either absorbed within communication service charges and so appear to the end user as metered phone calls, or provided as a flat rate service. Conversely in the USA, the situation is largely reversed with flat rate phone call charges for access to local Internet service providers appearing as unmetered access with a separate and comparatively large additional ISP facility fee on top. With the reality that if customers pay nothing then they can only reasonably expect nothing in the end, VoIP continues to evolve into much more than just another way of making telephone calls. In particular, VoIP can be viewed as the enabling technology behind many communications-rich applications and the first of many potential network applications that will emerge as the computing and communications industries finally converge. In this context, as shown in Fig 13.1, it is likely to be followed by a raft of integrated applications that will deliver new and exciting services based around IP network infrastructures. These applications are likely to offer rich multimedia conferencing, messaging and customer relationship management features. As considered in Chapter 1, a number of drivers associated with the use of IP make it increasingly attractive to develop and deploy such applications.

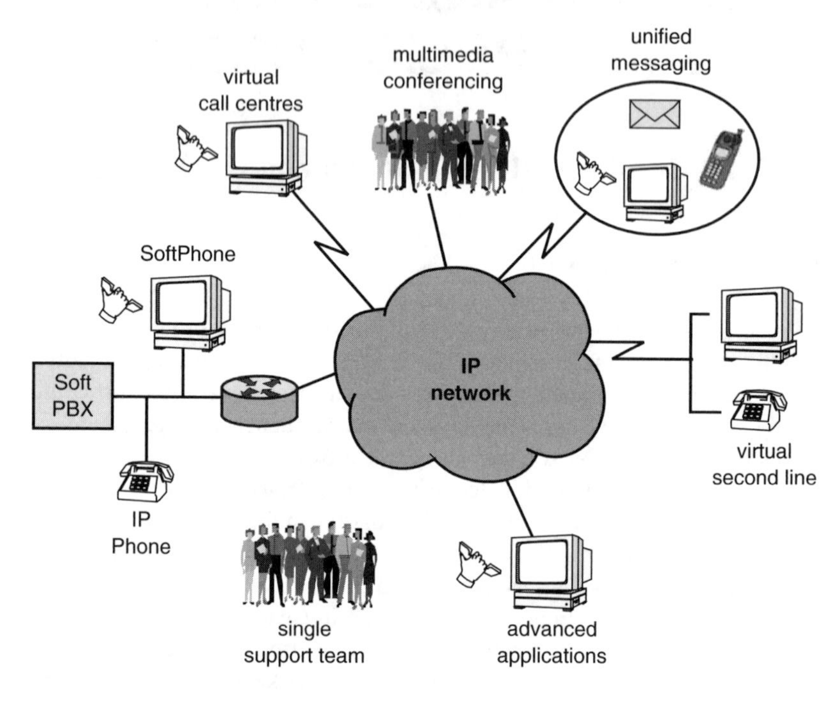

Fig 13.1 IP applications environment.

While the capabilities that may be afforded by VoIP in the future may seem attractive from a number of perspectives, there remains a certain amount of apparently mundane network and service development that needs to be addressed in order to translate such aspirations into reality. It is often these aspects that are either overlooked or their complexity not fully appreciated. This chapter explores some of these issues by considering some early VoIP applications that have been deployed over the years, describes some that have been recently launched, and subsequently considers the potential inherent in IP to produce a degree of seamless service integration not easily achievable in the existing PSTN. It concludes with a consideration of some of the important challenges still to be addressed if a feature-rich future is to be fully realised.

13.2 Basic Application Scenarios

There are a number of simple call usage scenarios that typify many VoIP applications. These scenarios complement or in certain cases supplant PSTN-based services. They are commonly classified according to the type of devices originating and terminating a call. As there may be either a telephone or an IP terminal on each end of a call, this gives four generic cases.

- PC — PC

  This involves the simple connection of two or more users communicating directly via IP clients as shown in Fig 13.2. The PC-to-PC scenario was probably the earliest area to be exploited as a VoIP application. Unfortunately early VoIP implementations relied upon proprietary technologies that did not readily inter-work with other solutions in the market and in doing so constrained their further deployment. However, the availability of Microsoft's NetMeeting application, which is an H.323-based VoIP Soft-Client distributed with the Windows operating system, has made this a popular and easy scenario with which people can experiment. Indeed NetMeeting forms the base functional component of many VoIP applications that have been deployed — particularly some of the early ones described in this chapter.

Fig 13.2 PC — PC calling scenario.

- PC — telephone

  In the simplest case depicted in Fig 13.3, the PC-to-telephone scenario involves a VoIP client communicating with an existing PSTN terminal via an intermediate PSTN-to-IP gateway. Further aspects of gateways are discussed in Chapters 1 and 7. However, an essential element of this scenario is the identification by the VoIP client of the egress gateway and the ability to be able to set up the IP-to-PSTN call. Once communication with the gateway has been established, the telephone number of the destination telephone must be transferred to the gateway to enable it to establish the PSTN segment of the call. This may be local, national or international depending on the specific circumstances. This degree of complexity, involving discrete call segments in both IP and PSTN domains, highlights the fact that a simple terminal centric analysis of application scenarios is incomplete for considering the design of large-scale solutions.

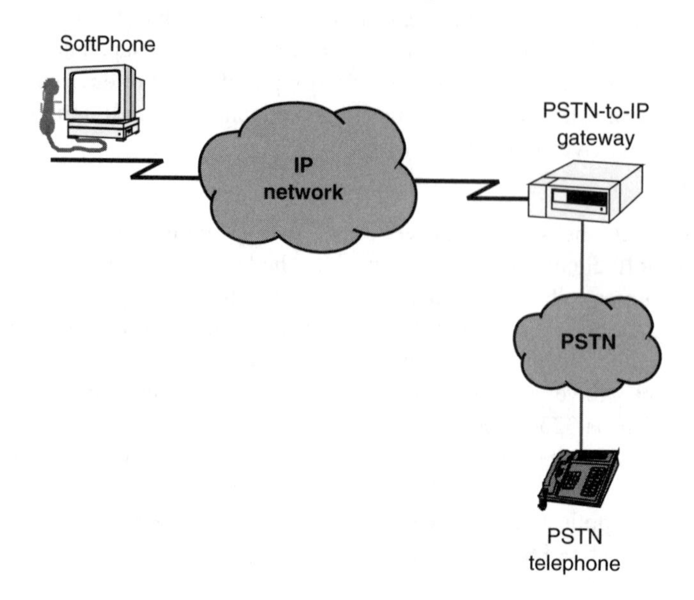

Fig 13.3 PC — telephone calling scenario.

- Telephone — PC

  This scenario utilises the same elements shown in Fig 13.3, with the difference being that the call is initiated from the PSTN side and terminated as VoIP on the IP side. The main issues in this case are the location of the ingress gateway to the IP network and the handling of VoIP call routing in the IP segment. This is somewhat constrained in that the naming and addressing scheme will be largely confined to the numerical system used in existing telephone networks, as discussed in Chapter 9.

- Telephone — telephone

  This scenario is the familiar case found in conventional telephone networks; however, it may involve VoIP if that is used as a network transport technology.

These four basic scenarios provide a useful starting point to view VoIP applications. However, they do not reflect the nature of the network infrastructure between the terminals — this was seen as a serious flaw in the early work on interconnecting VoIP networks undertaken by ETSI's TIPHON project (see Chapters 5 and 16). By expanding a view of the network infrastructure to embrace both IP and switched-circuit networks, the interworking points become exposed. This aids the understanding of the true end-to-end picture and exposes a more representative set of design issues. Unfortunately this also expands the core set of generic call scenarios that need to be considered to eight as shown in Fig 13.4.

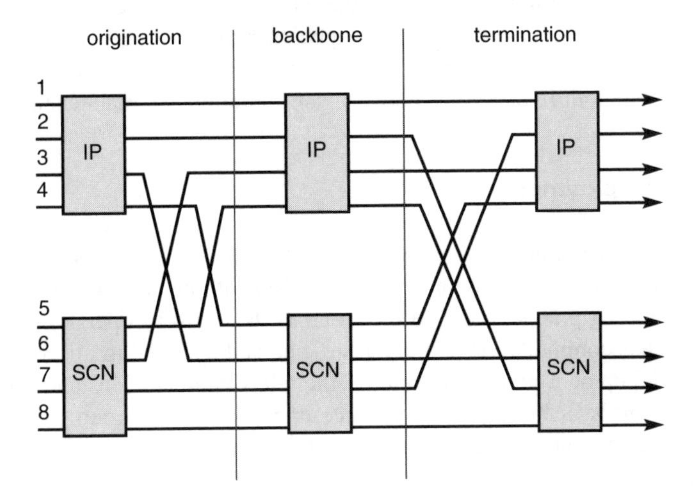

Fig 13.4 TIPHON scenarios.

13.3 VoIP Service Design Considerations

Although the technology is available to build VoIP applications, some of the restrictions in being able to launch an actual service exist in the service surround. In fact, the end-customer experience can be greatly impaired without some of the fundamental service surround being put in place. This in turn has a knock-on effect on the take-up of services and so the profitability and viability of a VoIP solution in general. To prevent this adverse perception of VoIP, many services that could have been launched much earlier in terms of the technology have been slow to market because of the time required to get cost-effective service surround solutions into place.

13.3.1 Maintenance

Historically there has been a fairly distinct line drawn between where the roles of the telecommunications technician and IT technician start and finish. For example SCN switches and voice networking equipment have typically been exclusively maintained by telecommunications technicians, while IP routers, servers and data communications equipment have been maintained by an IT technician. However, VoIP cuts across these historical boundaries. In particular, with the advent of devices such as PSTN-to-IP media gateways and signalling gateways, as discussed in Chapters 6 and 7, the line where the PSTN finishes and the IP network begins is somewhat blurred. From the telecommunications technician's point of view, the configuration and maintenance of IP network interfaces presents a new challenge. Similarly from the point of view of the IT technician, the configuration and maintenance of voice networks have their peculiarities needing to be mastered. For these reasons, the need has emerged for a whole new breed of support technicians that know about both IP and voice network configuration and maintenance.

13.3.2 Deployment

Traditionally PSTN equipment is housed in physical surroundings commensurate with very high reliability operation — in terms of power supplies, equipment practice, operating practices and so on. In contrast, many IT operations are run in general office accommodation that is constructed and operated to different policies. Since VoIP equipment potentially bridges the telecommunications and information technology worlds, it brings a convergence of these two environments. Problems may therefore arise concerning the choice of accommodation used to house VoIP equipment. In particular, this raises a number of questions that influence the deployment of operational equipment. For example, if the equipment is essentially PSTN based, such as a large-scale media gateway, should it be housed in an exchange type building or a data centre type building? Similarly if the equipment is essentially an IP data network element, such as a call server, should it be housed in a data centre type building? In either of these cases the reliability concerns for an end service are generally driven by the expectations of the customers using the applications provided and the commercial model involved. All of these factors must therefore be taken into account when designing the deployment of a new service.

13.3.3 Customer Service

Customer service for VoIP is a whole new area. In the past, customer service desks have been there to support service provision, cessation, and fault reporting, with very little technical knowledge of the products themselves. With new and innovative services being launched, the customer service desks are increasingly becoming a technical 'helpdesk' too.

The difficulty in doing this is that the 'helpdesk' now has to have a much greater knowledge of how the product works to identify whether a customer fault lies in the PSTN, the IP network, the VoIP application or the customer's terminal equipment. For applications running as software packages on general-purpose personal computers, there are also a near-infinite variety of computer manufacturers and suppliers in the market-place to contend with, in addition to machines that are custom built by their owners.

The 'helpdesk' customer service agents therefore have to be able to diagnose the problems a customer is having and be able to talk them through any modifications that are required to the customer's hardware or software to get the application working successfully.

13.3.4 Billing

There are many different challenges associated with billing customers for services. These range from identifying which customer to bill, through to what type of billing strategy to use. Also, whenever a financial transaction takes place, there will always be people who try to exploit it for fraudulent purposes. The security and integrity of charging and billing information are therefore essential service design considerations — all of which are difficult to achieve in open IP networks.

13.3.4.1 Customer Identification

One of the difficulties in billing customers in a VoIP environment is identifying exactly which customer is making a call. When a customer dials up their ISP they are usually allocated an IP address from a pool of IP addresses available at the time using DHCP. This means that they do not get allocated the same address every time they dial up. This makes billing against IP addresses more difficult and drives the need to be able to associate a name rather than necessarily an address for billing purposes. Further details concerning names and addresses can be found in Chapter 9.

13.3.4.2 Revenue Generation

Even once the problem of identifying the customer has been resolved, the decision has to be made concerning the payment strategy that is to be adopted. This essentially breaks down into the charging and tariff strategy and the billing strategy. There are a veritable array of different charging and tariff strategies which could be adopted, each with its own merits, but none that should be discounted out of hand when designing a new service. These include, for example:

- flat rate subscription with inclusive usage charges;
- flat rate subscription with usage charges;
- usage-based time and distance tariffs, for example different rates for local, long-distance and international calls;
- usage-based with differential rates depending upon media type, such as different rates for voice or multimedia calls.

In contrast, the different billing strategies provide the basis for the billing method and revenue collection processes. The most common billing strategies are in general:

- prepayment, which is very popular in the mobile phone and VoIP toll-by pass sectors,
- post-payment, as a telephone bill, such as the BT Blue Bill, for example, that is traditionally used for PSTN telephone service.

In either case there are potentially varying operational costs depending upon the specific implementation of the approach. These need to be carefully balanced against the risk of bad debt and fraud. In particular, it is important to appreciate the infrastructure requirements needed to support a specific revenue strategy since some solutions scale with demand much better than others.

These issues are considered further in section 13.5.5.

13.4 Early Services

13.4.1 Web-Based Services

13.4.1.1 Call Centres

Through WWW technology, the Internet offers opportunities for businesses to provide a more attractive and useful interface to customers for applications such as electronic commerce in the form of Web-based shopping. An obvious example would be a customer browsing a company's Web site who wishes to speak with a sales agent to clarify some details. Merely by clicking a button on the Web page, the customer could be connected via their PC to the company's call centre — along with any data the customer had entered on the Web page. To gain experience with this type of application, BT deployed a VoIP call centre application in the form of a SoftACD (automated call distribution) solution.

As shown in Fig 13.5, this comprised a special VoIP call-handling server that was deployed in conjunction with BT's ClickFree mass-market dial-IP service. By including a hot link on the support page, callers who had NetMeeting clients running could call the helpdesk using VoIP.

The incoming call from the customer was routed on to the SoftACD server for allocation to the next available agent. This demonstrated a number of practical issues with using this type of technology — the main one being that at the time, while the concept was very attractive, the QoS issues encountered with running over the public Internet meant it was difficult to consistently meet customer expectations of voice quality (see Chapters 2 and 3).

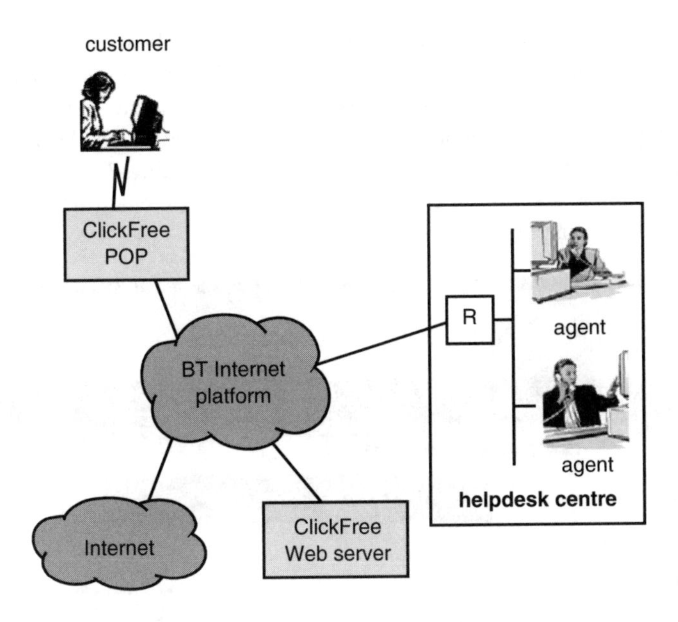

Fig 13.5 Web-based call centre application.

13.4.1.2 Communications Portals

Extending the concepts of the Web front end, BT conducted a trial to evaluate a communications portal concept comprising technology providing unified messaging, calendar, personal directories, and other applications. BTexact Technologies integrated these capabilities in collaboration with various partners to provide a focal point for access to communications services over the Internet. The communications portal is included here as it gives a good insight into the degree of integration of services and applications that can be achieved and went a long way toward trying to achieve the degrees of integration discussed above.

In this example, the communications applications provided included unified messaging, Internet voice, Internet fax, SMS messaging to mobile telephones, e-mail, business and residential directories, a personal calendar and an address book. The services were integrated under a BT Internet-style toolbar as shown in Fig 13.6. The desired services could be invoked by 'clicking' on the toolbar. Of particular interest here is the enhancement of the communications portal to support VoIP breakout to selected PSTN numbers via a PSTN-to-IP gateway. A development of BTexact's CTI-based 'Click-Dial' application was used to achieve the required functionality. To simplify the service surround, the deployment was restricted to calls to 0800 numbers and use of BT's Chargecard to avoid any potential usage-based billing issues that might arise as a consequence of making calls via the VoIP gateway. Figure 13.6 illustrates a business directory application being used, with 0800 numbers enabled with a 'Call...' button. To use the call button feature in the communications portal, the user must first download a helper application (developed by BTexact Technologies) from the BT Comms Portal Web server. This helper application is used to request Microsoft NetMeeting 3.01 to make a call, using the parameters set in a configuration file that is executed from the CPClick server on the Comms Portal platform.

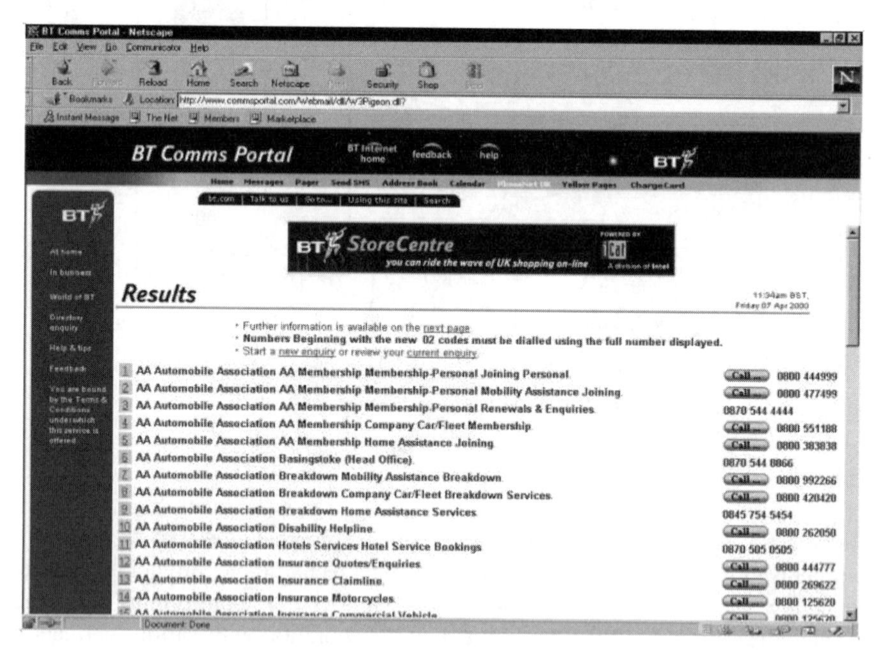

Fig 13.6 BT Comms Portal user view.

Figure 13.7 shows a call from the NetMeeting client being routed by the Comms Portal platform and a PSTN-to-IP gateway to the desired 0800 number, which in this context was a call centre agent. When a user clicks on the residential directory link on the toolbar, a message is passed to the CPClick server rather than the underlying residential directory Web server. A process on the CPClick server then sends out the original request to the actual Web server providing the directory service. When the resulting HTML page is passed back to the CPClick server, the server does a 'screen-scrape', placing a 'Call...' button next to any 0800 numbers, and returns this modified page to the user originating the request. When the user presses a 'Call' button, the associated configuration file on the CPClick server is executed by the user's PC using the helper application previously downloaded. The file contains the required 0800 number and the IP address of an appropriate PSTN-to-IP gateway. NetMeeting is initialised and sets up the call with the segment up to the PSTN-to-IP gateway being carried over IP, and the other segment over the PSTN. The service works similarly for a business directory scenario.

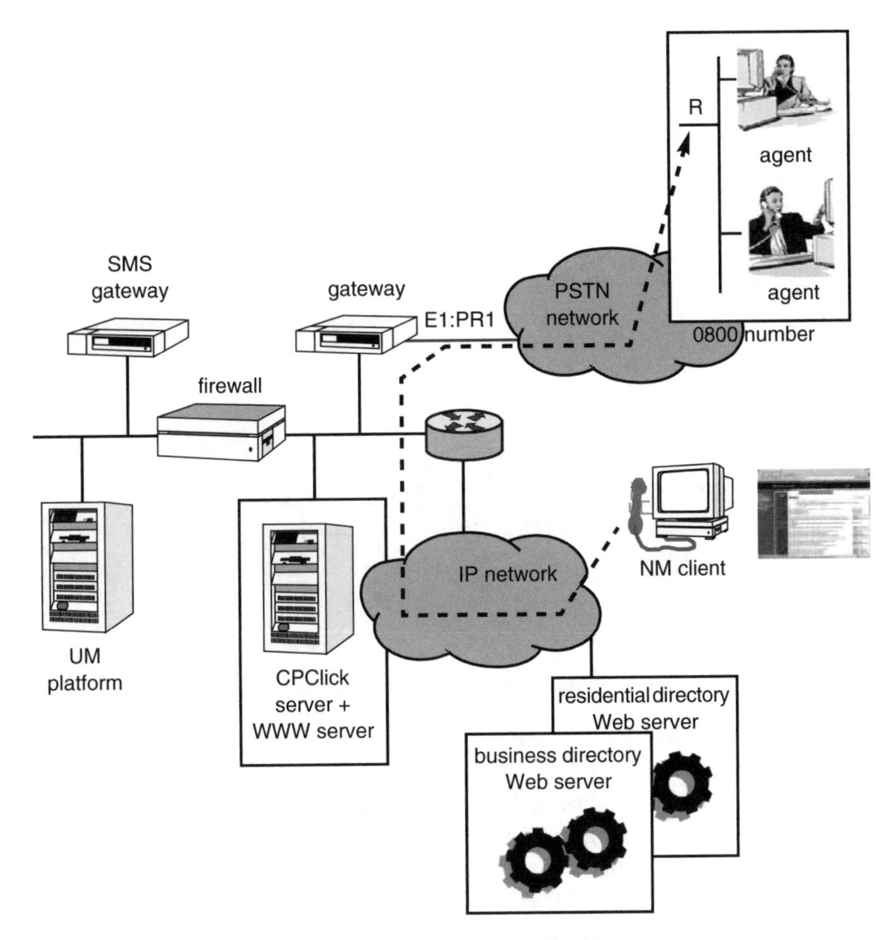

Fig 13.7 BT Comms Portal trial platform.

13.4.2 Consumer Services

13.4.2.1 Internet Locator Service

The BT Internet 'voice over the Internet' feature was first made available to the general public in the spring of 1998. BT had been evaluating and using voice over IP equipment in trials for a considerable period of time, however some of these early solutions were found to have quite detrimental effects on end-users' PCs! With the version 2 build of NetMeeting, it was felt that the NetMeeting/ILS solution, shown in Fig 13.8, was sufficiently stable to test market acceptance and explore end-user issues in a reliable and manageable fashion. To stimulate take up of the application, headsets were given away free to the first 2000 people to register. A helpdesk was equipped to provide users with advice on how to configure NetMeeting and resolve problems associated with sound cards. To provide further support, the development team were encouraged to use the application as an aid to both helpdesk operators and users by providing a known point for test calls.

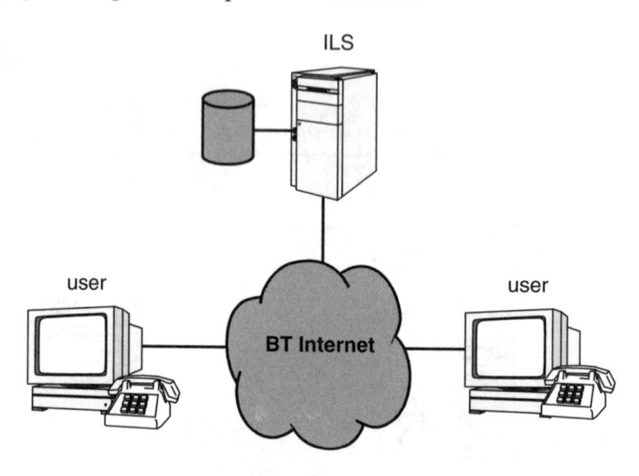

Fig 13.8 Simple user directory to assist VoIP calling.

Early versions of Microsoft NetMeeting enabled users to dynamically store their details in a central 'rendezvous' database known as an Internet locator service (ILS), shown in Fig 13.8. This is accessible through an LDAP mechanism included with NetMeeting and allows users to find and establish calls between each other by allowing addressing information to be made available, as shown in Fig 13.9.

An example of the look and feel of the ILS directory, as seen from NetMeeting, is shown in Fig 13.10. It indicates user-provided information such as e-mail address, first name, last name, location and comments (not shown). It also indicates video capability and whether they are currently engaged in a call.

BT Internet customers who have used the feature have been very satisfied with the experience — especially, or because, it was given as additional functionality provided free of charge. The deployment of this capability has been useful from the point of view of encouraging experimentation within the user community and has confirmed a number of operational concerns with such capabilities. It is still in use at the time of writing, consistently attracts interest from a large number of users, and regularly has a significant number of active users logged on. Although it has been supplanted by other VoIP applications it represents one of the first 'real' public voice over the Internet facilities in the UK.

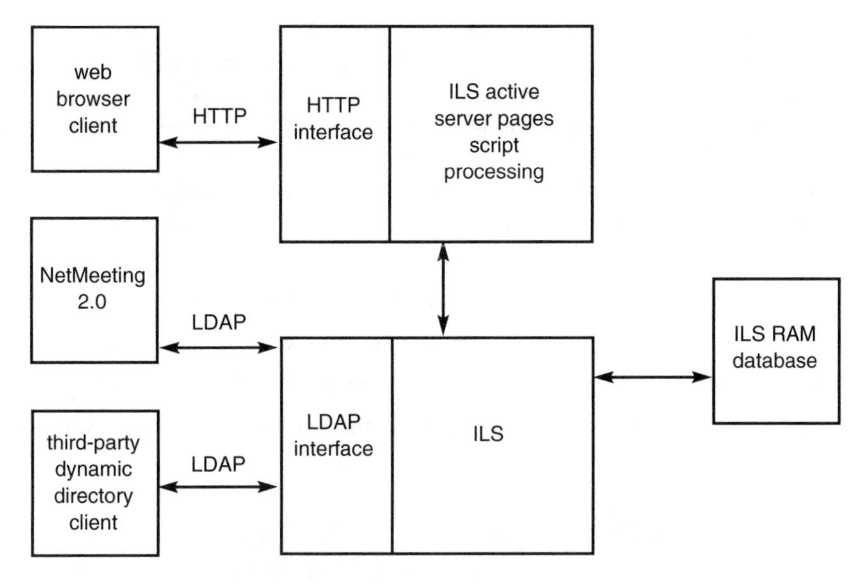

Fig 13.9 NetMeeting ILS server network architecture.

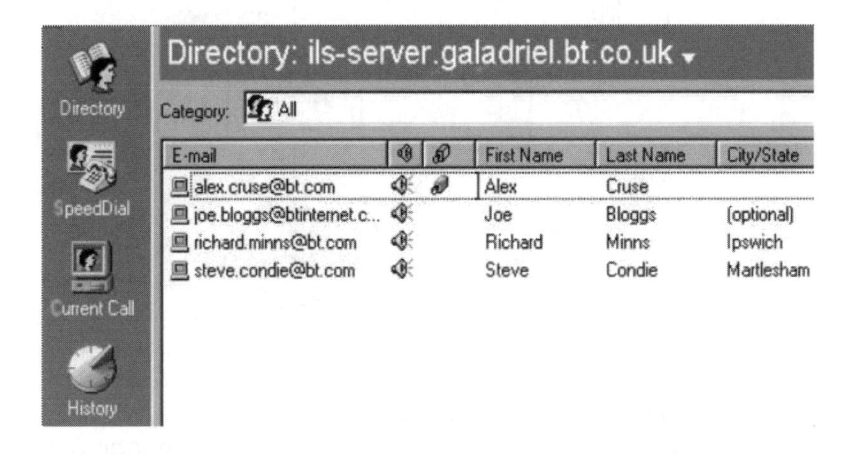

Fig 13.10 NetMeeting client view of an ILS directory.

13.4.2.2 Internet Call Waiting

Internet call waiting (ICW) is an application designed to provide a 'virtual second line'. It addresses the needs of customers with a single telephone line who still wish to receive incoming calls from the PSTN while on a dial-up link to the Internet, as shown in Fig 13.11. For the service to work, the customer or ICW service provider needs to arrange for the 'call divert on busy' network call feature to be enabled to divert incoming calls to the ICW gateway. The costs associated with this diverted PSTN leg can in some circumstances invalidate the business model for Internet call waiting; this is not surprising as intuitively there are effectively two telephone calls — one on the PSTN and one on the IP network — being priced as one. When a customer subscribes to ICW, they download a client, which replaces the standard Microsoft Windows dial-up connection. When used to initiate access to the Internet, the client:

- sets 'call divert on busy', redirecting subsequent calls to a gateway;
- dials the normal Internet service provider access code;
- registers with the ICW gateway via the IP connection established;
- deregisters from the gateway and removes the 'call divert on busy' when the session is ended.

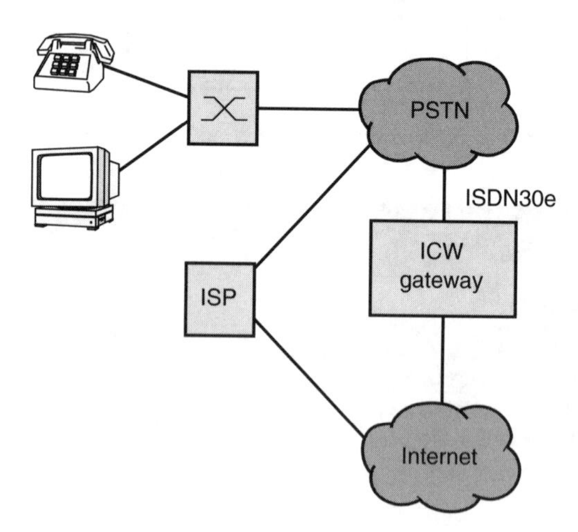

Fig 13.11 Network architecture for Internet call waiting.

When an incoming call is made to the ICW customer's line, the call is diverted to the gateway, which searches its list of current registrations to identify the IP address of the client. If it matches a registration to the called number, it establishes a call to the client using the IP address held in the registration. The customer is then presented with a pop-up window allowing them to:

- accept the call over the Internet;
- hang up the Internet session and take the call over the PSTN;
- reject the call.

If the call is accepted over the Internet, the originating call and the call made from the gateway to the client are bridged by the gateway. While Internet call waiting can work very effectively, there are QoS issues with running voice over such bandwidth-constrained links as dial-up Internet connections, some of which are discussed in Chapter 3. There are also a number of operational and service surround issues involved with deploying Internet call waiting services. These include the need to support customer-deployed software in the potentially hostile PC environment with the resulting, potentially complex, failure scenarios that this involves. In addition, the PSTN diverted leg presents both a commercial and an operational challenge since the cost and establishment of the PSTN call segment to the ICW VoIP gateway needs to be reliably and consistently controlled and paid for.

13.4.2.3 Internet Calling Services

BT SurfTalk, shown in Fig 13.12, is an example of a public Internet telephony application that provides Internet users with the ability to make voice calls over the Internet to other BT SurfTalk customers who are 'online', addressed by a telephone number, as discussed in Chapter 9. Since it is provided over the public Internet, there is no quality of service (QoS) guarantee associated with BT SurfTalk, and it is only possible to make or receive a single call at any one time.

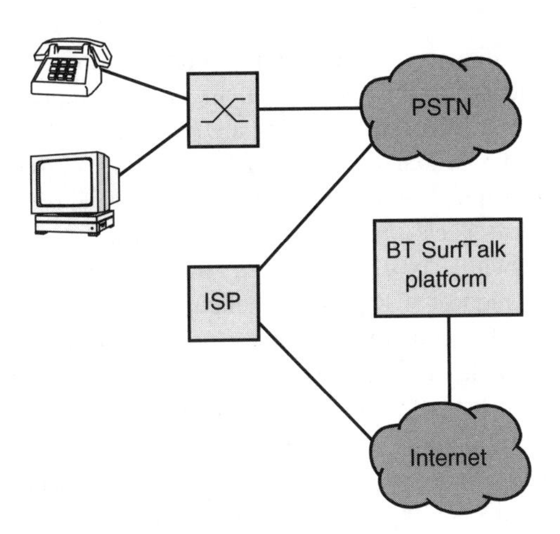

Fig 13.12 BT SurfTalk network architecture.

The core of the platform is a 'directory' gatekeeper and domain name server (DNS), as shown in Fig 13.13. These allow the distribution of incoming connections to a number of 'local area gatekeepers'.

This view of the design is very similar to that of the PSTN in that there are local area gatekeepers to cover customer connections in a specific region in the same way as a local (class 5) exchange would. Similarly the directory gatekeeper handles the signalling between local area gatekeepers in the same way that a trunk level exchange interconnects a number of local exchanges.

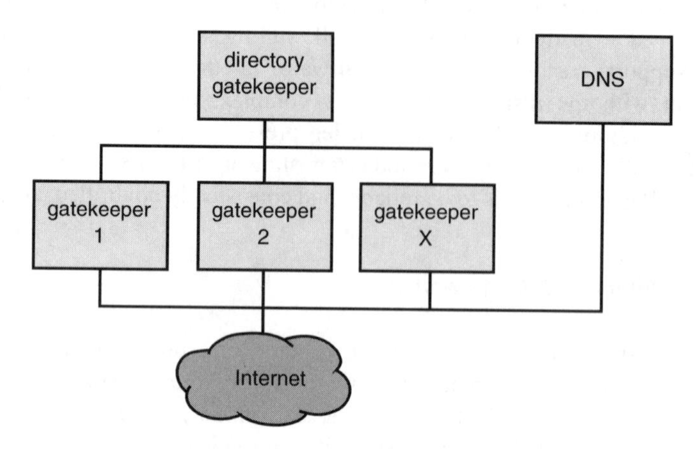

Fig 13.13 BT SurfTalk architecture

Registration

When using the SurfTalk service, the customer connects to the Internet using their existing ISP and launches the BT SurfTalk client. A DNS request is then sent to the BT SurfTalk DNS shown in Fig 13.13, using a URL made up from the STD code of the E.164 number with which the client is attempting to register and the BT SurfTalk domain name. Depending upon the STD code used by the client, the DNS then sends a response back that contains the appropriate IP address for the gatekeeper to register with. The client then uses the IP address returned by the DNS to send a registration request (RRQ) to a gatekeeper covering their local area. The H.323-ID field of the RRQ contains a username and password, while the E.164 field of the RRQ contains the home telephone number of the customer. The E.164 name is authenticated using a RADIUS server with the password contained in the H.323-ID field and registration with the gatekeeper is allowed if successful. If the authentication succeeds, then a registration confirm (RCF) is sent back to the client. If the authentication fails for any reason, a registration reject (RRJ) is sent back to the client (see Fig 13.14).

Making and Receiving Local Calls

Once registered with the platform, the user may access their e-mail and browse as normal. However, when an incoming call is received, the user is notified by a small pop-up window, that presents the name of the person calling, and allows the user to accept or reject the incoming call. If the user wishes to make a call, they enter the

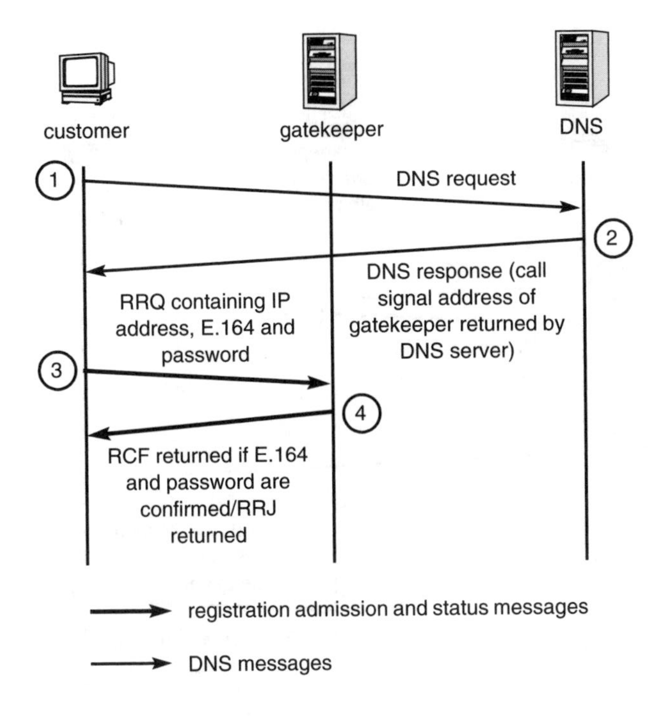

Fig 13.14 BT SurfTalk Registration.

full national telephone number of the person they wish to call and press the call button. The client then sends an admission request (ARQ) to the gatekeeper containing the E.164 number of the person they are trying to contact. The gatekeeper then searches its list of current registrations. If the person being contacted is registered on the gatekeeper, the gatekeeper sends back the IP address of that person in an admission confirm (ACF) message. The client then uses this address to set up a call directly between the two clients. In this case the H.225.0 RAS signalling is gatekeeper routed, but the H.225.0 call signalling and H.245 media signalling are direct routed. The call set-up is then initiated directly between the two clients. When a call set-up message is received, the client receiving the incoming call also has to get agreement from the gatekeeper to connect the call. It does this by sending an ARQ message to the gatekeeper in the same way as before, waiting for an ACF before completing the call set-up.

When the call ends, both end-points must signal the gatekeeper to inform it that the call has finished. This is achieved by sending a disengage request (DRQ) to the gatekeeper. The gatekeeper then confirms that the call has been cleared, by sending a disengage confirm (DCF) message back to each client as shown in Fig 13.15.

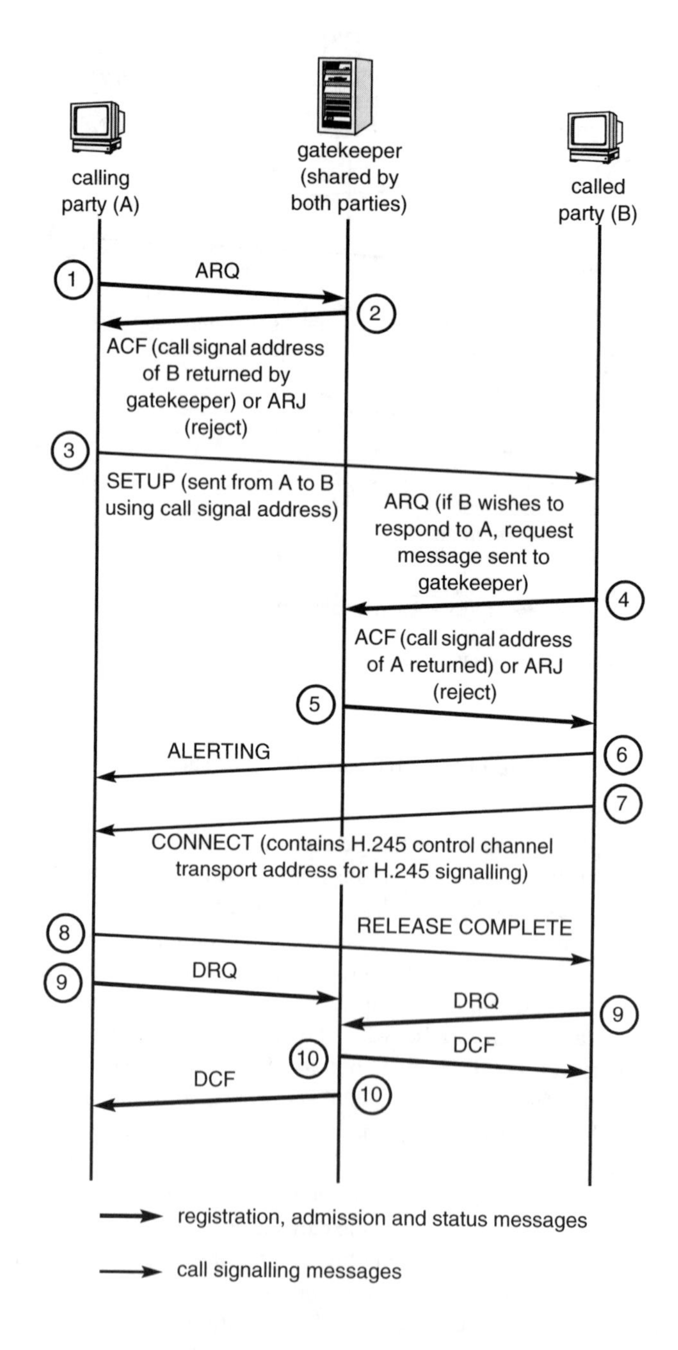

Fig 13.15 Local call signalling.

Making and Receiving Out-of-Area Calls

If the gatekeeper does not find a registration matching the E.164 number in the ARQ message, the gatekeeper sends a location request (LRQ) to the directory gatekeeper (see Fig 13.16). The directory gatekeeper acts like an LRQ proxy, searches its list of

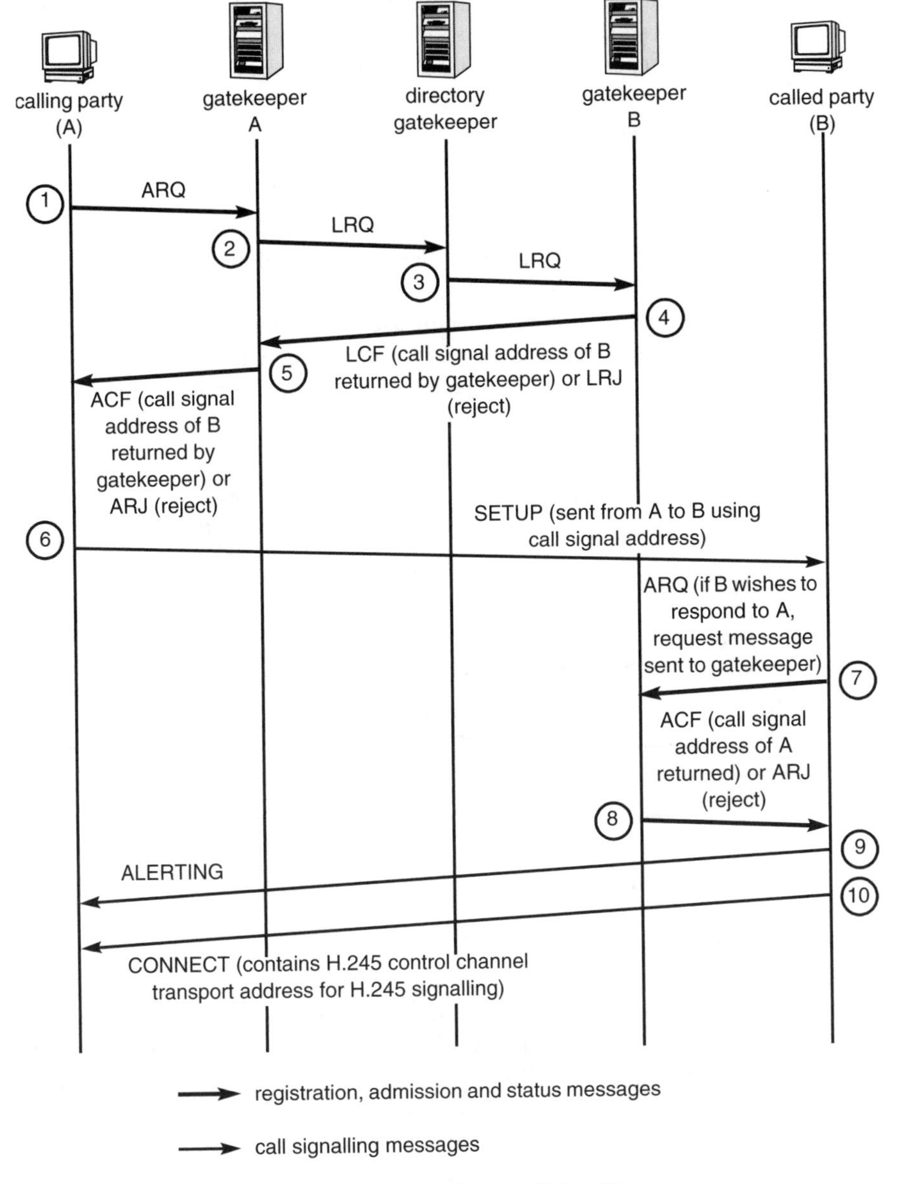

Fig 13.16 Out-of-area call signalling.

known prefixes/STD codes and forwards the LRQ to the gatekeeper associated with the particular STD code in question. The destination gatekeeper then searches its list of current registrations. If a registration with the E.164 number contained within the LRQ is found, then the IP address of the client is returned to the originating gatekeeper in a location confirm (LCF) message. The originating gatekeeper then sends back an ACF in the same way as before. However, if a registration is not found to match the E.164 number in the LRQ, a location reject (LRJ) is sent back to the originating gatekeeper. The originating gatekeeper then sends an admission reject (ARJ) back to the client, saying that the person they are trying to contact is not currently registered on the platform.

Unregistering

Before the user logs off from the Internet, they must shut down their BT SurfTalk client. When the client is shut down, it automatically sends an unregister request (URQ) to the gatekeeper, informing it that the user is no longer online and therefore that the gatekeeper should remove the current registration from its list of active clients. The gatekeeper then sends an unregister confirm (UCF) message back to the client to confirm the registration has been removed, as shown in Fig 13.17.

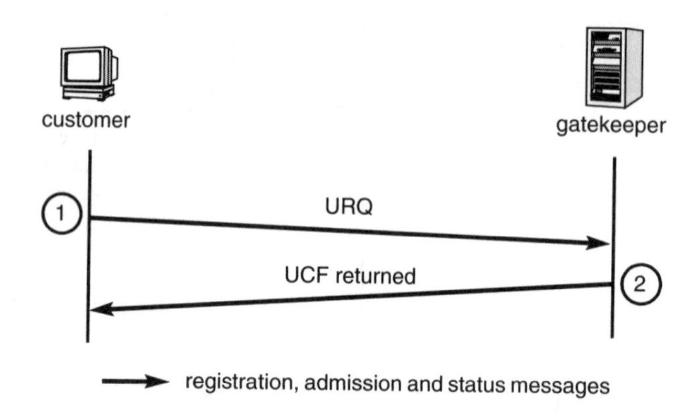

Fig 13.17 Unregistering signalling.

13.4.2.4 BT NetChat

Figure 13.18 shows the BT NetChat solution, which is the successor to BT SurfTalk. It has all of the functionality of the SurfTalk platform (not shown in Fig 13.18) but has the addition of PSTN break-in. The BT NetChat client registers with a gatekeeper in the same way as the BT SurfTalk client and customers are able to make and receive calls 'on-net' in the same way as BT SurfTalk. The main difference being that they are also able to receive calls from the PSTN.

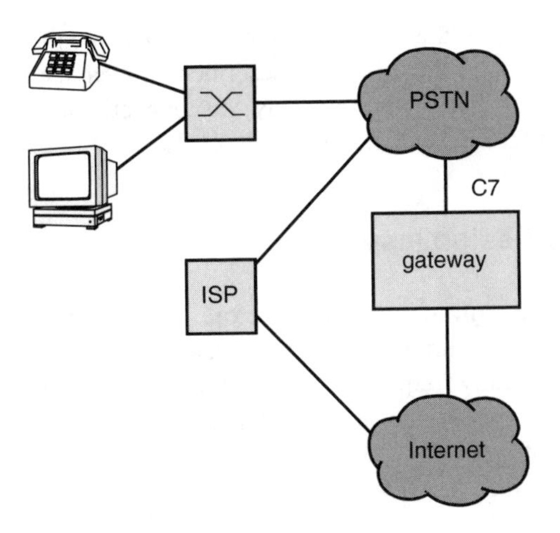

Fig 13.18 BT NetChat.

Unlike similar services, such as ICW, where the customer was required to pay for a call divert network feature together with the charge for the diverted leg of the incoming call, BT NetChat uses the 'administrative divert' feature used for other services such as BT CallMinder. This means that usage records are not generated for the diverted leg of the incoming call. The incoming call is trapped in the IN layer, and diverted to a PSTN-to-IP gateway, that is registered with an ingress gatekeeper in a similar way as a client. The call is then converted to IP packets and routed out over the Internet to the intended BT NetChat client recipient.

The call is presented to the gateway via a C7 link to the PSTN. When the call arrives at the gateway, the gateway sends an ARQ to the ingress gatekeeper containing the E.164 number of the incoming diverted call. The ingress gatekeeper then sends an LRQ to the directory gatekeeper. The directory gatekeeper then searches its list of known STD codes and forwards the LRQ to the gatekeeper responsible for handling calls with that STD code. The gatekeeper then searches its list of current registrations. If a registration matching the E.164 number contained in the LRQ is found, an LCF is returned to the ingress gatekeeper with the IP address of the associated BT NetChat client. The ingress gatekeeper then returns an ACF to the gateway, which then sets up a call directly to the client in the same way an 'on-net' call is set up. However, if a registration matching the E.164 number in the LRQ is not found, an LRJ is returned to the ingress gatekeeper. The ingress gatekeeper then returns an ARJ to the gateway. The gateway disconnects the incoming call, returning a disconnect cause of 'busy' and the call is rejected.

When the call arrives at the client, a pop-up window appears, giving the user the option to either accept or reject the incoming call. If the call is rejected or not answered, then the gateway disconnects the incoming call, returning a disconnect cause of 'no answer'. If the call is accepted, then a media path is established between the gateway and the client, as described in Chapter 1. When the call is terminated, by either party, the gateway disconnects the call and returns a disconnect cause of 'normal call clearing'.

13.5 Core Design Issues

There are many issues still to be resolved before large-scale VoIP networks can be fully deployed. In the interim, it is possible to deploy viable and working solutions; however, these generally require careful engineering. The following sections therefore describe some of the areas that require more careful attention when designing VoIP solutions.

13.5.1 Security

There are a number of fundamental network and service security issues that need to be overcome before VoIP can truly be deployed for high-volume commercial services on the open Internet. However, when deployed within an intranet environment, some of these hurdles largely disappear.

13.5.1.1 Firewalls

Firewalls are a major problem when it comes to deploying a VoIP service. Many corporate and large organisations now use firewalls to protect their internal networks from unwanted penetration through the Internet. Firewalls block any network traffic attempting to get into the intranet that is not specifically allowed by the firewall rule set.

User datagram protocol (UDP) packets are connectionless and can be used to infiltrate a network. As a result, UDP packets are generally blocked at firewalls to prevent malicious attacks.

The call signalling for H.323 uses a transmission control protocol (TCP) set-up message, which is both connection-oriented and on a specific port. This makes opening a firewall to the call signalling relatively easy. However, the media path is set up later, uses a random port number in most cases, and uses UDP as the transport protocol. The specific port number to be used for the media is contained within the payload of the call set-up messages and so cannot be easily accessed to set up a path through the firewall dynamically. As a result, if a firewall that was not capable of looking at the payload were used, a wide range of UDP ports would have to be opened up. Combined, these present significant security loopholes if opened up in a firewall.

Another problem associated with firewalls and VoIP is due to the fact that multiple firewalls cannot easily be run in parallel to spread load on connections in the same way that multiple Web servers can be load shared. The reason behind this is down to the routing of the different call set-up messages and the subsequent media stream. It is possible, even if you had a VoIP-aware firewall, i.e. capable of analysing the packets to ascertain which port to open up for the media stream, that the media stream could get routed to another firewall which in turn would not know about the call set-up signalling through the other.

13.5.1.2 Address Translation

Due to the speed with which the Internet has expanded over the past ten years, problems have arisen in that when IP was first conceived, a 32-bit address was considered to be more than enough to provide all of the IP addresses required. However, it was soon realised that there was a significant shortfall of available addresses.

To solve this problem network address translation (NAT) or port address translation (PAT), whereby a range of private IP addresses is used on a company/organisation's intranet, was introduced. A limited number of public IP addresses are then allocated to that company and either network or port address translation is used to send traffic between the intranet and Internet.

The problem arises as a result of the address translation. H.323 is an end-to-end connection-oriented protocol. When address translation is used, the H.323 end-point is unable to register its correct IP address with another end-point; therefore a call cannot be set-up successfully.

Another situation where this might occur is where two corporate customers wish to connect their voice over IP networks that use private address ranges. In this case there is the possibility that the same IP address could exist in both networks. To get around the problem, address translation could be used. However, there is the potential that a large number of public addresses would be required by both corporate customers to communicate across the border end-points.

13.5.2 Quality of Service

At this point in time, there is no kind of QoS supported over the public Internet. This poses a number of problems for service providers. These include the fact that without any QoS, service providers are unable to guarantee the speech quality delivered to the customer. As a result, the perception of the speech quality for voice over the Internet applications is that they could not possibly provide adequate QoS for speech services. However, in the short term, although no QoS is supported over the public Internet, infrastructure is being improved at such a rate that increased bandwidth is facilitating the deployment of such services without any QoS guarantees. Further discussion of these issues is given in Chapters 2 and 3.

In the future, however, it is possible that the public Internet will migrate to a QoS-enabled architecture, allowing the provision of different levels of service depending upon traffic type. To this end, a great deal of work is being done within the ETSI TIPHON project concerning QoS and how QoS can be applied to networks to give varying levels of service, as discussed further in Chapter 16.

13.5.3 Feature Support — SCN/IP Interworking

As voice over the Internet services are pushed more and more, the customers taking up the services want to have all the services they are used to enjoying on the SCN. For this reason, various standards bodies are looking for ways to standardise these SCN/IP interworking functions.

13.5.3.1 SCN/H.323 Interworking

As the legacy protocol, H.323 has had market dominance for a long time. However, to take account of the requirements for more SCN interworking, new protocols have been added to the H.323 protocol suite.

Primarily H.450 is used to provide the supplementary services available within the SCN. H.450 is split up into a number of sub-sections covering services such as call divert, call forward, hold, message waiting, etc. For more information on H.450.x services and how they interwork with the SCN, see Chapter 10.

13.5.3.2 SCN/SIP Interworking

Service provision for SIP users can be provided in a number of ways — call processing language (CPL) scripts, SIP-CGI, or Java SIP Servlet API. CPL scripts are a very powerful tool, allowing users to specify how their own calls are processed, rather than having to subscribe to specific services from a service provider. They allow the user to specify call divert, call forward, call waiting services, etc. For more information on SIP and SIP interworking, see Chapter 11.

13.5.3.3 SCN/BICC Interworking

Bearer-independent call control (BICC) is being designed to provide all the SCN (ISUP) services over a packet network. For more information on BICC, and the services that it can support, see Chapter 8.

13.5.4 Directory Services

Another problem associated with hosting a number of customers from the same equipment is that of privacy. From the point of view of directory services, one customer may not want to be visible in a directory, in the same way as customers within the PSTN can be ex-directory. From a corporate point of view, if multiple companies are being hosted by the same equipment, all members of the company may want to be listed in the corporate directory. However, they would not want their corporate directory to be visible to other companies being hosted by the equipment.

A solution to this is the use of proxy directories. A central LDAP database holds all of the information for all of the customers being hosted by the equipment. However, to access the directory, the users have to access a proxy directory, which does an LDAP look-up in the main database, but only has the access keys for the information for their specific company.

13.5.5 Billing

Billing of customers is difficult in more ways than one. As discussed earlier, there are difficulties in identifying which customer to bill, given that IP addresses are typically allocated using DCHP and therefore change between different log-on sessions. However, there are also moves to bill on both content and level of service. The aim being that, for example, a customer making a video call would use more bandwidth than a normal voice call and should therefore be charged a premium for being able to do so. On top of that, there may be a situation whereby two customers are making calls, one having 'Gold' service level, the other having 'Silver'. If the available bandwidth were not sufficient for both calls, there needs to be a method of identifying packets from 'Gold' customers, etc, to give them priority over lower service level customers. Without being able to do this, the ability to bill against service level agreements (SLAs) is somewhat dubious.

There are methods available today to easily identify from which IP address a call request has come, and, in turn, who is using that IP address at the precise moment the call was made. RADIUS is used extensively to log transactions. RADIUS log files can then be parsed to provide billing records. However, the difficulty comes when a customer makes a limited number of short calls within a charging period. These calls would generate very little revenue and could in some cases cost more to recover using a billing engine than the revenue they would generate.

Probably the simplest and most straightforward way around this problem is to use a billing strategy which does not require regular payments, which will typically be some form of prepayment solution. That way, customers can decide when they need to top up. If they have not got sufficient credit in their account, they will be unable to use the service.

There is also the question of how to bill calls that go between VoIP zones. For example, if a call starts and terminates within a single VoIP domain, the bill would most likely be generated by the billing engine servicing the domain.

However, if the call is terminated on someone else's VoIP platform, they are going to expect a share of the charges for that call, in the same way as SCN providers do. This leads on to the idea of a VoIP clearing house, which is discussed further in Chapter 14.

13.6 What the Future Holds

Simple bit transport and plain voice are no longer seen as the long-term high-value revenue generators. It is therefore thought that the major revenue streams of the future will be made from service provision and applications. For this reason, there is a need to liberate innovation in service and application development.

13.6.1 Instant Messaging

Simple, immediate, text messaging has been widely available for many years — most recently culminating in applications such as text chat. This type of application is very popular among the on-line networking user community. In addition, the short message service (SMS) can also be considered to be a form of 'instant messaging' (IM). These types of application generally work on the principle of transmitting short text messages that are transported across a network in near real time. The key functionality required for an instant messaging application is for the user to be able to make their presence known to the community with which they wish to communicate. They can then be contacted and an IM session established. This process is closely related to the registration process implicit within VoIP applications and it is not surprising therefore to find the industry focus is on adding this type of functionality to expand VoIP into a richer IP communications medium.

A number of 'instant messaging' services are available, although in the same way that VoIP in its early days was dogged by proprietary protocols restricting user groups, the existing services are limited to those customers who are with the same service provider. For example, an AOL Instant Messenger client cannot currently communicate with a Microsoft MSN Messenger client. For this reason, standards bodies have again picked up on the requirement for a standard protocol for messaging.

The instant messaging and presence protocol (IMPP) is being developed within the IETF (RFC 2778 and 2779). The IMPP working group is looking at a number of candidates — APEX (application exchange), PRIM (presence and instant messaging protocol) and SIMPLE (SIP extensions for instant messaging and presence). Once these interfaces have been standardised, the opportunity exists for service providers to be able to offer truly instant messaging. While text chat and instant messaging are envisaged to operate as peer-to-peer applications, the functionality it delivers is very similar, in terms of end-user experience, to fast e-mail. As Internet bandwidth increases, it is likely that there will be an increasing blur between e-mail and instant messaging unless more distinct functionality emerges in the IM area.

13.6.1.1 Providing Instant Messaging Using Existing Technologies

There are, however, opportunities for service providers already deploying VoIP services to offer instant messaging using existing technologies.

An H.323 gatekeeper is an ideal candidate for providing the address translation and location service required for doing instant messaging. The gatekeeper holds a translation table for E.164-to-IP address. In the same way as SMS is used to send short text messages to other telephones, if a customer is logged on to the VoIP platform, a valid current translation is available for E.164-to-IP address. So, using part of the existing H.323 protocol stack, the IP address of a destination party can be easily discovered. By using a TCP-based protocol for the actual message transaction, the service can also be guaranteed (see Fig 13.19).

Again, the difficulty with an approach such as this is that the actual message transaction protocol would again be proprietary. Alternatively, an extension to the H.323 version 4 protocol could be used to carry the message.

13.6.2 Presence and Location Based Services

As discussed above, the IMPP working group within the IETF is aiming to produce a standard protocol for presence indication. Presence is closely related to instant messaging since it involves making the availability of someone and their receptiveness to receive communications, available to a community with which they wish to communicate. It is then possible for others to see whether the person they are trying to contact is on-line at any given moment. This then leads to other more powerful features such as being able to filter calls and select different call dispositions for different groups of people attempting to make contact.

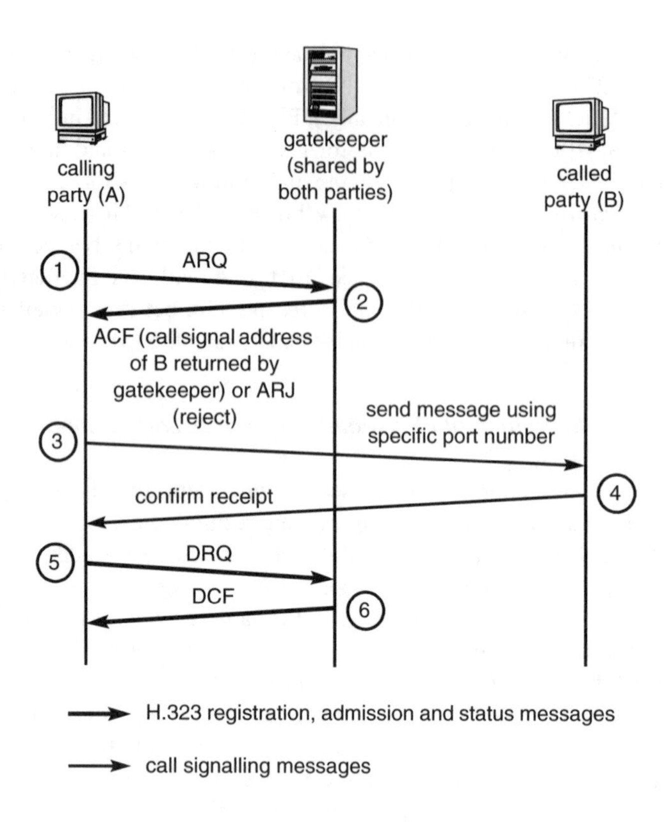

Fig 13.19 Instant messaging using H.323

Multiple call dispositions can allow someone to define how they want to receive their incoming calls. By splitting incoming numbers into specific groups, such as friends, family and work, it gives the customer the ability to route different types of call in different ways. For example, you might have it set up so that no matter where you are or what time it is, incoming calls from members of your family always get routed to your mobile, while at weekends work calls get diverted to a work voicemail service.

13.6.2.1 Possible Methods of Implementation

Some of these ideas can be implemented with current technology. The voice and multimedia services provided using the likes of H.323 and SIP require clients to register either with a call server of some description when wanting to make or receive calls.

When a client registers with a call server, that entity is then able to set a flag within a database to say that the person is on-line. This can then either be displayed as part of the client, or separately on a Web-based directory listing.

This idea can be extended further to display an 'in-call' indication as the gatekeeper or call server also has the intelligence to know about whether the client is in a call. This feature can be useful if someone wants to contact you urgently but does not want to leave it to chance for you check your voicemail — they can call as soon as you finish the call you are in.

13.6.3 Multimedia Services

Moving away from the idea of replicating SCN services, multimedia provides a whole new set of services that can be delivered. These include video, document sharing, white boarding and other forms of interactive collaboration. There are drivers to charge for such services based on their content and bandwidth. For example video is high bandwidth and also isochronous and should therefore be charged at a higher rate than e-mail and the like which are typically of a much lower priority.

There are applications that exist today that implement various aspects of such multimedia services. Most notably among these is Microsoft NetMeeting version 3.01 that is capable of these features as well as being capable of registering with an H.323 gatekeeper. This gives the user the ability to start a video call to another user by simply dialling an E.164 number, or alias, instead of having to identify an IP address which has been the case in the past.

13.6.4 Conferencing Management

Conferencing adds another dimension to voice and video calls. It gives service providers the opportunity to make conference calling a much richer experience. Combining conferencing with presence and multimedia applications spells the end of the 'So, who have we got on the conference bridge?' question. Being able to see a list of those present, or in the case of videoconferencing, a picture of those in the conference, makes it much simpler to figure out if everyone who is supposed to be there actually is — and that they are paying attention to what is being discussed.

13.6.5 VoIP for Mobile Applications

The mobile communications market has seen the evolution of second generation mobile networks such as GSM, with more than 250 million subscribers in more than 120 countries. Subscriber numbers have been growing at an ever-increasing rate in many countries, aided by new marketing strategies such as 'Pay as You Go' to the point where it is estimated that there will be 600 million cellular connections world-wide by 2002 [1]. The next step forward identified for mobile networks is the introduction of data with GPRS terminals being available in the UK in late 2001 and third generation UMTS terminals in 2002 and beyond [2].

In order to make UMTS more convergent with IP technologies, and hence more efficient in providing multimedia services such as those predicated on VoIP, 3GPP [3] has decided to adopt a VoIP approach to their future network architecture (R4/ R5 [4]). In particular, 3GPP have chosen to use SIP as the basis of call signalling and control (see Chapter 11). Since VoIP is oblivious to whether users are on fixed or mobile networks, it provides the capabilities to offer users the same services from any access technology. It does, however, enable negotiation between the end-points and intermediate filters, say, to adapt the content to the bandwidth and quality of the link. In this sense VoIP is an integral part of the IP-based realisation of the virtual home environment (VHE) — which was originally included within the UMTS vision — by enabling the provision of a common set of services for fixed, cellular and satellite access [5].

13.6.5.1 Other Services

VoIP is able to provide not only most classic IN services, but also a large range of more complex call-handling applications including:

- third-party call control — one party sets up a call between two other parties without necessarily participating in the call otherwise;
- time-dependent routing — calls receive different treatments depending on the time of day or the day of week;
- person-dependent routing — the call is routed to different end-points, depending on who is calling, e.g. users might require calls from their boss to be routed to their office desktop, and calls from their family routed to their home PC for example;
- media-dependent routing — the call is routed to different end-points, depending on the type of media requested, e.g. users might prefer to receive video on their desktop, instead of on their mobile device, where they may only have limited bandwidth;
- calling-name delivery — the name of the caller is displayed on the screen before answering the call;
- finding a party — as an example, when a user willing to play chess contacts the VoIP call server to request a partner, the call server then makes a look-up in the VHE database, discovers all the users with an interest in chess, and invites them to a session.

13.6.5.2 VoIP as a Component in an All-IP Fixed/Mobile Network

There are many initiatives [6, 7] that are attempting to define an all-IP fixed/mobile converged architecture (Fig 13.20). The key elements of these are a range of access technologies (ADSL, cellular, wireless LAN) allowing users to connect to a common IP core. Increasingly IP access networks will be equipped to provide QoS. They may also support mobility with IP mobility protocols such as HAWAII [8]. However, the common aspect of all these approaches is that the intelligence (security, call control, billing, etc) is pushed to the edge of the network [9]. This is

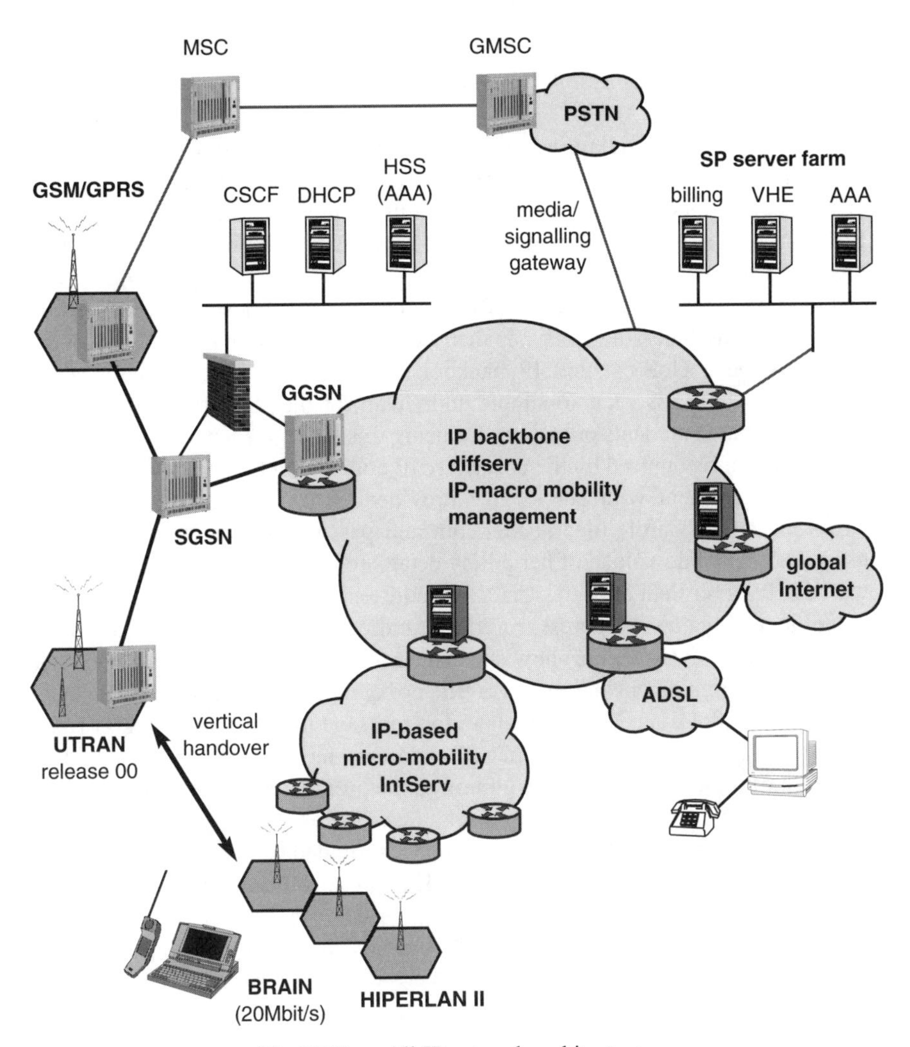

Fig 13.20 All-IP network architecture.

the network architecture that is emerging for wireless LANs from the BRAIN project [6] and UMTS (R5) from 3GPP.

With this approach, users connect to the IP core from a variety of terminals using the most convenient or cheapest access technology that is available. In Fig 13.20 the service provider is shown as providing a range of services such as billing, personal mobility (virtual home environment) and voice/video messaging. The network provider is shown as supplying network access security, call/session control and accounting. This architecture is by no means definitive but simply represents the BRAIN project business model and serves to illustrate some of the potential roles that VoIP could play in such a network.

13.6.5.3 Example of Mobile VoIP Application

The role of VoIP in providing session control for the network provider and personal mobility for the service provider is best illustrated by means of an example. Imagine a visitor, Carole Jones, to the University of Southampton. Carole has a laptop with a HIPERLAN II card and is informed that the University is running a HIPERLAN II network. In this example, the network has been created by simply adding wireless cards to PCs coupled to the University intranet at strategic hot spots (cafes, meeting rooms, etc). This access network has then been connected to a core WAN service, which may be a QoS-enabled IP backbone provided by BT. From the access network view, this has been so simple and cheap for the University to set up that they do not charge students or visitors for using this leg of the network.

Carole is not recognised by the university IP domain but the university does have a roaming agreement with her service provider (SP), Xtel, and so her network address identifier (Carole.Jones@Xtel.com and password) is referred to them for authentication and download of her policy details (e.g. charging arrangements). The local DHCP server then allocates Carole an address (192.123.65.23) and configures her terminal — gateway address, multicast address range and so on. Carole now needs to tell her SP (Xtel.com) how she can be reached for incoming sessions. One simple way to implement this is via a SIP proxy server with a location database. Carole, for example, sends a SIP registration message to a SIP proxy server run by her service provider to register her IP address, terminal capabilities and the bandwidth of her link. This information might be provided to the SIP component or a QoS broker by enhanced terminal stacks (see, for example, Kassler et al [10]). Carole's terminal opens with her personal portal provided by her service provider — adapted to her link, terminal and location. This presents her with a common set of services any time she gains IP access.

When Peter wishes to contact Carole he sends a message to Carole.Jones@Xtel.com. The message arrives at the SIP proxy server that in turn queries the SIP location database and finally the server selects possible destinations to forward the invite based on media type, terminal capabilities and user preferences. Carole receives the invite and exchanges further SIP messages to negotiate coding rate and codec type. Peter wishes to send a video stream at 128 kbit/s and so his application sends resource reservation protocol (RSVP [11]) messages to signal this across the various network elements. The backbone network operates a differentiated service for QoS [12] and simply uses the RSVP messages to establish how it marks and polices packets submitted at the ingress to the backbone, using charging-based policy to share bandwidth between competing users. The RSVP message may then perform a specific, per-flow, reservation within the micro-mobility domain — but this could also be a gigabit Ethernet with no QoS support. However, the reservation is definitely made across the HIPERLAN radio hop. Now, suppose that Carole moves out of the café and into a meeting room — HIPERLAN base-stations have continuous coverage and so a horizontal hand-over is possible. As the university routers are running a micro-mobility routing protocol, such as HAWAII [8], her IP address is valid throughout the domain. A temporary tunnel is set up between the base-stations to avoid loss of real-time packets until the routing protocol adapts to Carole's new location.

Carole now moves outdoors and beyond the range of the HIPERLAN base-stations. However, as she has a UMTS-capable card in her terminal it is able to find a UMTS network to continue her network connectivity. Carole authenticates herself by means of her smart card (this can act as a UMTS SIM card) and gains an IP address from this new domain. This may be part of Carole's deal with her service provider or she may be, for example, a BT Cellnet customer as well. Carole sends Peter a SIP invite with the new IP address but, importantly, the same call ID as was supplied by the proxy server. This enables Peter to identify this as part of the same call. Carole's application suggests a lower rate for video coding (64 kbit/s, for example). Peter's user agent accepts this invite and he now sends packets to Carole at her new IP address at the lower rate. Unlike the HIPERLAN scenario, some packets may be lost during this vertical hand-over.

When the call is over, Carole is anxious to access some travel information. Her UMTS terminal uses SIP to make a request for travel information to a modified SIP proxy server located within the domain of the UMTS network provider — this is known as the call server control function (CSCF). The role of the CSCF is rather like a multimedia call server. Carole could, of course, have simply browsed the Web, but finds the range of multimedia services provided by the UMTS operator to be both personalised and better focused, and to support, for example, free calls to merchants and help-lines from within sessions. For example, the SIP request for travel information might be modified by the CSCF — before being passed to the information provider who provides the travel service — to include Carole's location as determined by the UMTS radio network. The CSCF is stateful and can track the progress of a multimedia session — so that, when Carole decides she needs to speak to a representative of the train company, it can set up the appropriate voice session with the correct operator for her location. In this case the SIP server is acting as a call server; a complete description of the role of SIP within UMTS R4 can be found in Bale [4].

13.6.6 Virtual Home Environments

13.6.6.1 Centralised Services — User Profiles

There are currently a number of ways already proposed for managing services based upon user profiles. These include the IETF's call processing language (CPL) and Java Servlet APIs. CPL is interesting because it permits the user to construct their own services and upload them into their profile. CPL is considered safe as a scripting language because it does not permit loops — just conditional statements. Server processing therefore cannot be tied up indefinitely either accidentally, by a malformed script, or maliciously by a virus. When a request comes into the location server, the presence of one or more scripts is detected and executed accordingly, revealing just that subset of locations specified by the script.

In reality, the majority of users would probably not wish to write their own scripts from scratch. Rather they would most likely use some form of tool such as a number of options presented on a Web page. In this case details can be either selected or entered and scripts auto-generated behind the scenes. Auto-generated scripts can then be uploaded into the user's profile, potentially revolutionising service provision [9]. Figure 13.21 shows the prototype implementation of a simple HTML-based front end for setting up CPL scripts.

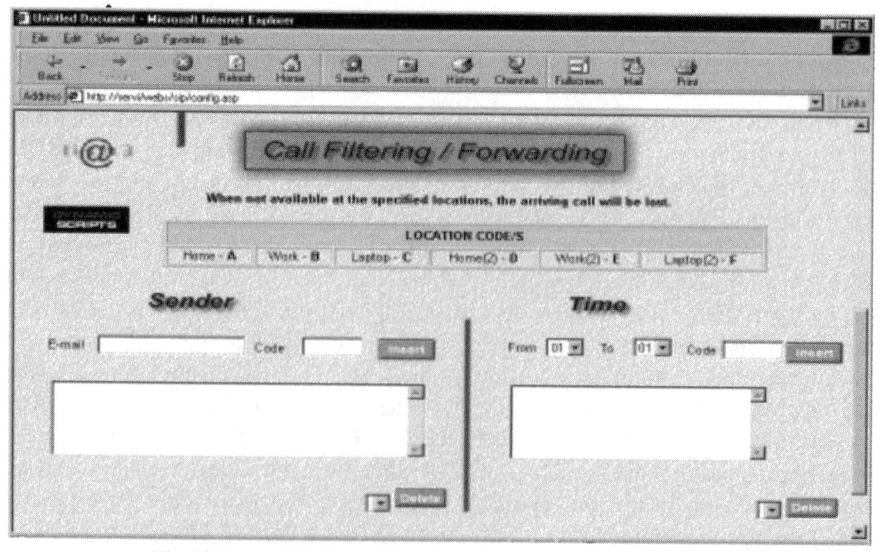

Fig 13.21 Web-based interface for setting up CPL scripts.

In contrast, Java Servlets and Web-CGI programs exemplify the traditional means of service provision, provided by the operator [9]. Small programs are executed to reveal locations according to the services for which a user has signed up in advance.

The flexibility of the SIP URL also facilitates some interesting service resolutions. SIP URLs can accommodate any sort of address, whether they be a user's PSTN telephone number (i.e. sip:E.164Number@gateway) or any other form of Internet URL. For example, during office hours, a location server could reveal the location of available operators to a company SIP URL, and out of office hours could reveal the company Web page URL that the SIP user agent would use to load into the Web browser of the terminal. Thus the user could learn the company's office hours, and perhaps even do some on-line shopping. Because user profiles are resident on the location server's database, effectively they are always available to the user, irrespective of the terminal currently being used. This represents one possible implementation of the virtual home environment (VHE).

An issue that also needs to be addressed is how resources are negotiated and supplied for sessions. For instance, if someone wanted to have a voice call over IP to their boss, they might send an INVITE message, negotiate a codec and receive a RINGING message. If, however, they could not subsequently reserve the resources necessary — including network bandwidth, CPU capacity or memory — then the call would fail. 3GPP is currently standardising the interaction of resource reservation and session initiation to address this problem. Bale [4] gives the detailed call flows being suggested for UMTS release 2000 (R4) while Kassler et al [10] discuss the whole question of resource reservation, QoS adaptation and the interaction between the session protocol (SIP/H.323) and the QoS components.

Another issue that is problematic for session-oriented protocols, such as those used for VoIP, is firewall traversal. Most firewalls do not currently pass VoIP messages. Future developments, such as those proposed by the IETF MIDCOM Working Group, allow micro holes to be created which will potentially solve this problem. Tunnelling SIP packets, possibly over a secure tunnel, or encapsulating them in HTTP packets are other possible solutions. Finally, network address translators (NATs), used, for example, to allow private address space to interface to the outside world by sharing a much smaller number of globally routable addresses, will potentially also cause difficulties for VoIP services. Any private addresses embedded within the VoIP messages, i.e. not in the headers that are dealt with by the NAT, will not be translated and therefore cause errors if used for routing in the rest of the Internet. However, one of the prime reasons for employing NATs — shortage of address space — will disappear as IPv6 is introduced. There are, however, potential advantages in using NAT and firewalls since this combination may enable message routing to be forced through proxy servers. There are also other advantages of using such network delineation devices in terms of managing QoS and enabling high integrity VoIP transmission to take place.

13.6.6.2 End-to-End Services

Potentially more revolutionary is the facility VoIP has to offer some of the services now associated with IN at the edge of the network. This may actually be on the user's terminal for more advanced terminals such as laptops. Since VoIP signalling and control appear as application-level protocols, they can be provided with information and configured accordingly to handle such conditions as call waiting, redirect on busy, call back when free, 1471, etc. However, a number of socio-economic factors will most likely determine whether this is a viable proposition.

Since in a pure IP environment VoIP solutions offer a full multimedia capability, they also provide an end-to-end means of doing things that are traditionally associated with client/server and store-and-forward. For example, e-mail is traditionally proxied between mail servers, before being stored on the recipient's mail server, and then forwarded when the mail client logs on to the network. Using VoIP, mail transfer could be sent directly to the mail client, effectively cutting out the messaging between mail servers, and storage within the network. The VoIP signalling negotiation to deliver the mail would also provide value added feedback upon the client's willingness to receive it, and confirmation that it had indeed been delivered.

Furthermore, should a secure mail delivery be required, VoIP signalling could be used to negotiate a secure channel between end-points, using tried and tested protocols, such as secure sockets (i.e. https), and tunnelling the mail message through it. Similarly Web browsing could be assisted by VoIP signalling. In this case the call server directs page requests to the nearest or best Web server mirror for that request. Web servers, too, could use VoIP signalling messages to push content to the user, for example, inviting the user's browser to be refreshed, when Web page content being browsed has been updated. Naturally the provision for pushing advertising can be addressed in the same manner, harnessing the wealth of user information provided by VoIP to make the targeting specific, e.g. location specific, gender specific, subject specific, etc.

13.7 Challenges for the Multimedia Future

A single IP-based infrastructure provides a ubiquitous network where more advanced applications can be built without the need for complex network infrastructure upgrades. In theory, a voice application can be considered to be simply another application running on an IP network.

It is therefore possible to link applications and media together to produce complex solutions highly customised to a specific need, controlled from anywhere in the world.

VoIP tends to be associated with an image of home DIY solutions — offering low quality speech, limited largely by the availability of bandwidth through modems and the capacity to process voice on a desktop PC. Additionally, the public Internet gives a very variable QoS when applied to speech performance, as discussed in Chapter 2. A further inhibitor to the development of applications has been the absence of appropriate standards. This situation has had the effect of limiting the scope of services seeking to exploit this emerging technology to functionally low and difficult-to-use proprietary applications, attractive only to those who are comfortable with the details of the technology. While this situation certainly persisted, technology has improved significantly over the past few years. PC telephony software has now improved, giving better, but still variable, performance over the public Internet. Together with the considerable improvements achieved with PSTN-to-IP gateway technology and the adoption of a standards-based approach, the environment is now right for new, innovative services to exploit the potential of IP. Indeed, it is quite appropriate now to contemplate the introduction of a new acronym, EoIP — everything over IP!

13.8 Summary

This chapter has provided a summary of the brief history of VoIP applications, an overview of the differences apparent in deploying VoIP applications on the public Internet compared with managed intranets, described some early and contemporary applications, and indicated where the direction of VoIP applications might be leading. VoIP, hyped up to offer 'almost-free' voice calls over the Internet, when faced with the reality of falling telephony charges, has failed to generate a large amount of interest from the business community. However, VoIP continues to promise a future of seamlessly integrated multimedia communications ranging from unified messaging to voice, video and presence that will potentially revolutionise the way we view and use communications in all aspects of our daily lives.

Current Internet applications, such as e-mail and Web-browsing, are client/server-based and do not require session set-up or control. They work very effectively on best-effort IP networks. In the future it is highly likely that both fixed and mobile services will be provided by a QoS-enabled all-IP core network. The next generation of peer-to-peer style applications, however, require advanced functionality in terms of user location by name, media negotiation and session renegotiation on handover. This functionality is analogous to call control and IN in the PSTN world, except that it is many times richer and potentially more complex. VoIP signalling and control offer an excellent candidate solution in this space. Given advances in technology and a greater business acceptance, it is expected that in the next few years IP-based voice applications are going to see continued penetration beyond the public Internet and into corporate intranets and extranets.

References

1 GSM Association — http://www.gsmworld.com/

2 Clapton, A. J. (Ed): '*Future mobile networks: 3G and beyond*', The Institution of Electrical Engineers, London (2001).

3 GPP: '*IP multimedia (IM) subsystem stage 2*', TS 23.228 (2000) — http://www.3gpp.org/

4 Bale, M. C.: '*Voice and Internet multimedia in UMTS networks*', in Clapton, A. J. (Ed): '*Future mobile networks: 3G and beyond*', The Institution of Electrical Engineers, London, pp 65-98 (2001).

5 Schwarz, J. da S., Arroyo-Fernandez, B., Barani, B., Pereira, J. and Ikonomou, D.: '*Evolution towards UMTS*', European Commission — DG XIII-B.4 (1996).

6 IST BRAIN project — http://www.ist-brain.org/

7 ETSI: '*TIPHON phase 1 scenarios*', — http://www.etsi.org/tiphon/

8 Ramjee, R., La Porta, T., Thuel, S. and Varadhan, K.: '*IP micro-mobility support using HAWAII*', Work in Progress (draft-ietf-mobileip-hawaii-01.txt) (June 1999).

9 Sage, C. J.: '*The service node — an advanced intelligent network services element*', in Dufour, I. G. (Ed): '*Network Intelligence*', Chapman & Hall, London, pp 95-108 (1997).

10 Kassler, A. et al.: '*BRENTA — supporting mobility and quality of service for adaptable multimedia communication*', Proc IST Mobile Communications Summit 2000, pp 403-408, Galway, Ireland (October 2000).

11 Braden, R. (Ed): '*Resource ReSerVation Protocol (RSVP) — Version 1 functional specification*', IETF RFC 2205 (September 1997).

12 Bernet, Y. et al: '*A framework for differentiated services*', IETF Work in Progress (draft-ietf-diffserv-framework-02.txt) (February 1999).

14

VoIP CLEARINGHOUSES AND THE OPEN SETTLEMENT PROTOCOL

R H Brennan and R P Swale

14.1 Introduction

The Internet is the largest public IP network today. From its earliest days, and as a consequence of its design and development, it has had to develop specific models for the interconnection and exchange of traffic between the physical networks from which it is constructed. From the point of view of engineering VoIP network solutions, it is important to consider these approaches and contrast them with the requirements of interconnecting existing voice network services. In particular, it is essential to appreciate how the technical and commercial options that have been used for the Internet may enable or inhibit particular VoIP services. This chapter considers these issues and describes how the VoIP clearinghouse approach has emerged as one particular solution. It concludes with a consideration of the technologies used to build clearinghouse solutions and highlights some of the design considerations that need to be taken into account.

14.1.1 Network Model of the Internet

The Internet comprises a hierarchical collection of IP networks; it is through the interconnection of such networks that the Internet is able to achieve both its global reach and scale.

From a service provider's point of view, Metcalfe's Law — which suggests that the value of a network is proportional to the square of the sum of the connected nodes — has become the core underlying principle that continues to fuel its growth. Put more simply, the more users you can attract on to your network, the more desirable it is for others to join your network rather than another one and so the more they will be willing to pay, within reason, for the opportunity to do so. As described more fully in Bartholomew [1], this has led to the formation of a small number of high performance backbone networks with a larger number of smaller networks providing aggregation of either end users or individual Internet service providers (ISPs) which in turn provide service to end users.

14.1.2 Interconnection of IP Networks in the Internet

14.1.2.1 Interconnection Models

From the early days of the Internet, the tariff arrangements used in its operation were entirely based upon largely non-commercial arrangements structured around the capacity of the transmission link used to connect the two parties involved in the network interconnection. This 'eat-as-much-as-you-like' model reflects neither the utilisation of the transmission link concerned nor the value of the traffic carried as perceived by the end user. As a consequence this has led to a structure where backbone service providers have been continually installing leading-edge high-performance transmission systems simply to keep up with the perceived traffic demand without being able to necessarily maintain profitability of their business. Signs emerged in the late 1990s that this model was difficult to sustain — in part due to the cost of procuring increasingly higher performance equipment and the decreasing time to its obsolescence. Service providers therefore contemplated the introduction of more complex tariff structures that reflect both the capacity of the connection offered and also the utilisation of the given link. However, these have not explicitly found their way into end-user tariffs other than providing an underlying influence when access contention ratios are being designed.

14.1.2.2 Peering

The method of interconnecting service providers IP networks is known as peering (Fig 14.1). While a more complete discussion of peering is provided in Bartholomew [1], the following analysis introduces peering from the point of view of VoIP networking. The act of establishing a peering relationship between two networks is usually enshrined within a peering agreement established between the parties concerned. The precise scope of a peering agreement is generally determined by whether it is a public or private peering arrangement. Generally, however, the commercial details behind peering agreements are not usually made publicly available.

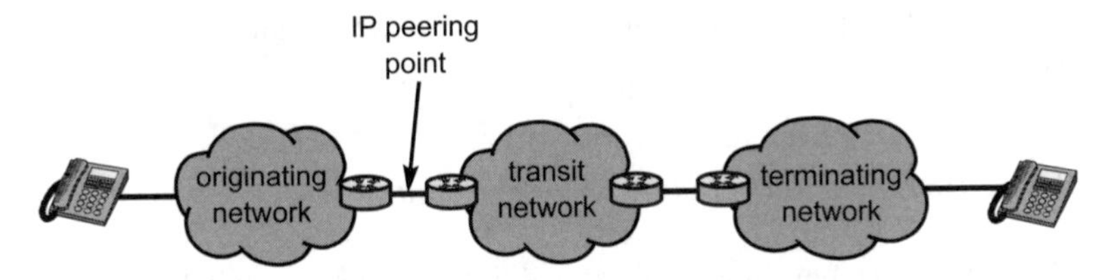

Fig 14.1 Typical IP peering implementation.

Public Peering

A key feature of the Internet is the inclusion of special nodes that enable service providers to exchange IP-based traffic. These nodes are known as metropolitan area exchanges (MAEs), of which there are three main sites — one on the west coast of the USA (MAE-West), one in the central area of the USA (MAE-Central) and one on the east coast of the USA (MAE-East). It is important to note that these arrangements only provide a simple IP-level interconnection. They do not provide any voice-related QoS support or traffic management as they only offer an entirely best-effort-based interconnection.

Private Peering

The alternative to public peering is to make special arrangements with another service provider to interconnect with their network at an agreed location or set of locations. The interconnection is usually planned to optimise traffic throughput and avoid network load balance problems. However, unlike public peering, private peering arrangements are not usually a practical option if the networks involved are of dissimilar size. This is because commercial concerns would tend towards the larger network, preferring that the smaller network simply purchase a normal connection from it.

14.1.3 Voice and Real-Time Services over IP

IP networks have emerged as the industry's dominant convergence layer for data-based applications within the information technology industry. Originally developed, and largely optimised, for the transport of variable-length 'bursty' data communications, the performance increases of IP technology have now made it possible for the support of real-time stream-oriented communications applications such as voice. A consideration of the interconnection of IP networks carrying real-time traffic therefore presents an essential feature of the design of large-scale VoIP systems.

14.1.3.1 Technical Implications of VoIP Peering

As VoIP and similar real-time applications depend upon the reliable transport of continuous streams of media packets within IP datagram payloads, the interconnection of two or more VoIP network domains represents a substantial challenge to ensure these requirements can be met. In particular, there are two main consequences of the current interconnect approaches for peering IP networks from a VoIP perspective. Firstly, QoS mechanisms are not sufficiently well standardised at present to allow the interconnection of IP networks while still supporting a stable QoS. Consequently, the network interconnect provides services offering only a best effort QoS level. Secondly, due to the potentially asymmetrical nature of data traffic, it is not uncommon for network interconnection points to have congested transmission paths in one direction while having relatively little traffic in the other.

14.1.3.2 Telephony Interconnection

In the current approaches to PSTN interconnect, specific points of interconnect are identified on a bilateral basis with the necessary inter-working functions being established and agreed between the parties concerned. The key aspect to note of this approach is that individual calls are generally distinguishable across the interconnection. The information gathered at the interconnection point therefore allows per-call or aggregated call billing between the interconnecting parties. In the case of VoIP, the IP peering interconnect arrangement can in theory be modified to encompass VoIP or associated real-time conversational services. However, there are really two options to be considered:

- one option is to consider each VoIP call and its associated signalling as a simple stream of packets;
- the other option is to engineer the solution to be able to associate the signalling and media packets across the point of interconnection.

The first of these is more closely aligned with the model of the public Internet. In contrast, the second option is more comparable with the existing PSTN model for network interconnection.

Beyond the aspects relating to how the VoIP traffic is passed over an interconnect, there are other aspects relating to service interworking that must be addressed. These include how the naming and addressing schemes are exposed across the interconnection. For example, there will be the need in the future to consider the type and structure of the naming scheme used since the current restrictions of a numeric naming scheme may give way to more user-friendly schemes in the future. For public services, however, there will be additional requirements such as the need to ensure that privacy and related requirements, including lawful interception, are also appropriately handled across the network interconnection.

14.1.3.3 Bilateral Interconnect Agreements

The oldest and simplest form of inter-domain traffic exchange is the use of a bilateral agreement between two service providers. Beginning in the earliest days of the telephone industry, bilateral agreements were necessary to hand-off traffic originating in one service provider's service area, and destined for termination in a geographic region served exclusively by another. Bilateral agreements have also become common within the VoIP industry, as service providers attempt to establish wider coverage for their IP-based voice services by gaining access to other service provider's VoIP gateways. In practice, bilateral agreements often appear to be quite desirable from the service provider's business perspective, but tend to become difficult to manage from a technical perspective. For example, a service provider implementing a single bilateral agreement is unlikely to have issues with the process due to the simplicity of the network involved. However, service providers attempting to manage more that one bilateral agreement, with overlapping service areas and dissimilar business terms, may experience operational difficulties when attempting to operate this model at scale.

An important issue in the implementation of bilateral agreements for Internet-based VoIP networks is that they tend to require technical implementations to follow a 'master/slave' model. This means that one service provider's network will usually give-up some portion of their operational autonomy, and possibly security, in order to function as a sub-network of the other partner. These types of relationships break down quickly when dealing with multiple network interconnections.

14.1.3.4 Clearinghouses

The primary alternative to implementing multiple bilateral agreements to achieve network service reach and scale is the clearinghouse model. The term 'clearinghouse' has been in existence for many years, and commercial facilities such as the central clearing function for credit and debit cards issued by financial institutions may be considered as a form of clearinghouse. Within the VoIP industry, however, the term has come to represent a very wide range of implementations that are concerned to a greater or lesser extent with route determination, authorisation, and charging for calls that need to be transported across domains of differing administrative policy, technology or ownership. This interpretation is assumed in this book. The term 'settlement', when used in the clearinghouse context, is the process of determining the value exchange between two or more service providers. A 'settled' call is one in which all available call detail records (CDRs) have been accounted for, verified, and normalised with appropriate wholesale costing (and/or retail pricing) elements being applied, according to the business agreements in place and the resulting values debited and credited appropriately to each service provider involved in a call session.

In its most simplistic form, a clearinghouse may be realised by as little as a post-processing solution for accounting resolution between the stakeholders concerned using CDR information provided by the underlying network technology. However, in a typical complex implementation, a clearinghouse can function as a trusted third-party service provider, handling policy administration for call and/or caller authorisation, pricing, quality of service, transit network determination and end-point address resolution. The clearinghouse may also act in a fiduciary role as the authoritative source of billable call detail records.

14.2 Approaches to Clearinghouses

There are essentially two models for clearinghouse operation, depending upon whether or not the clearinghouse function is directly involved in the call-processing path. In the simplest case, the clearinghouse exists as an off-line system that resolves call detail records generated within the end VoIP gateway systems. These are described in more detail below.

14.2.1 Off-Line Clearing

The essential principle behind off-line clearing is to collate historical billing-related data from a network as an aggregate. Once this has been priced accordingly, it can then be reconciled against the various originating or terminating domains concerned and an appropriate financial statement created. Although it is not largely advertised as such, Azure Solutions' INCA application [2] is an example of an off-line clearing approach which is used extensively between some of the largest communications companies in the world.

14.2.2 Real-Time Clearing

In contrast to the off-line clearing model, real-time clearing requires each call to be passed as a transaction in real-time from the originating domain to the clearinghouse to obtain the necessary credentials to complete the call. This may include pricing, charging, rating and onward routing information. As a consequence, it enables the clearinghouse to not only provide dynamic price-based routing but also participate in traffic management.

One advantage of the real-time clearing approach is that it enables the charges for a call to be made immediately available, usually via a Web page, once the call has cleared.

14.3 Clearinghouse and Billing-Related Protocols

There are a number of industry-accepted protocols that can assist in developing robust clearinghouse implementations. These protocols tend to divide themselves into three distinct categories:

- billing protocols — which always provide usage information, and may also provide other service metrics, including authentication and per user authorisation;
- location protocols — which provide network addresses for distant end-points;
- clearinghouse protocols, which aggregate billing and location functions along with other functions such as service pricing.

Table 14.1 illustrates the basic functionality of some of these protocols.

Table 14.1 Protocols for clearinghouse functions.

Protocol	Function						
	Authenticate	Authorise	Service location	Pricing	Token	Metrics	Usage (CDR)
Billing protocols							
RADIUS	**Y**	**Y**	N	N	N	N	**Y**
Diameter	**Y**	**Y**	N	N	N	N	**Y**
IPDR	N	N	N	N	N	**Y**	**Y**
Location protocols							
LRQ	N	N	**Y**	N	N	N	N
Trip	**Y**	N	**Y**	N	N	N	N
Clearinghouse protocols							
Annex G	N	**Y**	**Y**	**O**	**Y**	N	**Y**
OSP v1	N	**Y**	**Y**	**O**	**Y**	N	**Y**
OSP v2	**O**	**Y**	**Y**	**O**	**Y**	**Y** (Qos)	**Y**

Y = Yes N = No **O** = Optional

14.3.1 Billing Protocols

The advent of widespread dial-in access to the Internet in the early 1990s, led to the need for a network-based capability to provide customers with user name and password authentication. In addition, since at that time the network elements required for dial-in access, such as modem ports and telephone network connections, were expensive and scarce resources, it became a business imperative to include a method for accounting for 'minutes of usage' for on-line access. An early leader in this type of application was Livingston Networks, later acquired by Lucent Technologies, whose candidate protocol — RADIUS (remote authentication dial-in user service) was used as the basis for development of the protocol approved in 1997 as RFC 2058 in the AAA working group of the IETF. This was later superseded by RFC 2138 and RFC 2865.

14.3.1.1 Remote Authentication Dial-In User Service

RADIUS is defined in RFC 2865 [3] and provides the ability to manage a database of users, and allows user name and password authentication. RADIUS can also provide service configuration information for each user, which is passed to the specific network element providing access to network services. As shown in Fig 14.2, RADIUS operates in a client/server model, where the gatekeeper or VoIP gateway operates as a client of the RADIUS server. RADIUS servers can also act as proxy clients to other RADIUS servers or other kinds of authentication servers.

A key benefit to VoIP applications is the extensibility built in to RADIUS, since all RADIUS transactions are comprised of variable length 3-tuples — attribute, length and value. New vendor-specific attribute (VSA) values can be added without disturbing existing implementations of the protocol. This capability has been widely used as an interface between VoIP network elements, usually media gateways, and back-end servers, or billing systems providing implementation-specific authorisation services, such as for pre-paid calling cards. In these cases, VSAs are used to carry user identification, account balance information, call rating, and other related information. This may be required to trigger appropriate interactive voice response (IVR) interaction, as well as generate CDRs.

A secondary use of RADIUS is to provide an audit trail for both wholesale and retail billing reconciliation. While many first generation VoIP gateways were PC-chassis based, and contained hard disk drives suitable for locally archiving call detail records, most of the high-performance media gateways in the current market do not have this capability. RADIUS accounting is often configured to provide a complete record of all originating and terminating call attempts, successful or not, regardless of their status as 'billable'. In most implementations, the RADIUS archive server will need to be located in close proximity to the VoIP equipment which it serves, as any loss of IP network connectivity between the media gateway

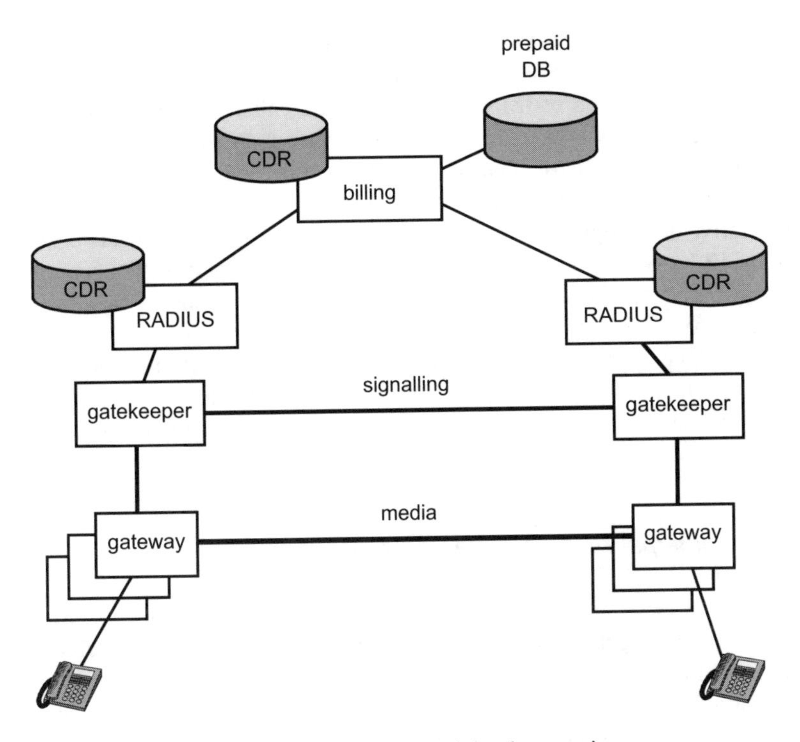

Fig 14.2 RADIUS network implementation.

and the RADIUS server can result in unreconciled, and ultimately un-billable, call minutes.

14.3.1.2 Diameter

Several IETF Working Groups, including AAA, ROAMOPS and MobileIP, have started work on the authentication, authorisation and accounting requirements for the next generation of network access servers. The proposed Diameter protocol [4], so called because it is the next generation protocol beyond RADIUS, addresses several known deficiencies in the RADIUS protocol. It is primarily intended for use within the NAS, roaming and mobile IP application arenas, but will also find a role within converged access and VoIP service networks as RADIUS has done. It is interesting to note that Diameter was used as the basis for some of the gateway control protocols discussed in Chapter 7.

An assessment made by the ROAMOPS Working Group in the IETF investigated whether RADIUS could be used in the context of a roaming network. This concluded that RADIUS was ill-suited for inter-domain purposes, because the implicit requirement that each ISP set up roaming agreements with all other ISPs did not scale. The Working Group therefore defined a 'broker', which acts as an intermediate server, whose sole purpose is to set up these roaming agreements. In this context, a collection of ISPs and a broker are called a 'roaming consortium'. There are many such brokers in existence today, many of which also provide clearinghouse financial settlement services for member ISPs.

The MobileIP Working Group has recently changed its focus to inter-administrative domain mobility, which is a requirement for cellular carriers wishing to deploy IP-based mobility protocols. The current cellular carriers' requirements are very similar to the ROAMOPS model, with the exception that the access protocol is MobileIP instead of PPP.

Since the Diameter proposal's capabilities are a super-set of the widely deployed RADIUS protocol, backwards compatibility issues should be minimised. It is therefore expected that Diameter will replace RADIUS for complex IP networks in the future.

14.3.1.3 Internet Protocol Detail Records (IPDRs)

Call detail records are the conventional unit of currency within traditional telecommunication networks. They are used to record information about usage activity within the telecommunications infrastructure, such as call completion. Recognising the role that CDRs have in helping to manage and charge for the use of network resources, the Internet Protocol Detail Record (IPDR) industry group has developed a suite of specifications centred around the network data management usage (NDM-U) specification [5] to extend this paradigm into the IP environment.

The IPDR specification is intended to address all potentially billable layers of the OSI stack [6]. It therefore aims to address the needs of a wide range of applications including VoIP [7], electronic mail, application services, and mobile access. It also aims to provide metrics useful in monitoring and managing IP-based services, such as quality of service, bandwidth, and latency.

A primary benefit of the IPDR approach to VoIP solutions is that it may be independent of VoIP network elements, and is often implemented using stand-alone network probes which function by observing the signalling and media flows on an IP link [7].

14.3.2 Location Protocols

As VoIP implementations moved out of their initial enterprise deployments and into larger corporate and service provider network implementations, the ability of a single network element to know or discover all of the other nodes within a network was found to be very limited. Most complex networks required every node to be manually provisioned with its location and with the location of every other node with which services were to be exchanged. In the most extreme cases, this required an FTP transfer of new city code or country code-based routing tables to each of several hundred globally dispersed VoIP gateways, each time that a route preference changed or new network nodes were added.

In both the H.323 and SIP environments, a second phase of protocol development took place, resulting in a 'location' function being added through which distant end-points could be discovered. In the case of H.323 v2, the protocol extension took the form of the location request (LRQ) message of H.225.0 (RAS), while it is intended that SIP-based networks will use the telephony routing over IP (TRIP) [8] which has been developed in the IETF's IPTel Working Group.

14.3.2.1 H.225.0 Location Request (LRQ) Message

The H.323 architecture described in Chapter 10 provides a mechanism, through the RAS LRQ message defined in H.225.0 [9], for a gatekeeper in one domain to query routing information from another. The use of this message follows the general H.225.0 RAS message syntax of a query associated with either a confirmation that the action has occurred, as illustrated in Fig 13.16, in Chapter 13, or a rejection of the request. While the gatekeeper receiving the query is able to provide routing information for the calling client using the LRQ message set, it has the limitation of only offering a simple routing query/response mechanism. The initial location request has space for a string of possible names for the called party, including e-mail and E.164 style names, along with the opportunity to include some proprietary information. However, as the LRQ mechanism is based on a simple transactional model, there is no opportunity to pass further information outside the initial request. A gatekeeper that is queried using the LRQ mechanism is therefore unable to apply subsequent call treatment to a call once it has responded, since it is not able to have any further interaction after the response to the query transaction has been sent. However, as indicated in Chapter 13, it is still possible to build some large and complex solutions even allowing for the restrictions of this simple mechanism.

14.3.2.2 Telephony Routing over IP (TRIP)

TRIP is an inter-domain gateway location and routing protocol modelled after the border gateway protocol 4 (BGP4). TRIP location servers exchange information with other location servers by multicast DISCOVER requests using BGP-like inter-domain transport mechanisms and formats. The information exchanged includes telephony destinations, routes towards these destinations, and information about gateways serving those telephony destinations residing in the PSTN.

TRIP can be used to exchange attributes between network elements, necessary for other BES applications to enforce policies and to select routes based on path or gateway characteristics, but the TRIP protocol does not define a specific route-selection algorithm. TRIP's operation is independent of any particular VoIP signalling protocol, and can be used as the routing mechanism for any protocol, including H.323 or SIP.

As a routing-only protocol, TRIP may be most often utilised where inter-domain clearing is not required, since there are no inherent mechanisms for admission control or CDR collection. These may be added by combining TRIP with other protocols, such as IPDR for the billing information. Additional work is proceeding within the IETF to utilise TRIP as the intra-domain SIP location server — SIP gateway discovery protocol.

14.3.3 Clearinghouse Protocols

Initial VoIP clearinghouse implementations were developed using proprietary protocols developed by either an equipment vendor, such as Clarent Corporation, or service providers, such as AT&T. As the industry has expanded, these solutions have been proved to be too restrictive since a clearinghouse needs to work with as wide a range of VoIP gateway and call server equipment as possible to avoid expensive one-off engineering for each new client of the clearinghouse. The early proprietary solutions have therefore given way to ones based on internationally agreed standards, such as the ITU-T H.225.0 Annex G solution for H.323 networks, or the ETSI TIPHON open settlement protocol (OSP) which is designed to work for any underlying VoIP network technology.

14.3.3.1 Proprietary — Clarent® Connect

The first clearinghouse solutions to appear on the market were developed as either bespoke applications or proprietary extensions to existing VoIP products. The Clarent Connect solution from Clarent Corporation is one such example of this generation of technology, and has been implemented by many customers, including the Concert, NTT, and Telia clearinghouses.

In order to add the Clarent clearinghouse capability to either a Clarent-based VoIP network, or to a stand-alone Clarent clearinghouse, an additional network element, the connect proxy server, is implemented in conjunction with a Clarent Command Center™ and Database (see Fig 14.3).

Clarent Connect is an IP-based network element equivalent to a gatekeeper proxy (or border element) that validates partner networks, and gives them call routing instructions based on location information within the database associated with the proxy server. Clarent Connect allows a VoIP network session to extend through multiple Clarent-based networks for greater coverage. Since each proxy server is a

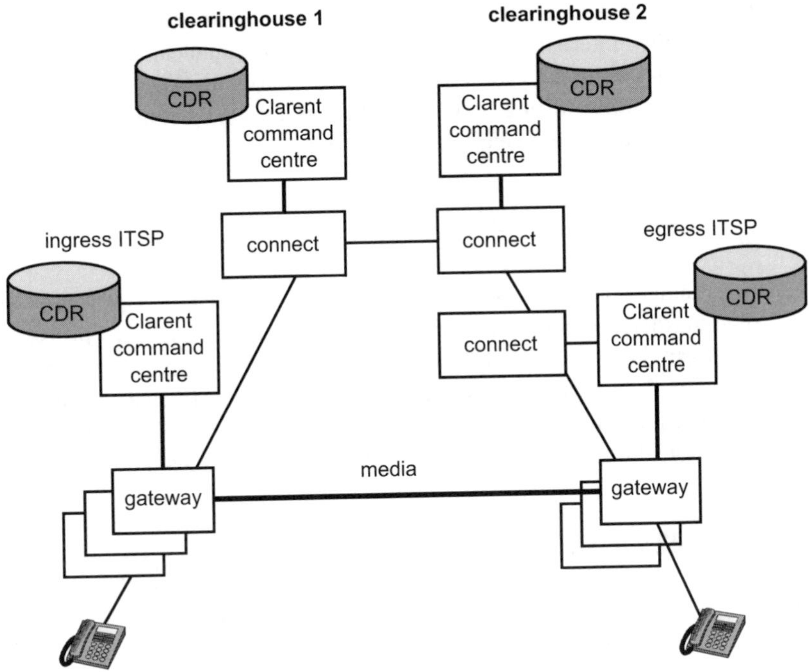

Fig 14.3 Clarent® Connect implementation.

state machine, all participating networks share call detail record information for every call transiting their network.

In a clearinghouse call, callers from the ingress network enter the telephone numbers they wish to call. The originating Clarent gateway checks the associated Clarent Command Center for a route. If the route is to a location outside the originating domain, the Command Center returns a route to the gateway which points to a connect proxy in another network. Upon receipt of the gateway route request, the neighbouring network connect proxy will query its own command centre for routing. The command centre will return either a local egress route for a terminating gateway, or, in the case of a multi-hop scenario, the address of another network's connect proxy, which in turn requests a local egress termination from its associated command centre. After call completion, billing records are created for each participating network, and the clearinghouse and partner networks utilise billing records to settle partner accounts in near real time.

14.3.3.2 H.225.0 Annex G

The use of H.225.0 Annex G was initially championed in 1999 by VoIP carrier ITXC, as a part of its 'iNOW!' initiative [10] to accelerate the deployment of interoperable inter-domain H.323 networks. In June of 2000, iNOW! was formally transferred to the International Multimedia Telecommunications Consortium (IMTC), which has continued to support development of iNOW! profiles, and to promote interoperability testing at forums such as their SuperOp event. H.225.0 Annex G is described in more detail in section 14.4.

14.3.3.3 Open Settlement Protocol (OSP)

The open settlement protocol was originally proposed by Thomas of TransNexus [11]. OSP is further described in section 14.5 below and is currently in use by a number of clearinghouse operators. United Telesis' Revolution global clearinghouse is one such example and was the first recorded deployment of Lucent Technologies' OSP-based Lucent Multivoice Settlement Engine. OSP has seen a growth in support by a number of leading manufacturers in addition to the case cited above.

14.4 H.225.0 Annex G

The initial H.323 architecture, as described in Chapter 10, provides a framework for delivering multimedia services within a single administrative domain. In reality, it is unlikely that large-scale networks can be created on this basis as it will be necessary to provide some form of interconnection between networks without having to resort to dropping calls back into a TDM network — which is largely the case with current implementations. The interconnection of H.323 networks at the IP level is therefore of major importance for the future of VoIP networks. However, as discussed in Chapter 12, this type of interconnection must satisfy a number of requirements including the authorisation of the call and associated media streams in addition to any service features that are to be invoked. This creates a number of points in the architecture where the interconnection needs to, or may, be created. These occur between:

- network elements constituting the point of interconnection between two networks (reference point A);
- a network element constituting a point of interconnection and an associated gatekeeper within the same domain (reference point B);
- two network gatekeepers within the same domain (reference point C);
- a network interconnection element and a surrounding service domain (reference point D).

H.225.0 Annex G [12] addresses only the first of these although its use in the context of reference point C is currently for further study. As it is assumed that network elements used for creating a network interconnection will require access to gatekeeper functionality, it is also assumed these functions will be collocated. By implication reference point B will therefore most likely remain internal to a specific device implementation.

14.4.1 Relationship to Other Standards

H.225.0 Annex G is part of the H.323 framework of specifications for providing multimedia communications over packet-based networks as described in Chapter 10. Since the core H.323 specification does not concern itself with the minute details of constructing large H.323-based networks that involve the interconnection of multiple administrative domains, H.225.0 Annex G has been developed to address the issues involved in these types of network scenario. Since multi-domain operation raises a large number of issues, Annex G differs from most other Annexes of H.225.0 by being published as a separately maintained document by the ITU-T SG16.

14.4.2 Development of Annex G

The ITU-T SG16 completed the first version of H.225.0 Annex G in December 1998, with it reaching approved status in May of the following year. A number of issues were identified with the first release of the recommendation that has led to proposals for the development of a new version. At the time of writing, it is expected that H.225.0 Annex G version 2 will be published in February 2002 and will consider performance optimisation of version 1 of the protocol and also the relationship between Annex G and the TRIP protocol being developed in the IETF. This chapter therefore relates to H.225.0 Annex G version 1 unless otherwise explicitly stated.

14.4.3 Functions Provided by H.225.0 Annex G

The core H.323 specification allows calls to be handled in an environment that includes multiple gatekeepers. However, it was never intended to provide a complete solution for routing calls efficiently in an environment that incorporates network elements deployed across multiple administrative domains. So while it can support calls in this type of environment it is not necessarily an efficient and scalable solution for achieving this goal. The basic H.323 functionality can therefore be more readily likened to the user network interface (UNI) of a switched-circuit network.

To address these concerns, H.225.0 Annex G provides functionality supporting address resolution, access authorisation and usage reporting between administrative domains for H.323 systems. It therefore reflects the type of functionality more commonly associated with network-to-network interfaces (NNI) found in switched-circuit networks.

14.4.4 Annex G Architecture

The H.225.0 Annex G solution extends the basic H.323 architecture, described in Chapter 10, by introducing back-end services (BES) and border elements (BEs), as shown in Fig 14.4. However, as H.225.0 Annex G specifically relates to the protocol across reference point B, it concerns the exchange of information between border elements rather than any specific interaction with the BES elements themselves.

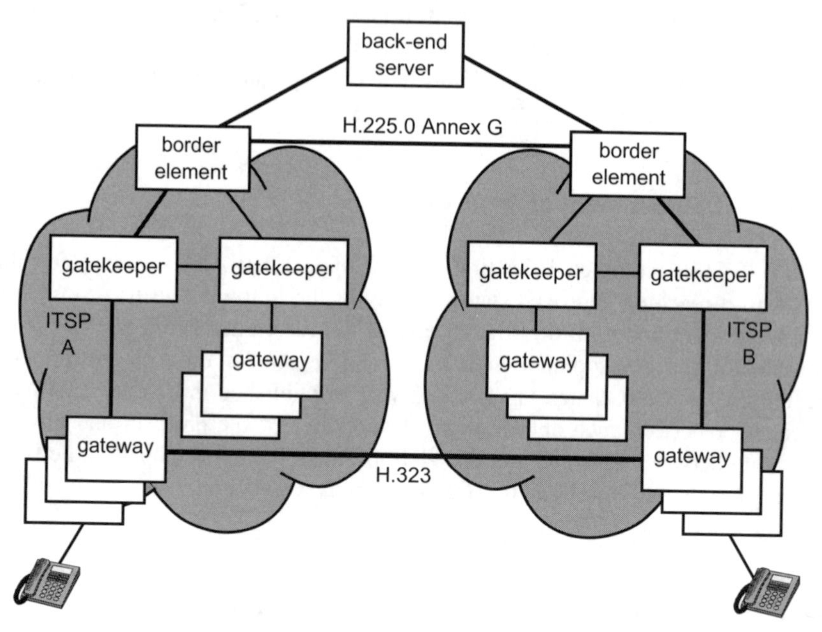

Fig 14.4 Annex G network architecture.

The core functionality provided by Annex G operates on the exchange and use of addressing and routing information held in a format known as an 'address template'. An address template comprises:

- a pattern which consists of an H.225.0 AliasAddress or range of AliasAddresses;
- RouteInfo which consists of a set of routing related information, including usage and pricing information associated with this template;
- TimeToLive that indicates the duration for which the template is valid.

Address templates may be grouped together into an aggregation known as a 'descriptor'. These entities constitute the unit of currency for the expression of routing information exchanged by the H.225.0 Annex G protocol. In this respect, they offer similar functionality to the 'routes' concept used in TRIP.

A border element will maintain templates for all the routes for which it is responsible. These templates may be either statically or dynamically configured on to a specific BE and the H.225.0 Annex G protocol provides functions that enable these descriptors to be managed. It also provides functionality through the DescriptorRequest message which enables one BE to discover the routing information available from another.

14.4.5 Network Implementation

In a typical configuration, shown in Fig 14.5, the H.225.0 Annex G protocol is exchanged between individual border elements via a clearinghouse BE that acts as a proxy between the two networks concerned.

In the case shown, normal H.323 gatekeeper signalling occurs between the originating and terminating end-points. It has also been assumed that the BE functionality within the calling and called domains has been collocated with a gatekeeper function.

14.4.6 IMTC iNow! Profiles

While the H.225.0 Annex G protocol offers an effective solution to interconnecting border elements at the reference point B previously discussed, it is insufficient for achieving a complete practical implementation.

Among other things, to achieve a workable solution, the semantics of the information transferred in the information flows that enable a given call flow scenario need to be clarified. This became the focus of the work undertaken by the iNOW! working group in the IMTC who have produced an interoperability profile [10] based upon the use of H.225.0 Annex G.

14.5 Open Settlement Protocol

As the clearinghouse model can be applied to a wide range of network-based services, there is the opportunity for a generic protocol to be designed to support it. The open settlement protocol, described in this section, has emerged as one candidate solution for such a protocol.

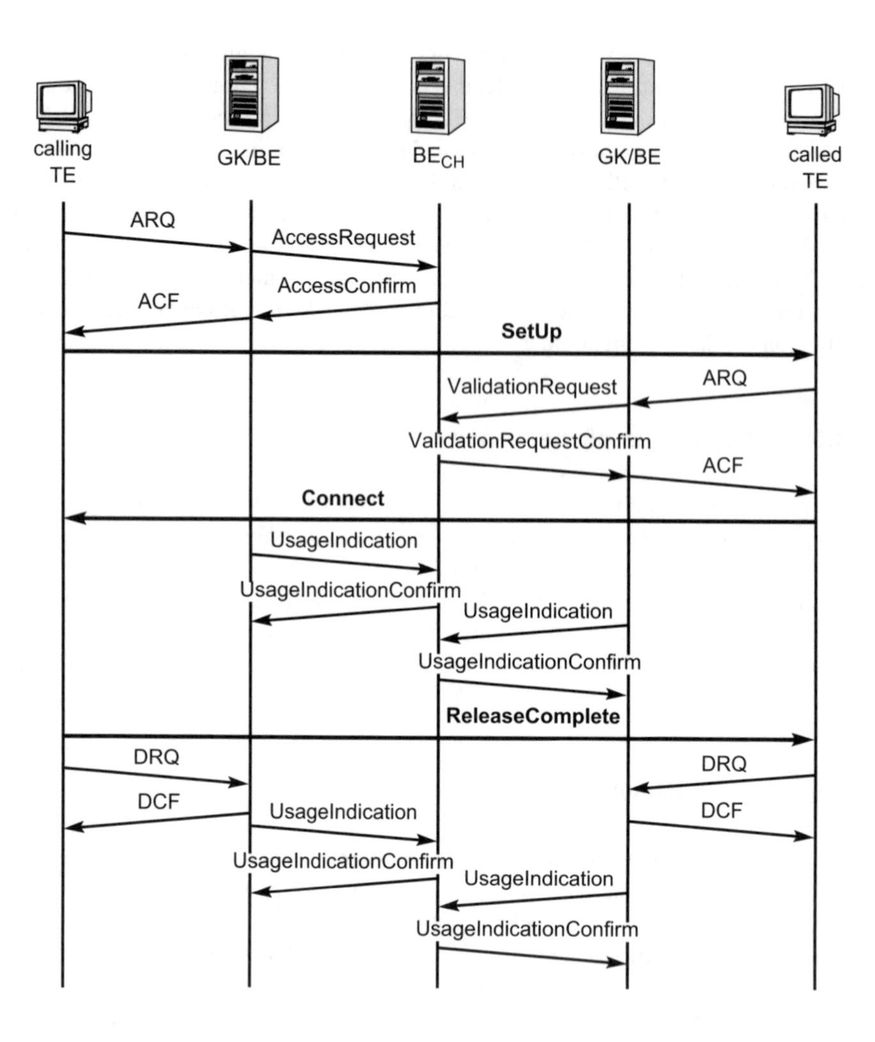

Fig 14.5 Annex G message exchange.

14.5.1 Relationship to Other Standards

Unlike H.225.0 Annex G, the open settlement protocol is an XML-based protocol that is independent of the specific VoIP protocol used. It can therefore be more easily deployed in a heterogeneous network deploying protocols in addition to, or instead of, H.323.

14.5.2 Development of OSP

In late 1997 the ETSI project TIPHON [13], which is further described in Chapters 5 and 16, began to consider issues involved in authorising, completing and obtaining billing records for inter-domain VoIP calls. Formal protocol development work began in early 1998 and with the co-operation of a number of ETSI members, including service providers, equipment vendors, and clearinghouses, a consensus was built around a protocol structure originally submitted by Thomas [11] who was then of TransNexus LLC. Version 1 of this protocol, later known as the open settlement protocol (OSP), was approved and published in December 1998 as ETSI TS 101 321 v1.4.2 (1998-12). In July 2000, OSP version 2 was published as TS 101 321 v2.1.1 (2000-07), which included a number of additional features, including the ability to provide subscriber authentication for VoIP roaming, and enhanced usage elements for reporting QoS metrics [14].

14.5.3 Functions Provided by OSP

OSP provides a method of obtaining inter-domain authorisation, route determination, and, optionally, pricing. It also includes functionality that enables the exchange of indications such as call detail records. OSP may be flexibly implemented to allow the use of any or all of its functions, which include:

- service authorisation;
- user authorisation (OSP v2);
- location determination;
- token issuance;
- pricing indication;
- QoS metrics (OSP v2);
- usage indication.

It is important to recognise that OSP is independent of any VoIP signalling or media-stream protocol, and has been demonstrated to work with a variety of implementations, including H.323, SIP and proprietary VoIP and FoIP (fax over IP) architectures.

14.5.4 OSP Architecture

The OSP protocol is an XML implementation, transported using HTTP [15], the standard protocol for Web-based communications. The OSP message is a MIME [16] message containing OSP primitives expressed in XML [17]. If optional SSL security is desired, the OSP message signature is included in the MIME multipart message transported using HTTPS. To achieve non-repudiation, usage information may be conveyed using a secure MIME (S/MIME) digital signature. Figure 14.6 shows the OSP protocol stack.

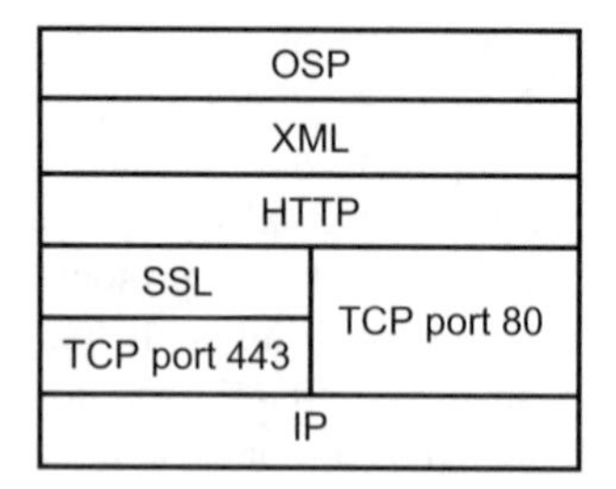

Fig 14.6 ETSI OSP protocol stack.

A key feature of OSP is the ability to deliver signed tokens for authorisation, call progress, and call completion. This capability allows 'blind' admission of a VoIP session by a terminating service provider, based on the trust relationship with the token issuer, whether this is a bilateral partner or clearinghouse.

14.5.4.1 Network Implementation

In a typical OSP network implementation (Fig 14.7), the OSP server, which implements the route determination policy, token generation, and CDR collection,

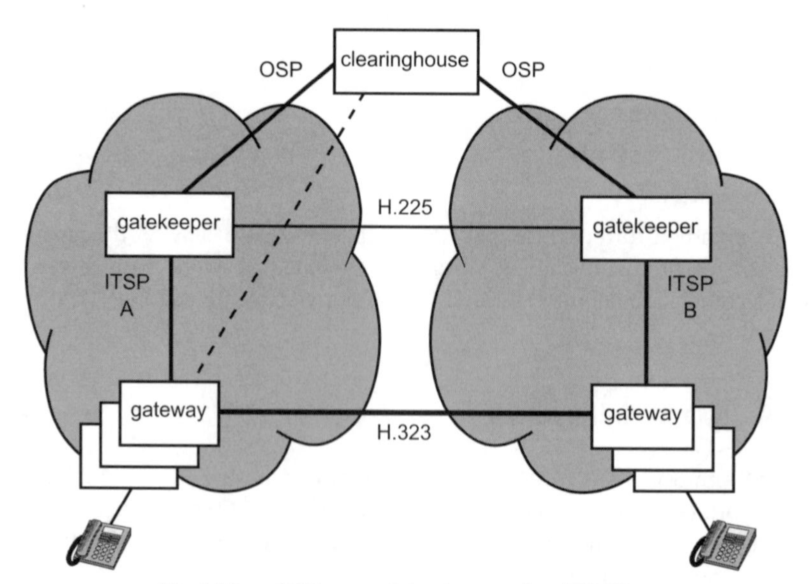

Fig 14.7 OSP network implementation (H.323).

will be located in a BES domain. The BES may be collocated within either the originating or terminating ITSP domain that provides VoIP services. This will be typical for a bilateral traffic agreement. Alternatively it may be implemented as a neutral 3rd-party service outside the ITSP networks concerned, such as in the clearinghouse service model.

Within an ITSP network, there may be one or more 'zones' of VoIP service, each managed by separate network elements, such as an H.323 gatekeeper or SIP proxy server. The specific implementation details of how each zone utilises OSP and/or other protocols such as TRIP or H.225.0 Annex G to obtain distant zone information, and from which network element the OSP query is launched, are left to the network architect.

OSP implementations are executed in a wide range of topologies, extending from multiple-node gateway-only implementations (running 'gatekeeperless'), to super-domains that are made up of concatenated sub-domains, implementing equipment from several vendors. OSP may be utilised to find only unknown distant termination locations, or to identify terminations within bilateral partner networks, or even within a single network to map terminations between sub-domains or unmanaged gateways.

14.5.4.2 Message Exchange

Prior to processing any call-related messages, any network element implementing an OSP client must be provisioned with the addresses of the appropriate OSP server(s), and must be issued with a digital crypto-certificate from the root certificate authority. The procedure is as described below.

- A call begins when a VoIP terminal or gateway requests authorisation to place a call to a destination number. In the H.323 example shown in Fig 14.8, an ARQ (admissionrequest) is sent to the gatekeeper managing the local ingress zone. If the gatekeeper cannot provide an appropriate termination location for the dialled number, an OSP client capability is invoked.
- When the OSP client is triggered to query for authorisation, it sends an OSP AuthorisationRequest message, containing a timestamp, unique call identifier, the source (calling number), requested destination (called number), service type, and the number of candidate destinations it wishes to receive.
- The OSP server replies with an AuthorisationResponse message, containing a timestamp, transaction identifier, crypto-token, an ordered list of candidate destinations, validity duration, and the original call identifier.
- The originating gatekeeper will then attempt a call set-up to the 1st candidate destination (followed if needed by the 2nd, 3rd, etc), using the crypto-token to identify this call session to the terminating network as an authorised transaction.

- Upon acceptance of the token, the call proceeds until released or otherwise ended.
- Both the originating and terminating network elements forward UsageIndication messages to the OSP server associated with the authorised token, and the OSP server replies with UsageConfirmation messages. The UsageIndication message contains the necessary information elements to construct a CDR.

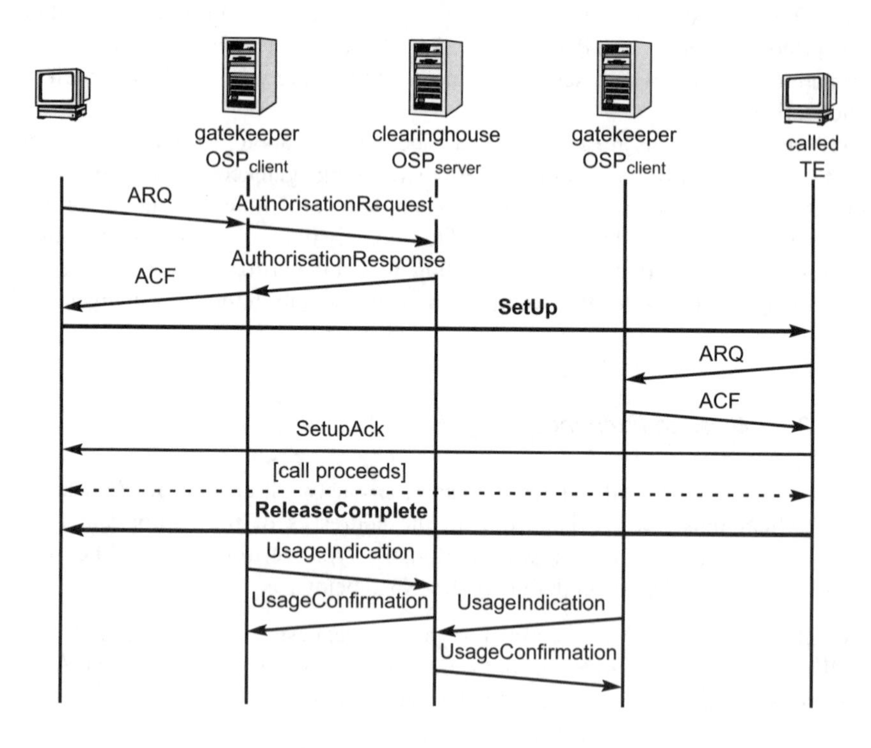

Fig 14.8 OSP message exchange.

14.5.5 Implementation

In a clearinghouse scenario, upon completion of the message exchange for a specific call (or group of calls) and receipt of the usage information, the CDR collection, normalisation, rating, and billing functions, collectively referred to as 'clearing', take place. In this process, all record tuples would be reviewed for accuracy, completeness, missing or mismatched records, and error conditions, including abnormal disconnects or zero-duration calls. Mismatched or missing half-call records are normalised according to the business agreements in place between the service providers, and the resulting valid CDRs may have call cost ratings applied to convert them from minute transactions to value transactions. The call values are typically expressed in either currency units (e.g. Euro (€) or USD ($)), or in weighted 'units'. The clearinghouse will then debit the originating ITSP account, and credit the terminating ITSP account, and make available a bulk record of the normalised and rated CDRs. If the clearing process is achieved in near real-time, a Web-based interface may be generated, often with the addition of retail pricing factors, to allow authorised customers to view their call records on-line.

14.5.6 ETSI TIPHON Profiles

Towards the end of 2001, new Work Items were created within ETSI-TIPHON Working Group 6 to develop OSP test suites and a protocol implementation extra information for testing (PIXIT) pro-forma. These documents, as well as the protocol implementation conformance specification (PICS), contained within the base OSP specification (ETSI-TIPHON TS 101 321 Annex J), provide useful tools for both equipment developers and network architects to compare implementation parameters, and to determine specific interoperability profiles. Careful attention to these implementation details has resulted in the ability to create multi-vendor, inter-domain VoIP traffic models.

14.6 Core Design Issues

In the implementation of the clearinghouse model, or any other inter-domain VoIP business model, there are several areas of design that must be carefully thought through well in advance. Three of the design aspects specific to the VoIP clearinghouse application are security [18], performance, and accuracy.

14.6.1 Security

A primary issue, which must be agreed by all parties involved in an inter-domain VoIP traffic exchange, is determining the level of security desired, and the specific security implementations to be used. Specific areas of concern include:

- source of the root certificate authority;
- use and implementation of SSL or TLS;
- use of signed tokens for admission control;
- use of S/MIME for messages (especially call details).

14.6.1.1 Multiple Root Certificate Authorities

Root certificate authorities (CAs) may be required to issue crypto-certificates for each of several security applications, including the use of optional SSL or TLS message encryption, and separate certificates are usually required for each BES serving as a source of authorisation tokens. This has been shown to cause conflict with the design implementations of certain network elements, which may allow only a limited number of root authorities to be registered on a device. Most vendors are aware of this limitation, and are working to expand the number of simultaneous active certificate authorities.

Since the certificate authority is most often associated with the source of fiduciary trust, in multiple bilateral network arrangements there may be business issues involved in effectively determining which partner's CA shall act as the root. While clearinghouses have tended to act as the CA for members of their clearing consortium, there are often issues due to ITSP membership in other bilateral or clearinghouse relationships.

14.6.1.2 SSL and TLS

The need for encryption technology on clearinghouse messages is highly dependent on the network topology over which the messages are being transmitted. In a highly localised, multi-zone ITSP domain, intra-domain messages may be considered to be within a trusted network, and less susceptible to external threats. In these cases, SSL or TLS encryption might not be implemented for communications between network elements and intra-domain BES components.

In the traditional inter-domain clearinghouse model, however, the need for encryption on clearinghouse messages is evident, since it would be highly undesirable to allow sensitive information such as individual user's calling and called numbers to flow across public or other unsecured IP network connections in plain text.

However, implementation of encryption technologies such as SSL carry with them inherent additional complexity, and, more importantly, a severe performance impact due to the reallocation of scarce hardware and software resources to cryptographic processing.

14.6.1.3 Signed Crypto-Tokens

Where VoIP networks are tightly coupled and 'well-known' to each other, the use of signed crypto-tokens may not be required for admission control. This is certainly the case for sub-domains within an ITSP's extended network, and may also hold true for some bilateral or clearinghouse relationships. The extent of actual implementation of this capability is often constrained primarily by the ability of network elements to originate and accept calls with a mixture of both signed tokens 'off' and signed tokens 'on'.

As with SSL, the additional resources required for cryptographic processing, both creation and processing of crypto-tokens, can tax the available processing power of smaller VoIP network elements. Embedded processor devices are particularly susceptible to these impacts, and a typical network implementation might avoid this problem by off-loading the OSP interface to a larger, server-based, zone-management node.

14.6.1.4 S/MIME

The use of an S/MIME [19] digital signature is optional on all OSP messages, and has primarily been used to create a UsageIndication message (CDR) with non-repudiation characteristics. Once again, the processing overhead to generate digital signatures for large numbers of messages is considerable, and two major strategies have emerged to de-load network elements from this burden; limiting S/MIME to only the UsageIndication messages, and grouping multiple UsageIndication messages into time or quantity blocks, such as 5 minutes or 100 records.

14.6.2 Performance

Clearinghouse implementations must be deployed with dimensioning and topology suited to handling the traffic being carried on the underlying VoIP ingress and egress networks. The primary traffic measurements that characterise the load on BES or clearinghouse servers are:

- busy-hour call attempts (BHCA);
- maximum simultaneous call attempts;
- total call volume (quantity of completed and incomplete calls).

Back-end or clearinghouse servers must be sized for each of four VoIP call phases, including:

- set-up phase — authorisation, route determination, token issuance;
- call proceeding phase — call state monitoring (if required);
- call ended phase — delivery of CDR;
- post call processing phase — CDR rating and settlement.

14.6.2.1 Message Throughput

The proportion of total busy-hour call attempts (BHCA) in the originating ITSP networks which are requesting inter-domain service will give an accurate indication of the number of sustained BES AuthorisationRequests (or AccessRequests) which will be handled by the clearinghouse or BES.

The maximum simultaneous call attempts is usually a percentage of the total number of available ingress ports, de-rated by the ability of the originating gateways to process multiple simultaneous PSTN-IP signalling conversions. This calculation provides an approximation to the maximum rate of AuthorisationRequests (or AccessRequests) which need to be processed by the clearinghouse or BES server(s).

In addition to the actual offered load volume, the level of security required for each transaction often has the most impact on BES capacity. The performance of an internal BES server running without SSL and without signed crypto-tokens, may be several times greater than the same processing platform implemented as an inter-domain clearinghouse, and using full SSL and signed tokens.

The total call volume, represented as the total quantity of both completed and incomplete calls, is a direct indication of the volume of UsageIndication messages that can be expected from a given implementation. In fact, many implementations will collect usage from both the ingress and egress network nodes, and the number of records will be at least double the number of calls.

14.6.2.2 Crypto-Acceleration

A primary method of off-loading server-processing platforms from the burden of handling various cryptographic tasks, including SSL and signed authorisation tokens, is to enhance the server with adjunct crypto-processing capability, optimised for acceleration of the public-key cryptographic functions of SSL, TLS and other security protocols. Generally implemented on a BES processing platform as one or more add-in board-level products (e.g. PCI or C-PCI), these devices off-load and speed up the CPU-intensive secure operations that would otherwise saturate the server's main processor. In high-throughput implementations, server response time can be expected to improve by as much as ten times.

14.6.2.3 Distributed Processing

In order to ensure sufficient network capacity and non-stop availability, the clearinghouse application is usually distributed across globally dispersed servers. A robust implementation of BES components will usually involve multiple fault-tolerant processing and storage nodes, located in strategically placed geographic points-of-presence, each provided with more than one independent IP access facility. Careful consideration should be given to the need for fault-tolerant and or hot-swappable CPU arrangements, as well as disk mirroring and local cache storage of processing records.

Software tools are often used to facilitate the even distribution of messages across the distributed clearinghouse servers, and these tools, originally designed to spread consumer traffic across multiple Web servers, are quite effective in performing the same function for HTTP-based inter-domain clearinghouse protocols. A typical implementation of this type of load-balancing software will provide automated policy-based clearinghouse services allocation, spreading regional high-volume requests over multiple nearby servers, as well as providing a disaster recovery capability by re-routing messages away from failed or overloaded BES nodes.

14.6.2.4 Transaction Throughput

Once messages have reached the BES clearinghouse platform, the most critical element for an effective implementation is the processing time required for the BES platform to complete the necessary transaction. The two primary instances of transaction demanding high throughput levels are:

- authorisation (location) requests;
- call detail record collection.

Authorisation requests are a contributor towards undesirable VoIP post-dial delay (PDD), and, given the real-time processing requirements of the policy rules-engine applied to a highly complex routing database, management of the average transaction time is crucial to achieving suitable network call-completion rates. BES platform design must both optimise the efficiency of the rules-engine code, and also ensure that a high-performance database solution is implemented.

Processing of collected UsageIndication messages into settled call detail records is the primary business of a clearinghouse, and with potentially millions of individual records involved, the burden of handling this volume must be specifically addressed. In many BES implementations, the actual BES function is divided into two separate server types — highly distributed servers performing authorisation and collection, and more centralised servers processing the resultant CDRs. In this environment, the task of uploading the cached CDRs from multiple distributed nodes to a central processing site(s) must be carefully engineered, to make sure that the periodic transfer of records is accomplished without overloading either of the servers, or the network between them.

The decision on how to provide settlement processing is usually dictated by the business model of the clearinghouse and its affiliated service providers. Processing can range from bulk periodic process runs (i.e. overnight or hourly), through near real-time processing of individual or small groups of records. Each of these methods has implementation-specific issues related to processing throughput. In the periodic process, the concern is usually the raw quantity of records, and the CPU capacity required to complete the settlement processing in a timely manner. For near real-time processing, there may be a substantial over-provisioning of CPU processing capacity required to handle peak busy-hour loads, as well as the burden of collecting multiple CDRs from disparate sources in a timely manner.

14.6.3 Accuracy

In order for a clearinghouse to uphold its fiduciary role as the authoritative source of billable CDRs, an accurate, auditable set of records must be first collected, then processed, and finally retained. Issues which can affect the ability of the BES to perform the settlement function accurately are: inaccurate call timing; incomplete records, or improper processing.

14.6.3.1 Call Timing

While it is obvious that correct call timing and accurate record generation are essential to an effective settlement process, unfortunately, in many cases, either the networks or the individual network elements tend to disagree on the exact timing of a call session. Discrepancies can be as minor as the difference between expressing call start in local time instead of GMT, or can be much more serious, such as inaccurate answer supervision on analogue terminations.

In the case of network implementations operating without end-to-end SS7 PSTN signalling, there is a strong likelihood that timing discrepancies may occur between the originating network's call 'start' and 'end', and the CDR reported at the egress VoIP gateway to the terminating PSTN. A typical scenario is that the VoIP network will report a 'zero duration' call attempt, while the PSTN network is billing the terminating service provider for a short call completion. These cases need to be both mitigated by careful network design, and carefully considered in the business arrangements between service providers.

14.6.3.2 Record Completeness

One of the most vexing operational problems with the clearinghouse process is the normalisation of missing, mismatched, or incomplete records. In most cases, a business policy or decision will have to be made to determine if certain record types should be treated as a *bona fide* billable call.

A typical example is the situation where only one of the half-call records is present. The business arrangements between the clearinghouse and service providers should provide a policy regarding the status of using either originating or terminating half-call records for settlement. In general, there is usually a higher incentive to bill a terminating call record, since, in many cases, the terminating VoIP service provider will be billed by the local PSTN operator for every network completion.

In many cases, additional audit information may be required to accurately determine whether groups of certain call types are valid. These records most often take the form of VoIP RADIUS records from an associated AAA server, or raw PSTN switch call records. An effective BES implementation may utilise statistical data analysis tools to provide bulk comparison between these dissimilar record formats.

14.6.3.3 Processing Accuracy

With the increased competitive pressure in today's telecommunications market-place, service providers will attempt to implement highly granular, non-traditional, and rapidly changing pricing schemes for their service offerings. It is the mission of the clearinghouse to accurately apply the myriad of pricing elements to the millions of transactions flowing across their interconnected networks. In addition to the expected time-of-day and day-of-week pricing, there are a seemingly infinite number of customised arrangements between carriers, such as bulk volume discounts, threshold discounts (i.e. up to 1M calls, up to 5M calls, etc), and guaranteed volume commitments. Complying with these requirements, and creating an audit trail through which it may be shown that the rate in effect at the call start timestamp was appropriately applied to the entire call record, is an imperative for a successful clearinghouse service.

14.7 Building Systems with OSP

A minimal OSP inter-domain clearinghouse could consist of only three components:

- an originating VoIP zone OSP client;
- a BES domain OSP clearinghouse server;
- a terminating VoIP zone OSP client.

Although clearinghouse networks that approximate this simplicity do exist, using small integrated VoIP gateways with OSP client interfaces, the more interesting cases utilise a mixture of network elements in more complex configurations.

14.7.1 H.323 Inter-Domain

The initial implementations of OSP clearinghouses have been focused on the exchange of telephony minutes between wholesale carriers. These early networks were, by necessity, vendor-specific sub-networks, limited by the ability of the network elements to interoperate across their media and signalling protocol implementations.

These limitations impose severe operational restrictions on the ability of clearinghouses to build effective least-cost-routing tables, and exchange traffic efficiently. However, as each pair of vendors demonstrates basic technology interoperability, the sub-network may be simply concatenated within the OSP server's routing tables, and the network can smoothly transition out of the sub-network model.

In setting-up an inter-domain implementation, the network elements containing OSP clients must be configured to communicate with the appropriate OSP server or servers, and issued a digital certificate from the CA server. At this point, the network is ready, but the BES is not.

In most BES clearinghouse implementations, each service-provider member is responsible for submitting the information necessary both to populate the destination location database, and to provide the input to the policy engine which creates a fully implemented least-cost route. Once the required information has been received (usually including at a minimum IP address, E.164 numbers served, and pricing elements for each E.164 numbering range), the clearinghouse BES creates the least-cost routing (LCR) tables, and is ready to respond to OSP requests.

Depending on the clearinghouse policy agreed with the service provider, the OSP server may offer one or multiple destination targets (most commonly three), and the originating network will use these addresses to attempt call set-up.

A minimal OSP inter-domain clearinghouse could consist of only three components:

- an Originating VoIP zone OSP client;
- a BES domain OSP Clearinghouse server;
- a terminating VoIP zone OSP client.

Although clearinghouse networks that approximate this simplicity do exist, using small integrated VoIP gateways with OSP client interfaces, the more interesting cases utilise a mixture of network elements in complex configurations.

14.7.2 Enterprise — Managed Network Services

A variation of the H.323 inter-domain model can be created by service providers offering VoIP services to enterprise customers, by using an OSP-enabled gateway at the customer's location. The enterprise gateway may be owned and managed by either the customer, or by the service provider, but in either case, the gateway is configured to communicate with an OSP server located in the BES domain of the service provider, or an associated clearinghouse.

Since enterprise gateways ranging from as few as two voice ports up to multiple T1/E1 capacities are available with OSP support built-in, enterprise networks of any size are able to target global VoIP destinations directly through the OSP location service, without the burden of maintaining global routing tables within the enterprise device.

When incremental VoIP capacity is obtained simply by adding DSP and telephony interface ports to an existing IP router platform already being provided as part of a managed data network offering, this allows the service provider to easily add a portfolio of voice services to their previously data-centric offering.

14.7.3 Asymmetric Inter-Domain InterOp

The stated goal of Inter-Domain InterOp efforts over the last several years has been to allow voice traffic to not only originate and terminate in different operational domains, but also to allow the use of network elements from multiple vendors, implemented in differing topologies.

By early 2001, it was possible to build inter-domain networks with elements from multiple OSP-compliant vendors, utilising a variety of network elements. These inter-domain building blocks include OSP-enabled integrated VoIP access gateways, traditional gateway/gatekeeper implementations, and monolithic switch/gateway/gatekeeper devices. Building on the work of InterOp events such as those hosted by ETSI, or the IMTC SuperOp, the base gateway interoperability achieved in these events has been extended into the interdomain space by utilising OSP as the common denominator between dissimilar networks.

14.8 Enhanced Applications

The ability to act as a server for OSP does not make a clearinghouse — the policy-based processing software behind the interface does. One of the intentional omissions of OSP was built-in constraints on either the technical or business implementations possible. The following represent a sampling of the extended capabilities possible.

14.8.1 Enhanced Applications of Clearinghouses

Having established a basic level of functionality for providing VoIP call clearing, it is possible to add more complex call-handling features and billing options. Three examples of such functionality are discussed below.

14.8.1.1 Least-Cost Routing

In a simple network with no service provider overlap, there would be one route, and one pricing policy for each dialled destination. In practice, however, there are always multiple service providers who are able to provide call completion from a calling user to a called location. Unfortunately, no one service provider ever has unlimited capacity to any specific destination, nor will one service provider ever have the best wholesale rate to every destination. This is the source of the complexity in building an effective least cost routing table.

In order to service OSP AuthorizationRequests, the BES complex must be able to take in rate and route changes (monthly, daily, or sometimes at will) from their associated service providers, and reformulate this information into an ordered list of potential destination locations, ranked by lowest cost first. If network capacity were unlimited this would be all that is required, but in many cases, the service provider with the lowest rate is also the provider with the least capacity, sometimes only a small fraction of the total required route sizing.

Rather than forcing the originating network to iteratively search down the ordered list of destinations in the AuthorizationResponse for an available high-capacity termination at a marginally higher price, it would be far more effective to temporarily remove congested networks or network nodes from the LCR. OSP v2 has provided a mechanism for this capability, with a 'high-water' indication — the AlmostOutOf Resources element — within the normal usage indication. Setting this element 'true' indicates that the network node should be temporarily removed from LCR rotation, and setting it 'false' indicates that there is sufficient capacity to reinstate traffic.

14.8.1.2 Real-Time Pricing

One of the more desirable end-user applications for a clearinghouse service is for the originating service provider to offer a real-time Web interface to their customer. Typically, this service could supplement or replace the call history portion of a monthly statement on a post-paid account, or provide a call history and value remaining for a pre-paid account. The operational problem for most clearinghouse applications is that they deal with wholesale cost, not the user's retail price.

There are several potential approaches to this problem, the first, simplest, and least accurate, is to simply ignore the settlement process, and perform real-time retail rating on the originating CDR. This capability can often be provided by VoIP-capable billing systems.

A more sophisticated approach, which requires tight integration between the clearinghouse and the originating service provider's retail billing system, is to perform both the clearinghouse settlement and then the retail rating in real-time, with the resulting Web call history display providing an accurate, auditable record of the call session.

14.8.1.3 Virtual Bilateral (Preferential) Routing

The per-minute pricing elements provided by service providers may not merely be applied to post-call CDR processing, but utilised to play a very large part in determining the location determination function for LCR tables, and create virtual bilateral (preferential) routing arrangements. As an example, if an originating service provider has a 5 million minute per-month bulk commitment to a terminating service provider, the clearinghouse might be asked to modify their standard least-cost routing algorithm to accommodate this business arrangement.

A policy decision would then need to made as to the method of implementation, i.e. should the first 5 million minutes go to the terminating partner, or should the total traffic volume be split with the standard LCR table, in such a way that the 5 million minute commitment is met by the end of the month. In either case the complicating factor arises from the fact that the LCR policy engine will need to acquire periodic minutes-of-use history from the settlement system in order to accurately steer traffic to the preferred route.

14.8.2 Non-Voice Applications

The clearinghouse model is applicable to a wide range of applications beyond voice, three examples of which are the support of fax services, middlebox location and roaming services. These are discussed below.

14.8.2.1 Fax Services

Because OSP is call signalling, media-stream, and device independent, it has found use in networks outside the traditional VoIP application. A Canadian company has utilised the OSP protocol to provide location determination for fax store-and-forward nodes within the networks of their global affiliates. Rather than provide static route tables within their fax nodes, an OSP client within each node is used to contact an OSP server, which provides a single, easily updatable source of inter-domain fax node location information.

14.8.2.2 Middlebox Location Determination

Middleboxes are a class of VoIP network element that typically provide functions such as QoS management and/or traffic aggregation at the VoIP network edge. OSP has been proposed as an easily implemented mechanism for one middlebox to determine the location of the next downstream middlebox with compatible capabilities.

In this type of implementation, the policy-based routing engine within an OSP server is used to create a topology-aware mapping of multiple networks. A middlebox OSP client interacts with the OSP server, and allows the middlebox to take actions based on aggregate traffic destination (e.g. country or city code, terminating service provider), to assure optimal transit of the network based on QoS or route capacity criteria.

14.8.2.3 ISP/ITSP Roaming Services

The addition of user authentication capabilities in OSP v2 was primarily intended to facilitate VoIP users from one service provider (the 'home' ITSP) to originate and be authorised for pre-paid or post-paid calls from a distant originating service provider. It is a logical extension of this capability that OSP could be extended to service any type of service request from a distant domain, including remote dial-in Internet access (IP roaming), paid content services, or network-based media services, such as conferencing, collaboration, voice-mail, text-to-speech or speech-to-text. OSP, or ultimately an OSP-like protocol, is likely to play a role in next-generation eCommerce services, and preliminary discussions have taken place within ETSI regarding this possibility.

14.9 Summary

As VoIP networks become more widely deployed, there is the increasing need to interconnect them. In the early phases of deployment, such interconnections have been dominated by relying upon the existing TDM infrastructure of the PSTN. However, this is unlikely to remain a stable position as the TDM infrastructure itself migrates through natural network evolution away from TDM towards packet-based solutions.

Achieving the interconnection of VoIP-based networks using IP technology is therefore an essential element of the VoIP environment in the future. This chapter has discussed a number of issues surrounding such interconnections and has highlighted current protocol developments envisaged to address these needs. In particular, the chapter has presented an overview of both the H.225.0 Annex G and OSP solutions.

References

1 Bartholomew, S.: '*The Art of Peering*', in Willis, P. J. (Ed): '*IP carrier-scale networks: designing and operating Internet networks*', The Institution of Electrical Engineers, London, pp 43-54 (2001).

2 Azure Solutions — http://www.azure-solutions.com/solutions/interconnect.htm

3 Rigney, C., Willens, S., Rubens, A., and Simpson, W.: '*Remote Authentication Dial In User Service (RADIUS)*', IETF RFC 2865 (2000).

4 Calhoun, P. R., Zorn, G., Pan, P., and Akhtar, H.: '*Diameter Framework Document*', draft-ietf-aaa-diameter-framework-01.txt (2001).

5 IPDR: '*Network Data Management — Usage (NDM-U) For IP-Based Services Version 2.6*', (2001) — http://www.ipdr.org

6 ITU-T Recommendation X.509: '*Information technology — Open Systems Interconnection — The Directory: Public-key and attribute certificate frameworks*', (2000).

7 IPDR: '*Network Data Management — Usage (NDM-U) For IP-Based Services Service Specification — Voice over IP (VoIP) Version 2.5-A.0*', (2001) — http://www.ipdr.org

8 Rosenberg, J., Salama, H., and Squire, M.: '*Telephony Routing over IP (TRIP)*', draft-ietf-iptel-trip-09.txt (2001).

9 ITU-T Recommendation H.225.0: '*Call signalling protocols and media stream packetization for packet-based multimedia communication systems*', (1998).

10 '*iNOW! Inter-domain telephony profile, version 2.1*', International Multimedia Teleconferencing Consortium (November 1999).

11 Thomas, S.: '*Application of ETSI TS 101 321*', TransNexus, 16TD125, ETSI TIPOHON, San Diego (December 1999).

12 ITU-T Recommendation H.225.0 version 2, Annex G: '*Call signalling protocols and media stream packetization for packet-based multimedia communication systems — Communication between administrative domains*', (2000).

13 ETSI Technical Specification TS 101 321 v1.4.2: '*Telecommunications and Internet Protocol Harmonization Over Networks (TIPHON); Inter-domain pricing, authorization, and usage exchange*', (1998).

14 ETSI Technical Specification TS 101 321 v2.1.1: '*Telecommunications and Internet Protocol Harmonization Over Networks (TIPHON); Open Settlement Protocol (OSP) for Inter-Domain pricing, authorization, and usage exchange*', (2000).

15 Fielding, R. J., Gettys, J., Mogul, H., Frystyk, H., and Berners-Lee, T.: '*Hypertext Transfer Protocol — HTTP/1.1*', IETF RFC 2068 (1997).

16 Freed, N., and Borenstein, N.: '*Multipurpose Internet Mail Extensions (MIME) Part One: Format of Internet Message Bodies*', IETF RFC 2045 (1996).

17 Bray, T., Jean, P., and Sperberg-McQueen, C. M.: '*Extensible Markup Language (XML) 1.0*', World Wide Web Consortium (February 1998).

18 ITU-T Recommendation H.235: '*Security and encryption for H-Series (H.323 and other H.245-based) multimedia terminals*', (1998).

19 Hoffman, S. P., Ramsdell, B., Lundblade, L. and Repka, L.: '*S/MIME Version 2 Message Specification*', IETF RFC 2311 (1998).

15

CASE STUDY: RIDE REPLACEMENT — A HYBRID PSTN/VoIP APPLICATION SOLUTION

J C Parr and A J Heron

15.1 Introduction

This chapter considers the design and deployment of an application solution that spans both existing TDM networks and emerging VoIP networks. The first Recorded Information Distribution Equipment (RIDE) system was deployed by BT in the late 1980s [1] to provide economic mass access to recorded announcements. The system was designed for excellent availability for applications with a high customer profile, generating a very large number of calls. Services deployed on RIDE include announcements for national code changes, timeline (the 'speaking clock') and televoting.

By the end of the 1990s it was recognised that the system needed to be replaced as key hardware components were becoming obsolete, the system could not support the more sophisticated voice processing capabilities required, and there was a need to deliver similar services (e.g. a 'televote') into Internet-hosted applications. These requirements are typical of many TDM-based network solutions currently deployed. The chosen replacement system, supplied by Telspec plc, is based on a 'distributed service node' architecture that has elements of the original RIDE architecture, together with features drawn from intelligent network and computer telephony integration (CTI) architectures [2].

The use of VoIP is particularly suitable for distributing live audio of the kind used in the RIDE solution. This is because these unidirectional audio feeds are not conversational. The usual problem of transmission delay in a VoIP network (see Chapters 2 and 3) is therefore not generally an issue. As the amount of VoIP traffic is also fixed, and relatively small, the IP network can be easily optimised.

The RIDE replacement solution takes account of current progress in the design and standardisation of protocols for supporting enhanced services in VoIP networks. It is planned that the network will be upgraded to support new standards in this area as they are ratified (see Chapter 5 for details of the various standards bodies).

In the new RIDE architecture, services are provided to customers from service nodes located around the UK. Each service node provides application and voice-processing capabilities that are independent of the underlying network — whether circuit-switched or packet-switched (VoIP). It therefore represents an example of a hybrid PSTN/VoIP application.

High throughput and optimum use of the data network are achieved by making extensive use of IP multicasting techniques, both for the distribution of live audio feeds using multicast VoIP and for the distribution of service data from the management system to the service nodes.

15.2 The First RIDE Network Solution

The original RIDE network architecture is shown in Fig 15.1. The recorded announcements fed to the RIDEs were carried in 64 kbit/s channels of 2 Mbit/s pulse code modulation (PCM) systems, connected to form rings linking all the RIDE switching nodes. The switching nodes contained custom-designed switches. Those announcements that did not have to start at the beginning were continually broadcast around the rings. Calls not requiring a 'start at beginning' (non-SAB) announcement were connected to the appropriate channel on the ring by the switch. Start-at-beginning (SAB) announcements were stored by the switches in local memory. A sophisticated system management function was developed to control and manage the network of RIDE switches [1].

The original RIDE system proved highly effective, handling announcements for a number of major UK code changes and achieving record UK call volumes for televotes such as those associated with Channel 4's television programme 'Big Brother' in September 2000.

15.3 Core Design Issues

Over time, a number of new requirements have arisen for the RIDE system as a consequence of the evolution of both the televoting market and network technology. The new requirements fall into three main areas:

- more flexible and feature-rich services;
- support for VoIP users in addition to PSTN-based users;
- automated, Web-based service provisioning.

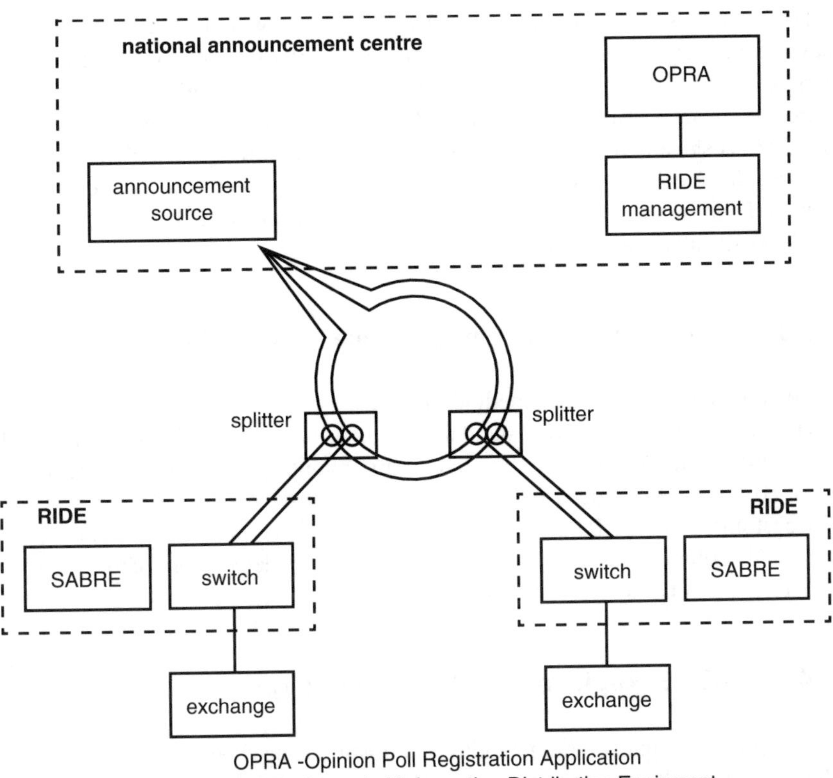

Fig 15.1 Original RIDE network — a simplified view.

15.3.1 Flexible and Feature-Rich Services

Services have become more sophisticated and personalised. For instance, in a televote, menu-style dialogues are required. Instead of callers having to remember a separate number for each choice, a single access number is used and the caller invited to make a choice by pressing the appropriate key on the telephone.

15.3.2 Support for VoIP

The 1990s saw the phenomenal growth of the Internet and related technologies, leading to the recognition of the benefits of having a single network for carrying both voice and data traffic and the advent of 'voice over IP' (VoIP) (see Chapter 1).

Today, BT is implementing new VoIP-based networks. The users of these networks may expect to have access to services similar to those provided for PSTN users. Clearly it is better to provide services to both networks from a common service platform — minimising costs and ensuring consistency of service. To avoid increased costs and to maintain the overall quality of service delivered, it is desirable for VoIP services to be provided natively, rather than converting them back to TDM via a VoIP gateway. The new platform must therefore be able to handle VoIP calls in addition to those from the PSTN.

15.3.3 Web-Based Automated Service Provisioning

The Web is now the preferred method of accessing service management systems. It provides a consistent, low-cost, standardised interface with flexible connectivity for all potential users. At the same time, the user community for the RIDE platform has increased and now includes independent service providers who deliver service to their customers via the BT network. To simplify service provision, the aim for the new RIDE solution has been to enable service providers to provision their own services without BT's direct intervention.

15.4 RIDE Replacement System Design

Telspec plc, a UK telecommunications manufacturer, has been selected to deliver a solution for replacing the existing RIDE platform that meets the existing requirements for traditional telephony services and supports the new requirements for delivering those services via various forms of IP access.

15.4.1 Network Architecture

The new RIDE system is based on a distributed service node architecture, with centralised service and network management. The initial deployment calls for 40 service nodes covering the whole of the UK. The main components of the new RIDE architecture are shown in Fig 15.2.

The service node [2] can be viewed as a combined service switch point (SSP), service control point (SCP) and intelligent peripheral (IP). Each service node has all the service logic, voice processing resources and data required to service each call. The aim in this architecture has been to provide services as close as possible to the origin of the call in order to minimise the traffic that has to be carried in the transit layer and transmission network. This approach differs from the traditional intelligent network (IN) [2] approach in which voice processing is distributed but the application and data are centralised. In particular, it has been chosen over the centralised approach which cannot cost effectively provide the performance required in mass-calling applications of the type required to be supported by RIDE.

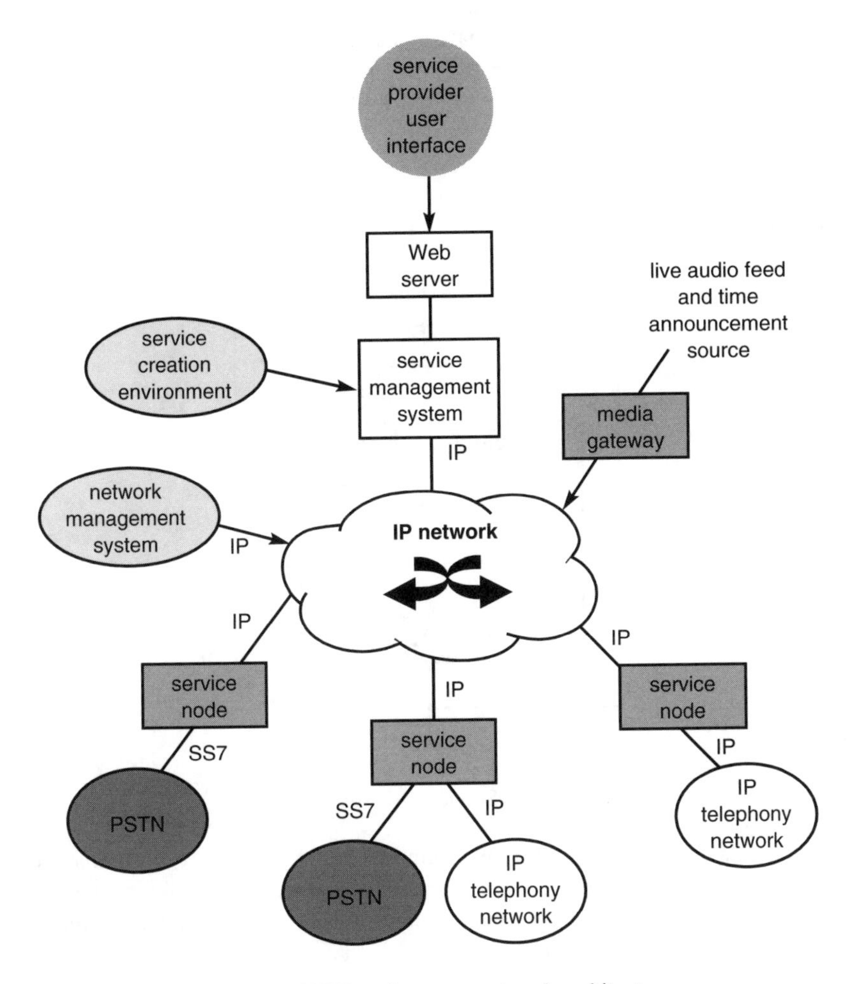

Fig 15.2 RIDE replacement network architecture.

Live audio distribution is implemented by converting the incoming audio feed to VoIP. Multicasting techniques are then used to distribute it to all service nodes. To increase service reliability, each audio stream is delivered as two channels to each service node where it is converted back to PCM and then broadcast to multiple callers.

Service provisioning is controlled from the central service management system (SMS). At the SMS, the service logic, service parameters and voice files that make up an individual service instance are combined and distributed to the service nodes.

15.4.2 Service Node

The service node (see Fig 15.3) provides the platform for delivering service applications to customers. It can handle calls received from either the PSTN (circuit-switched PCM) or a VoIP network. Each service node consists of three logical components, implemented as a single system:

- narrowband switch (IPS);
- intelligent media server (IMS);
- media gateway (MG).

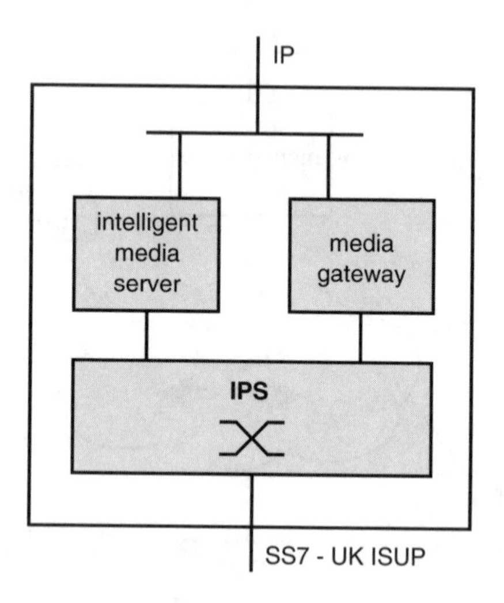

Fig 15.3 Service node sub-systems.

PSTN calls arrive over PCM trunks from BT switches (next generation switches in the transit layer of the UK network) or over interconnect trunks from other public telephone operators and are terminated at the service node. UK ISUP SS7 signalling [3] is used on these trunks. For VoIP calls, the service node will collaborate with a call server (gatekeeper or SIP proxy server) in the VoIP network via a control interface.

15.4.3 Narrowband Switch Subsystem — Architecture

The IPS follows the VoIP server-routed signalling approach (described in Chapter 1) in which the signalling and call control is separated from the physical switching (see Fig 15.4). The key components are the IPS720 PCM shelves, duplicated host computers, known as system control units (SCUs), and a mediation element manager (MEM).

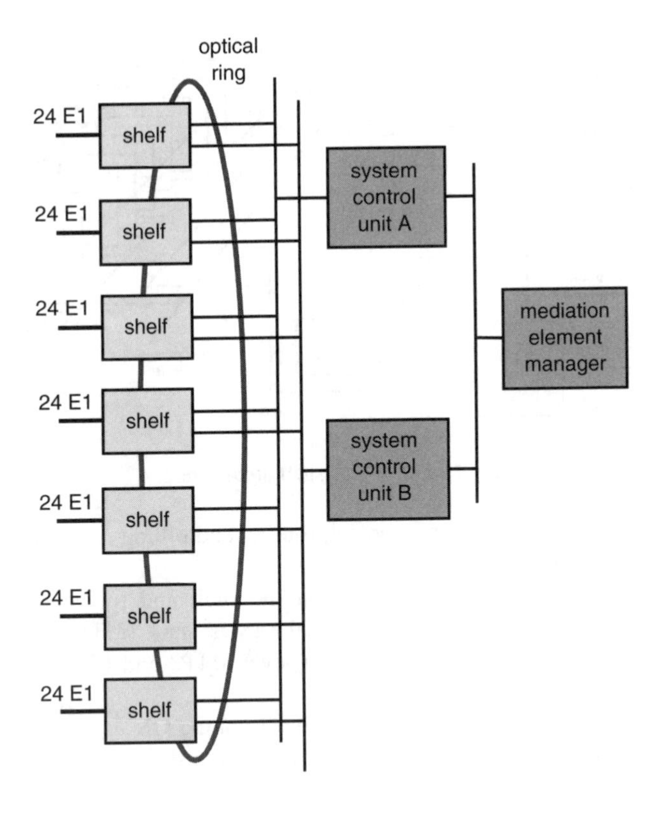

Fig 15.4 Switch architecture.

Each PCM shelf can contain up to 12 dual primary-rate-interface (PRI) cards, giving a capacity of 24 E1s per shelf. Shelves are linked together via a dual counter-rotating optical ring. Up to 10 shelves can be linked in this way.

Each PCM shelf also contains duplicated CPU/Fast Ethernet cards, power supply units (PSUs), and optical cards. Thus all common control equipment is duplicated for resilience. The cards are all hot-swappable, and there is no loss of service on a switch-over. Secure communications are maintained to all the switch shelves and the MEM in the multi-shelf switch, by using dual redundant hubs and connecting one of the CPU/FE combinations on each shelf to each hub. As each CPU is a 720-channel cross-point matrix, calls going in and out on the same shelf do not need to be routed via the ring. Inter-shelf routes are set under the control of the service control unit (SCU) via the optical ring, with traffic being appropriately routed through a particular shelf CPU to the correct E1 port (see Fig 15.5).

The SCUs are based on rack-mounted industrial computers with Intel Pentium processors running the QNX operating system. QNX is a real-time POSIX-compliant Unix. An additional layer of software has been implemented to provide a

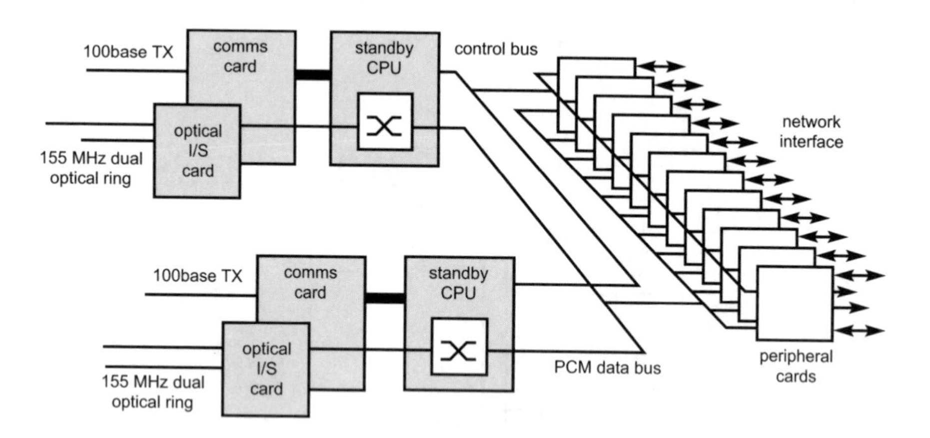

Fig 15.5 Voice shelf architecture.

resilient duplicated platform. Call control, routing and billing record generation are all handled by the SCU.

SS7 signalling channels are groomed at the shelf and routed over dedicated channels to SS7 cards. The lower levels of the ISUP stack (MTP1 and MTP2) are processed on the SS7 cards while the upper level MTP3 and ISUP are handled on the SCU.

The MEM is also a rack-mounted computer running QNX. It is responsible for statistics processing, for securing billing data on disk, and providing network management functions.

15.4.4 Key Features of the IPS

15.4.4.1 Signalling

In the initial deployment each service node will be interconnected with at least two Ericsson next generation switches (NGSs) using UK ISUP signalling [3]. UK ISUP is based on ETSI ISUP version 2 with extensions to support UK-specific features. The IPS will support STP working if required. The signalling between the IPS and IMS subsystems is primary-rate ISDN.

15.4.4.2 Special Features

A number of special features that are of particular use in providing RIDE services are available on the IPS, including:

- broadcast;
- wait-on-start;
- mid-call diversion;
- call deflection.

The broadcast feature is a unidirectional broadcast connection, allowing many callers to hear audio from a single source external to the IPS, such as from an interactive voice response (IVR) or 'live feed'.

Wait-on-start (WOS) is a unidirectional broadcast connection, with the addition that SAB announcement services can be provided. With WOS activated, an incoming call can be put on hold, with ring tone being returned to the caller by the IPS. The call will then be connected when further calls requiring the same outgoing route arrive (i.e. the queue size exceeds a preset limit) or a time out expires. Mid-call diversion allows the terminal equipment (typically an IVR) to divert an incoming call to a third party — analogous to the call transfer facility on a PABX.

The mid-call diversion feature is defined in BT SIN 316 (Enhanced Digital Volume Call Access) [4].

Call deflection (CD) is an ISDN supplementary service. Call deflection enables an ISDN user (e.g. an IVR) to respond to an incoming call by requesting redirection of the call to another user prior to answering the call. The ISDN user has access to information such as the calling line identity on which to make the re-routing decision. The distinction between the CD service and call forwarding is that the ISDN user re-routes the call on an individual basis.

15.4.5 Routing

The routing features are extremely flexible. The system has a routing table with the ability to choose a user-selectable routing function (USRF) from a fixed pool and apply it to one of the user-selectable levels within the routing table. Four user-selectable levels are allowed, with the order of route determination as follows:

1. Incoming group
2. USRF 1
3. USRF 2
4. DDI
5. CLI
6. USRF 3
7. USRF 4
8. Route determined

The USRFs include calling party category, clearing cause, nature of address, time of day, and day of week. Up to 16 routing tables are held on the switch — but only one will be active at any given moment, controlled by a schedule.

The use of wild cards and masks for digit matching and manipulation significantly reduces the size of the routing tables, for example:

'*' = any sequence of one or more digits

'?' = any one digit

'B+' = second and all subsequent digits

15.4.6 Billing

One of the key design challenges for VoIP and hybrid VoIP-TDM applications is to establish an appropriate approach to billing and event recording. In the case of the RIDE replacement solution, call detail records (CDRs) are produced for all calls, whether they are successful or unsuccessful. Although this may appear curious at first glance, the collection of unsuccessful call details is, however, useful for network management purposes. The CDRs are output in batches via TCP/IP to an external collector and are also stored on duplicated RAID drives for collection by a billing mediation system. The integration of new network solutions can have a wide impact on existing service surround solutions unless carefully designed. In this case, the CDR format used in the solution is the BT Platform-Independent Usage Record (PIUR) format, which enables the CDRs to be processed by BT's existing billing systems.

15.4.7 Intelligent Media Server

The intelligent media server consists of two parts — the application server and the voice processing resource cards (see Fig 15.6).

The application server contains the service execution environment to run the service logic for each call.

The voice processing resource cards provide the interface to the outside world (PCM or IP interfaces) and digital signalling processors to process the voice. The basic voice processing functionality required includes:

- play a voice file;
- record voice;
- detect a fax call;
- provide echo cancellation for automatic speech recognition (ASR).

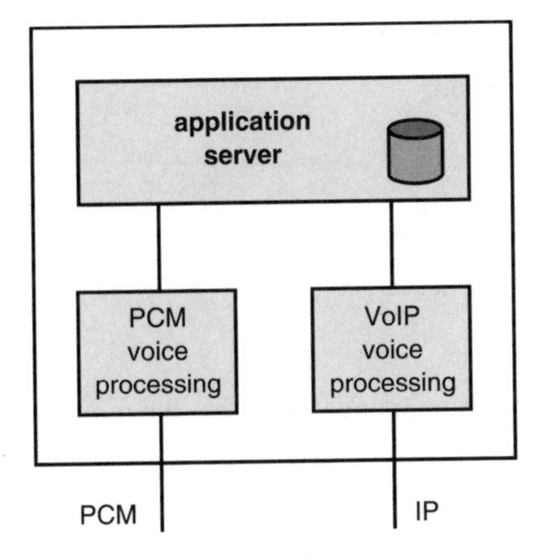

Fig 15.6 IMS sub-system.

The cards used are Intel Dialogic Quadspan™ for PCM, and IPLink™ for IP. As the voice-processing capability of the cards is the same, applications can be written independently of the card type.

15.4.8 Media Gateway

The media gateway (MG) sub-system provides the conversion of media streams between packet-switched and circuit-switched networks for the distribution of real-time audio feeds. Chapter 7 provides more detailed information on MGs.

The MG at the central node multicasts the audio streams. The MG in the service nodes converts the incoming streams back to PCM and presents them to the IPS sub-system for connection to incoming calls.

15.5 Network Management System

The network management system allows centralised alarm surveillance, and operations and maintenance, of the entire network of service nodes. The network management system views each service node as a single system — alarms from the IPS, IMS and media gateway subsystems are combined and presented together. Key features of the management system are:

- integrated centralised management system;
- graphical display of all service nodes with colour codes to represent operational status;
- acknowledgement of active alarms;

- current and historical alarm logs;
- remote software download;
- centralised back-up;
- remote diagnostics;
- traffic statistics;
- call trace;
- activity display.

Under normal conditions no manual intervention at the service node sites will be required.

15.6 Service Management System

15.6.1 Service Creation Environment (SCE)

The SCE provides the capability for rapid application development. For most applications this means that conventional programming is eliminated and system design is reduced to the creation of a flow chart, using graphical action cell objects. The SCE architecture is flexible and open, providing for access to other applications. Custom extensions of the development environment can be written in any standard programming language.

Within the SCE, the developer uses graphic representations to decide the logical flow of the application, by connecting action cells. The design philosophy is object oriented, with each action cell encapsulating a set of properties and events. Properties allow the developer to customise the personality and performance of an action cell. Events allow the developer to dictate the next programming step, based on the results of executing the action cell. These action cells are divided into groups of functionality, comprising:

- call control;
- connection control;
- database control;
- file management;
- housekeeping;
- programmatic;
- serial port control;
- string manipulation;
- voice processing.

Action cells can be combined together to produce subroutines of generic functions such as number translation, queuing, playing an announcement, and recording speech. These functions are common to many different services. It is this reusability that leads to a reduction in development time-scales. Once the application has been created, it is compiled into a small, efficient, run file.

Customer or service provider control is made possible by allocating particular DDI numbers which invoke applications that can access data in the central database that controls characteristics of the service.

The SCE supports open database connectivity (ODBC) to access external databases, and optimised drivers are available for SQL Server, Oracle and Microsoft ODBC. However, for most mass-calling applications where throughput is paramount, the use of conventional database access techniques would impose a very high overhead and therefore such applications are optimised by handling all data elements in memory.

The service creation environment is a powerful tool that enables new applications to be developed quickly. However, as with any kind of software development, it requires disciplined testing and configuration management processes.

15.6.2 Service Creation and Provisioning

The role of the service management system is to automate and co-ordinate the provisioning of services in the network of service nodes.

There are two aspects to this process:

- controlling the downloading of files containing service logic and service data, including announcement files, to the service nodes;
- service scheduling, which includes bringing them into service, so that they can be accessed by callers, and de-scheduling them when no longer required — this is achieved by means of a routing table, which initiates the appropriate application depending on the DDI number dialled.

ColdFusion is used to develop the Web interface that provides the screens by which the service provider defines and schedules a particular instance of a service.

15.6.3 Data Management

The data required by a service node to execute a particular service falls into three categories:

- service scripts;
- voice files for announcements;
- service control data.

The master copy of this data is held within the service management system. To ensure correct operation, this data must be available and correct at all service nodes in the network. The SMS is therefore responsible for the distribution of data to the service nodes. In order to provision a new service, all that is required is to make the necessary data available to the SMS. The SMS is then responsible for replicating the data across the appropriate service nodes.

The service management system maintains a central library of service logic files, announcement files, routing tables, and other global files, such as tables mapping CLIs to advertising regions. These files originate from a number of sources. For example, the service logic files are generated from the service creation files, whereas the announcement files may have been created externally in a recording studio.

In a large distributed network, the efficiency of the file download process becomes critical and it becomes more difficult to ensure consistency across the nodes [5].

The simplest approach is to store all data to be replicated in files. The distribution system will then ensure that any changes to the master file will also be applied to all of the replicated files (Fig 15.7).

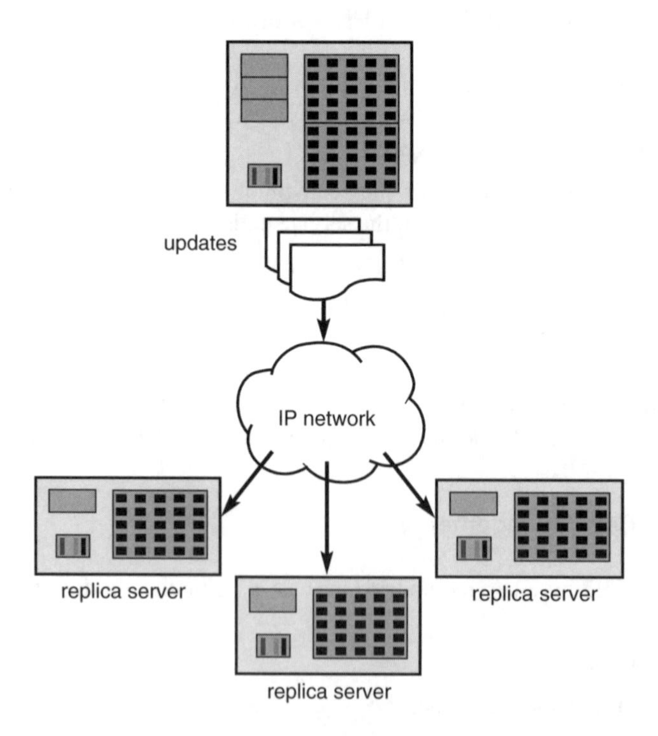

Fig 15.7 File replication.

Key requirements for the file replication process are:

- to centralise management of server synchronisation;
- to maintain availability during updates, since the performance of the network must not be degraded when an update is in progress;
- to ensure automatic synchronisation, e.g. if a service node is 'down' when an update occurs, it must be automatically updated with any changes when it is brought back into service.

15.6.4 Service Scheduling

The routing tables maintained on the IPS component are very simple. All calls received from the PSTN are passed through to the IMS, providing a spare channel is available. If all channels are busy, the call can be either re-routed through the PSTN to another service node, or given 'busy' indication. The IMS routing table is consulted to see which service logic should be invoked to handle the call. The IMS routing table therefore effectively controls the scheduling of services, and scheduling is achieved by updating the IMS routing table.

15.6.5 Statistics

The demand to collect a vast amount of data during a peak period, while at the same time providing statistics in virtually real time, presents a major design issue. In order to overcome this, the platform uses two types of data and a different approach for the collection of each.

Data from the CDRs and the IVR system is transferred in batches to the data warehouse where it is processed into a relational database management system (RDBMS). This data can then be analysed using RDBMS queries and used to create active Web pages, reports or downloadable data sets (see Fig 15.8).

Summary data, such as the total number of calls between a start and end time or total votes cast via dialled digits, is collected in real time by the real-time statistics summariser. The data to be summarised is controlled by rules defined by the management system. The master summariser collates the summary data from all service nodes and presents it in near-real time via the active Web pages. The management of statistics collection is performed through the Web server interface.

All statistical data is collated and processed by the bulk data collector. This process is responsible for parsing the CDRs from all service nodes and importing the data into the RDBMS. It is also responsible for retrieving all data from the IMS systems and assimilating this into the RDBMS. As this component performs as a single unit, it is not replicated. The RDBMS therefore needs to be implemented on a high-end server with very high levels of dual redundancy.

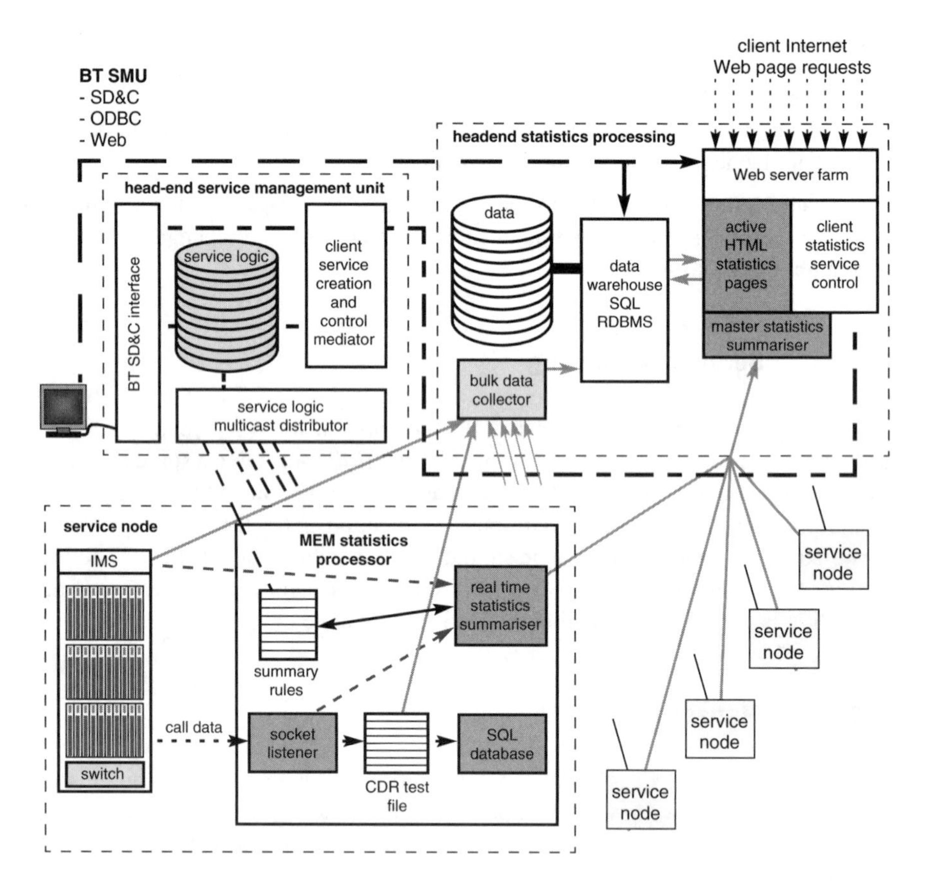

Fig 15.8 Data warehouse.

All data extraction is achieved through active Web pages. This includes reports, graphs and data downloads in CSV format. The data from the warehouse is stored for two months before being discarded.

15.7 IP Network Design

The key challenge in implementing a data network carrying VoIP is in providing an acceptable quality of service against a background of varying traffic types. The RIDE IP network will carry four kinds of traffic:

- multicast VoIP for distribution of real-time audio from the central control site to the service nodes;
- distribution of files containing service provisioning data from the service management system to the service nodes;

- transfer of billing data and statistics information from the service nodes to the service management system;
- network management data traffic between the service nodes and the network management system for reporting equipment alarms, etc.

The characteristics and quality of service (QoS) requirements of each of these kinds of traffic are different. VoIP traffic is most efficiently carried in small packets. The main requirement is that the packets are delivered quickly, though some packet loss is acceptable without affecting the perceived voice quality (see Chapter 2 for further discussion of QoS). In contrast, the data traffic is most efficiently sent as large packets. The main requirement is accuracy. In other words, the data must not be corrupted and delivery must be guaranteed — although the speed of delivery is not so critical. As is apparent from the discussion in Chapter 3, the requirements for these two types of traffic are clearly conflicting. For example, the small voice packets can be susceptible to delay caused by a queue of larger data packets. Striking the right balance between these conflicting requirements is a key challenge facing the widespread introduction of VoIP. However, the voice services being provided over the RIDE network are unidirectional. This has the great benefit that the receiver will not perceive any delay in receipt of the broadcast. So long as voice packets are received at a constant rate, the receiver will not notice any degradation in voice quality compared with a circuit-switched call.

One notable exception to this general rule relates to the delivery of the TimeLine service. This is considered to be a live-feed multicast service since the time announcement is derived from a clock device with an audio output fed into a media gateway at the central site for distribution to all service nodes via multicasting. Delivery of the announcement to the end caller is therefore subject to the delays of the IP network. In this instance the delays are considered to be service affecting since they detract from the accuracy of the information delivered to the caller (i.e. the time).

In the new RIDE platform this problem has been overcome by characterising the performance of the IP network to identify the constant end-to-end delay. The time of the clock source, which is synchronised to a national standard source, is then advanced by the characterised delay to compensate for delaying the announcement to the point of termination.

15.7.1 VoIP Network Engineering

The QoS requirement to support RIDE replacement has been addressed by:

- using IP multicast to minimise the bandwidth required;
- using network equipment supporting QoS standards, namely diffserv;
- applying traffic engineering techniques for optimal network design.

Figure 15.9 illustrates the architecture of the IP network that will be implemented to support the RIDE replacement platform. The implementation of the network engineering requirements identified above is discussed in the following sections.

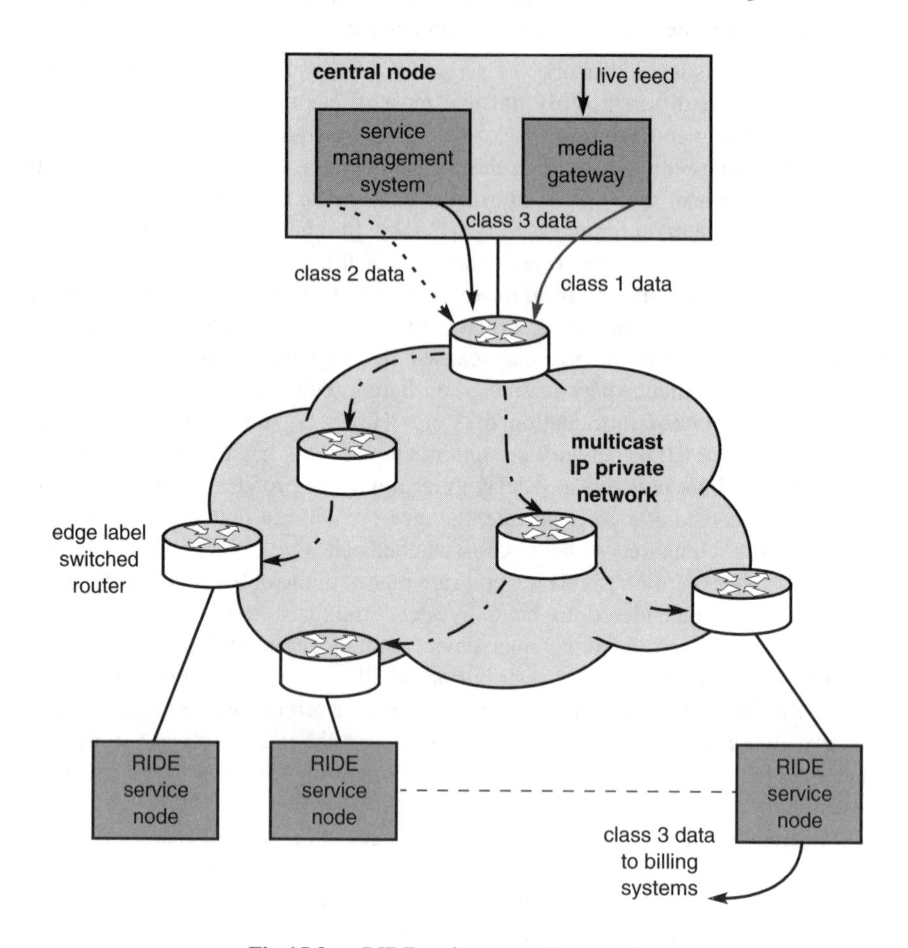

Fig 15.9 RIDE replacement IP network.

15.7.2 IP Multicast

Internet protocol (IP) multicast [6] is a routing technique that allows IP traffic to be sent from one source and delivered to multiple destinations. Instead of sending individual packets to each destination, a single packet is sent to a multicast group, which is identified by a single IP destination group address.

In a multicast environment, the source server needs to transmit a single media stream for each multicast group, regardless of the number of clients that will receive it. The media stream is then replicated as required by the network's multicast routers and switches to allow any number of clients to subscribe to the multicast address and receive the broadcast. In the router network, replication only occurs at branches of the distribution tree.

Where there are a large number of recipients of a replicated transmission, multicast technology makes a tremendous difference to both server and network load, even in a small network with a small number of router and switch hops.

Other advantages of multicast are:

- multicast transmissions are delivered nearly simultaneously to all members of the recipient group — the variability in delivery time is limited to differences in end-to-end network delay;
- the server does not need to know the address of any particular recipient of the transmission — all recipients share the same multicast network address.

Over and above the multicast capability in the network, a multicast transport protocol is required to implement reliable multicasts, e.g. there must be a feedback mechanism in case packets do not get delivered.

A principal component of IP multicast is the Internet group membership protocol (IGMP). IGMP relies on class D IP addresses for the creation of multicast groups and is defined in RFC 1112 [7]. IGMP is used to register individual hosts dynamically in a multicast group with class D addresses. Hosts identify group memberships by sending IGMP messages, and traffic is sent to all members of that multicast group.

Pragmatic general multicast (PGM) is a reliable multicast transport protocol for applications that require ordered or unordered, duplicate-free, multicast delivery from multiple sources to multiple receivers.

A fundamental characteristic of the distributed nodes comprising the RIDE platform is that they are all identical in terms of the services delivered and therefore the audio announcements stored.

The central node is the 'master repository' of both service logic and audio data, and is responsible for maintaining the content and configuration of all distributed nodes. This is achieved by using IP multicasting (SmartPGM) to distribute data to each of the distributed nodes from the central node. The use of multicasting here realises two important objectives:

- ensuring that, at a given point in time, all nodes are guaranteed to contain the required services and audio announcements to support those services;
- reducing significantly the bandwidth required to distribute audio announcements (and live-feed audio) to all forty distributed nodes — using non-multicasting techniques would require a bandwidth of 'number of nodes × data volume', while multicasting requires a bandwidth close to the 'data volume' (i.e. plus a small overhead for using multicasting protocols).

15.7.3 Diffserv

Diffserv operates at layer 3. It allows an application to set the relative priority of different types of traffic by setting bits in the IP type of service (ToS) field. The ToS field describes one entire byte (eight bits) of an IP packet (defined in RFC 791 [7]). Precedence refers to the three most significant bits in the ToS field. These bit fields are used by routers to control routing across the IP network. The new diffserv standard (RFCs 2474 and 2475 [7]) extends the number of bits used for setting priority to six, i.e. 64 classes. A more detailed discussion on diffserv, and related techniques, is given in Spraggs [8].

The RIDE IP network uses diffserv to classify three 'types' of traffic between the central nodes and the distributed nodes.

- Class 1

  Class 1 data has the highest priority and will always be transported — used for all voice traffic (e.g. live feed audio).

- Class 2

  Class 2 data is subordinate to Class 1, but can use 100% of the available bandwidth if it is not being used by any other service, and is used for high priority data (e.g. 'command and control' data from the central node).

- Class 3

  Class 3 has the lowest priority and is the first to be discarded when insufficient bandwidth is available to meet demand. This is used for lower priority (in terms of time of delivery) data such as billing data and other records used for statistical analysis.

15.7.4 Bandwidth Engineering

Typical of all current QoS-engineered VoIP solutions, the RIDE IP network is a 'closed' network with predictable traffic flows (unlike the public Internet), and the use of multicasting helps to minimise this traffic. Additionally, bandwidth economy resulting from the use of multicast allows G.711 encoding to be used, thereby avoiding loss of voice quality due to compression.

As a consequence, a network can be provided that is relatively easy to dimension for 'worst-case' traffic scenarios, without resorting to the inefficiency of bandwidth over-subscription. The deployed network to support RIDE will provide a bandwidth of ~2.25 Mbit/s into the central node and ~0.85 Mbit/s into each of the distributed nodes.

15.8 VoIP Enhanced Services Framework

VoIP standards are still emerging and evolving as practical experience tempers idealism. For VoIP to realise its full market potential it must provide at least the same range and quality of service as the existing PSTN. The focus of effort is now moving beyond the establishment of a basic point-to-point call and towards the provision of enhanced services. Among other groups, the International Softswitch Consortium [9] is working in this area (see Chapter 5 for more details), and the approach taken in the design of the RIDE replacement solution accords with this functional architecture, as shown in Fig 15.10.

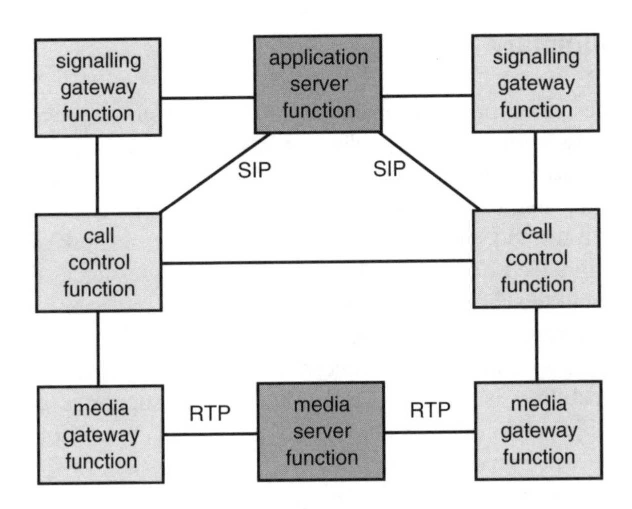

Fig 15.10 Functional services architecture.

Within this architecture, the application server function provides the execution and management of enhanced services, handling the signalling interface to a call control function and APIs for creating and deploying services. It is becoming increasingly accepted that SIP will be the protocol used between call control and application server functions. A more detailed discussion of SIP can be found in Chapter 11.

The media server function provides for specialised media resources (e.g. IVR, conferencing, fax, announcements, speech recognition) and handling the bearer interface to a media gateway function.

Within a RIDE service node the application server and media server functions are provided by the intelligent media server (IMS) sub-system.

Calls requiring announcements will be terminated on a telephony server providing call control in the IP telephony network, which will use SIP to initiate and communicate with the particular application to provide the required service. Where voice processing is required, the media stream will be provided via an RTP interface on the media server.

The call control function, media gateway function and signalling gateway function provide the mechanism for setting up calls across the IP network.

In the RIDE implementation, where a strictly limited number of channels is required, these functions are all implemented in a single board (the media gateway sub-system) for efficiency.

15.9 Summary

The solution described in this chapter solves the problem of providing increased capacity and more flexible services today, while providing a bridge and future-proof platform to new-style networks based on distributed intelligence and IP transport. The RIDE replacement system is designed as a generic, enhanced-services platform that will initially provide the original RIDE services of network announcements, televoting and live feeds. However, the voice processing resources deployed in the service nodes, and the service creation and provisioning mechanisms implemented in the service management system, may be utilised in the future to provide other advanced services — using, for example, the conferencing, messaging and speech recognition capability, as a voice portal. The services supported are any kind of mass-calling application, for example, providing the control of entertainment services such as video on demand.

The RIDE replacement network incorporates several state-of-the-art technologies in its design, including file replication, multicasting and the use of voice over IP. However, it uses these techniques in a controlled and carefully engineered way. This ensures that the benefits are realised without compromising the fundamental requirements of reliability and quality of service implicit within the public telecommunications network.

References

1 Allard, E. and Warren, A.: '*RIDE: Recorded Information Distribution Equipment*', British Telecommunications Eng J (April 1992).

2 Dufour, I. G. (Ed): '*Network Intelligence*', Chapman and Hall, London (1997).

3 BSI: '*UK ISUPv2.2 PD6623:1999 PNO-ISC Specification 007*', (1999).

4 BT SIN 316 Enhanced Digital Value Call Access — www.sinet.bt.com

5 Marchant, M.: '*Reliable replicated databases using the ISIS toolkit*', BT Alliance Engineering Symposium (May 1996).

6 Challinor, S.: '*An introduction to IP networks*', in Willis, P. J. (Ed): '*Carrier-scale IP networks: designing and operating Internet networks*', The Institution of Electrical Engineers, London, pp 9-23 (2001).

7 IETF Web site — http://www.ietf.org/rfc/

8 Spraggs, S.: '*Traffic engineering*', in Willis, P. J. (Ed): '*Carrier-scale IP networks: designing and operating Internet networks*', The Institution of Electrical Engineers, London, pp 235-260 (2001).

9 International Softswitch Consortium — http://www.softswitch.org/

16

TIPHON — PSTN SUBSTITUTION AND BEYOND

P G A Sijben, P A Mart and R P Swale

16.1 Introduction

As will be appreciated from the previous chapters in this book, it is unrealistic to expect the large-scale deployment of VoIP solutions to take place completely independently of existing networks and services. Recognising this problem, the Telecommunications and IP Harmonisation Over Networks (TIPHON) project has been created under the umbrella of the European Telecommunications Standards Institute (ETSI) with the aim of addressing service level interworking between traditional switched-circuit networks (SCNs) and the emerging next generation networks based on VoIP technology.

TIPHON's objective is to support the market for real-time telecommunications services between users, including voice and related voice-band communication, such as facsimile, over multiple network technologies. So telephony, multimedia conferencing, instant messaging and eCommerce may all be examples of applications and services within the scope of, and enabled by, TIPHON. However, it is important to recognise that the capabilities of such services may go beyond the common services provided by either basic Internet connectivity or the PSTN today.

TIPHON's principal goal is to enable users connected to IP-based networks to communicate between themselves and also with users in SCNs, especially those served by PSTN, ISDN or GSM networks. In this regard TIPHON addresses the difficult, yet extremely important, area of multi-network interworking across multiple administrative and technology domains. Since it emerged as the first and most dominant VoIP technology in the late 1990s, the initial focus of the TIPHON project has been concerned with issues relating to the deployment of H.323-based networks (see Chapter 10 for more details). However, as the early implementations of VoIP technology available when TIPHON started could not deliver a telephony service directly equivalent with that obtainable from existing networks, defining interworking between such networks emerged as a rather daunting task. To overcome this problem, the TIPHON project has had to develop a detailed understanding of the provision of conversational services over packet-based network infrastructures to advance the development of its specifications. For example, it has had to develop a detailed appreciation of the challenges of providing a universal communications service offering across disparate networks and technologies while simultaneously enabling the possibility of innovation and differentiation in equipment and services. In doing so, it has had to consider a broad and diverse set of requirements including security, quality of service, numbering and the very approach to communications standardisation itself. This chapter presents an overview of TIPHON from the point of view of the most recent release of specifications from the project, which is referred to as TIPHON Release 3.

In Release 3, TIPHON addresses the problem of providing public communications services in a heterogeneous environment by defining a generic means of creating services that is independent of any specific underlying network technology — irrespective of whether it is switched circuit or packet based. To achieve this objective, TIPHON has identified the need for an overarching technology and domain-independent protocol framework — known as the TIPHON meta-protocol. This is used to generate profiles for the protocols associated with any given communications network technology. By mapping this 'meta-protocol' to individual network technologies, TIPHON aims to ensure a higher degree of end-to-end capability than would otherwise be possible.

16.1.1 In the Beginning

The TIPHON project started in May 1997 with the intention of developing specifications for interworking between the existing PSTN, ISDN and GSM switched-circuit networks and the H.323 VoIP-based networks that were emerging at that time. The approach adopted for this initial work was based upon a physical view of the problem space, as shown in Fig 16.1.

The TIPHON Release 1 work was also predicated on the study of a simplistic set of numbered scenarios comprising calls from:

- an IP device to a telephone (scenario 1);
- a telephone to an IP device (scenario 2);
- a telephone to another telephone via an IP network (scenario 3);
- an IP device to another IP device via a switched-circuit network (scenario 4).

These scenarios were further grouped into 'phases' of activity within the TIPHON project with Phase 1 relating to scenario 1 and so on. However, as the work progressed within the project, it rapidly became clear that a simple physical architecture was insufficient to achieve the project's aims. This is because it left little room for equipment manufacturers and service providers to create multi-vendor, multi-service provider, network solutions that could scale reliably. In other words, this generation of solution is completely dependent upon the existence of a switched circuit, or similar, based PSTN and therefore unsuitable for consideration as a useful basis of future public network evolution.

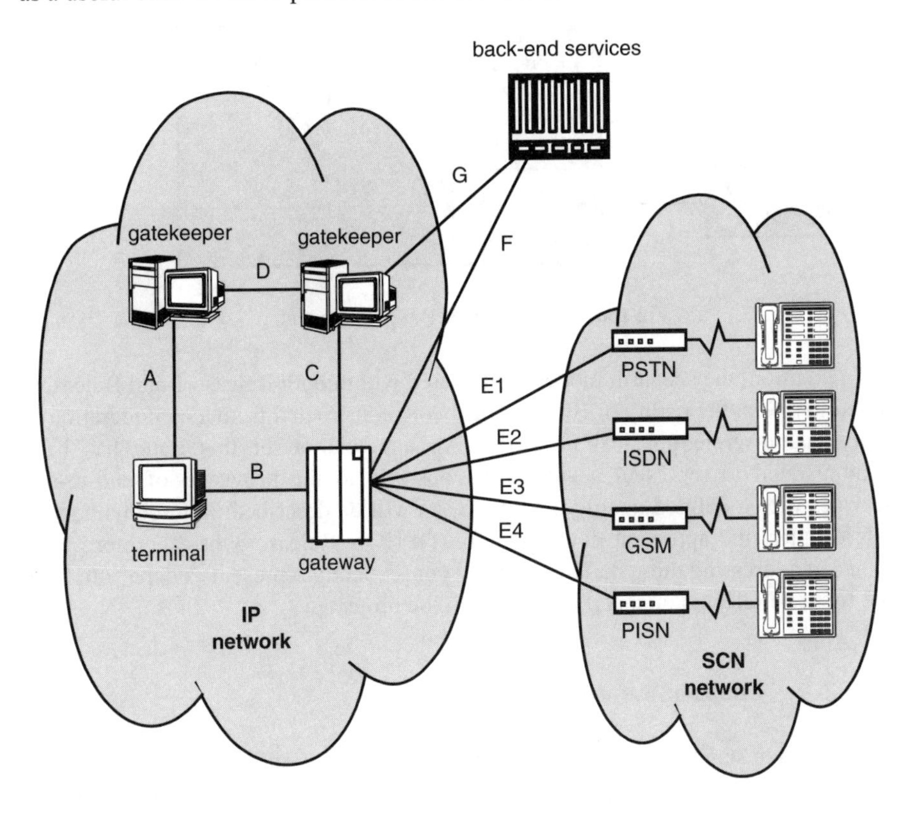

Fig 16.1 TIPHON Release 1 Architecture.

The next generation of activity within the TIPHON project, known as Release 2, attempted to address some of these limitations through the development of a more functionally oriented architecture, shown in Fig 16.2. This work exposed a greater understanding of gateway devices required to interconnect IP-based terminals with switched-circuit networks.

However, the harder problems concerning how the TIPHON Release 2 architecture enabled new services across IP and switched-circuit networks remained intractable. This was due to the architecture maintaining a physical device-oriented view and, in doing so, avoiding a number of fundamental issues — for example concerning whether the actual call control was implemented by the gatekeeper or the media gateway controller, both of which are physical entities from this point of view.

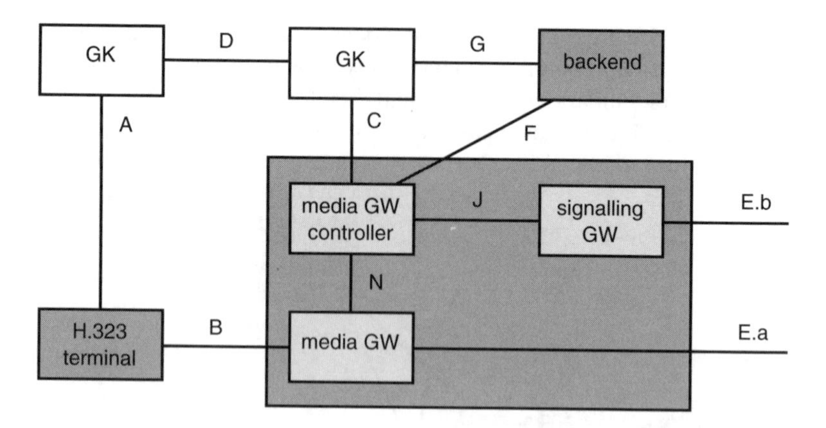

Fig 16.2 TIPHON Release 2 Architecture.

In addition, the scenario models considered within both Release 1 and Release 2 failed to reflect the reality of IP networks being constructed from a confederation of individual discrete physical networks. An appreciation of this issue has been demonstrated to represent a substantial obstacle to the provision of end-to-end services with a defined quality of service. As will be described in the remainder of this chapter, the approach developed for TIPHON Release 3 has therefore been aimed at addressing these deficiencies and consequently represents a departure from the first two releases of specifications from the project.

16.1.2 Lessons Learned

In the earlier stages of the TIPHON project, there was an over-emphasis on one implementation technology, namely H.323, and in the profiling of it for use to interconnect VoIP with SCN-based networks. While this was closely aligned with the original aims of TIPHON, as the project progressed it became clear that the competing work of other standards bodies was both progressing at an increasing rate and often being driven by different objectives. Generally, however, the work of other groups has mainly focused on developing specific types of VoIP technology rather than defining how they may be deployed in a practical environment. This has caused considerable debate concerning whether specific interfaces should or should not be subject to standardisation. There has also been the additional complication in that there have often been differing interpretations of the functions performed by a given entity in various bodies. For example, the debate within the TIPHON project over the split of functionality between the media gateway controller, media gateway and gatekeeper started the requirements capture process for the benefit of both the IETF Megaco working group and ITU-T SG-16. As such, the work in TIPHON at this stage may be categorised by a focus on particular implementations or 'boxes'.

These 'boxes' were being targeted at particular signalling systems, which are the real instruments of competition. The battles between SIP, H.323, MGCP and the rest embodied this problem space. It seemed that there had to be a better way than expecting one signalling system to win so that universal service provision could be obtained.

16.1.2.1 A New Way of Making Toast

With the exception of trying to maintain network reach and connectivity, most attempts to forge new telephone technologies have been founded on a single service — the support of 3.1 kHz audio telephony. With the possible exceptions of DECT and TETRA, all new base technologies in telephony have therefore done little more than support the same narrowband voice communications service. However, second generation cellular systems, and in particular the GSM system, were a much bigger success than any industry analyst could have envisaged. They provide a convenient way of supplying personal communication wherever the user is, with the added bonus of text messaging, using the short messaging service, enhancing the attractiveness of the solution and making the service even more popular. Unfortunately a disadvantage of the universal availability of the GSM system is that the service offering it enables is fully standardised. Service providers are not therefore empowered to compete on their service offering; they can only largely compete on price. It is therefore attractive to contemplate a new approach to standardisation for the communications industry that is able to provide the basis for fuelling the next generation of telecommunications networks by enabling true competition in service features.

16.1.2.2 Security

One of the underlying design goals for the Internet is the achievement of end-to-end transparency. This requires that network behaviour is independent of distance, and that addresses are globally routable and independent of location. Therefore to a machine on the Internet, the whole world looks like a big local area network. This works well for the service it supports since the network itself only transports packets and is not concerned with the content. Network transparency has aided the growth of the Internet as it has allowed new services to be deployed quickly in a peer-to-peer fashion, with users being connected directly to the server of the new service provider to access the new application. Unfortunately, public direct peer-to-peer communication has also presented an open invitation to hackers to try to disable servers and users' terminals. Network transparency has also created a situation in which all users can be, and often are, monitored from well beyond their immediate network connection. As an example, the design of modern Web browsers and their protocols allow Webmasters to establish where users have come from and to analyse their browsing habits. In some cases the links the users follow on leaving the site may also be monitored. This capability is partly created by the design of the hypertext transfer protocol [1] and partly by the transparent peer-to-peer relationship users have with sites. Therefore, while these developments are clearly useful, they are also potentially very intrusive to user privacy. As a consequence, it is highly probable that the design of the Internet of the 21st century will have to address these issues and therefore need to allow users to control the release of potentially sensitive personal information such as network address, equipment type and physical location.

16.1.2.3 Quality of Service (QoS)

For a number of reasons, and contrary to popular opinion, the Internet in its current guise cannot be considered as a viable way forward for the next-generation converged public network. These include the absence of a scalable, commercial and technical model for delivering QoS on an end-to-end basis. As a result ISPs and backbone network providers are not empowered, nor given the incentives, by today's Internet technology to provide a better service than their competitors and obtain a better price for it. In particular, despite considerable activity over recent years in the area of QoS mechanisms by the IETF [2], there is currently an absence of a complete concensus on how it should be achieved. Further, even though several QoS mechanisms have been specified by the IETF, none have seen wide-scale deployment. However, even when a technically viable solution does finally emerge, commercial models supported by appropriate charging and billing mechanisms will still be required. Fundamentally though, irrespective of the commercial or technical reason, the absence of QoS support is hampering the deployment of wide-scale real-time applications, such as VoIP on the Internet.

16.2 TIPHON Release 3

Since TIPHON's principal aim is to interwork voice-band telecommunications services across packet-based and existing switched-network technologies, it needs to maintain a focus that encompasses key target technologies rather than just any one specific technology. Therefore, while TIPHON does address the topic, it is not just about voice over the Internet. Similarly, it is not purely voice over ISDN or voice over GSM or voice over ATM either. Rather, TIPHON Release 3 is about offering voice services over a plurality of network types and the harmonisation of all types of voice telephony services over these different network technologies. More succinctly, the project is about the broader requirements of providing a multiservice environment that is transparent across commonly deployed network technologies.

As identified from the deficiencies in its earlier work, TIPHON has found it necessary to consider a very broad view of communications network interworking. This has resulted in the identification of a suite of tools that enables the problem space to be appropriately addressed. As described in the remainder of this chapter, Release 3 of the TIPHON specifications constitutes the first attempt at defining these tools and applying them to real-world networks.

16.2.1 Extensibility and Service Interworking

The development of network technology and services for the PSTN, and more recently ISDN and GSM networks, has until relatively recently been carried out in an environment largely determined by the relationships between nation states at a global level, as described in Chapter 5. This has been a contributory factor in determining the approach adopted for the development of standards and specifications across the industry. The result of this has been the use of a three-stage process which begins with the specification of a communications service, then the development of a set of supporting information flows, and ultimately the definition of the communications protocol which realises those flows. The basic premise at the service specification stage is to specify completely the behaviour of a service from an end user's point of view. This creates the problem that when a second service is specified, this specification has to consider not only the behaviour of the second service but also how this second service interacts with the first, and so on. While this has enabled the successful creation of a global communications infrastructure it has some notable and significant weaknesses for the contemporary environment, chiefly:

- it removes the ability for equipment manufacturers, network operators and service providers to rapidly innovate while maintaining the ability of achieving end-to-end service delivery;
- since the approach of specifying the interaction between services is complex, it is never completed in depth, and this approach therefore allows the consideration of the difficult problem of service feature interaction to be exposed without really solving it in any practical or meaningful way;
- it does not readily support or enable reuse of constituent functionality across different services outside the immediate one in mind, which potentially increases costs;
- there are potentially few, if any, other modern technology-based industries that adopt the cost and complexity of this approach of minute specification of end-user behaviour.

It is therefore attractive to contemplate an alternative approach that is more supportive of the needs of the contemporary communications environment, reflecting a multiplicity of equipment manufacturers, systems integrators, service providers and network operators. In particular, since the effects of deploying VoIP networks are very much to create this environment, it ultimately becomes inescapable to embrace[1].

Since TIPHON is fundamentally concerned with the interworking between different network types, including the addition of VoIP elements, it has had to consider these challenges. In doing so, TIPHON has identified an alternative approach that, although related to the existing three-stage process, represents a significant departure from the established practice. The goals for this approach are to:

- enable innovation in network equipment and services while achieving a high degree of end-to-end interworking across a heterogeneous network environment;
- achieve an approach that is no worse at handling service feature interaction than the current three-stage approach;
- recognise that the constituent functions can be distributed across a concatenation of networks.

A key tenet of TIPHON's approach is that an end service can be decomposed into its constituent elements and that these elements can be expressed as functionality in a reusable form — this can enable a high degree of interworking between disparate network technologies.

16.2.1.1 Service Applications and Service Capabilities

The view of services that TIPHON has developed [3, 4] considers an end service to comprise one or more service applications set in a commercial context. A service application is considered to represent the technical functionality that is the essence of, and realises the functionality for, a specific communications service proposition. Useful as this model may be for structuring a consideration of a service from an engineering point of view, it is the reusable constituent elements of a service application — which are known as service capabilities — that are the core of the approach. Service capabilities are, from a specification point of view, statements of self-contained functionality that can be reused across a number of service applications. A service capability definition comprises:

- an identifier or label for the capability;
- a declaration of any attributes essential to the capability;
- a declaration of the set of normal behaviours essential to the capability;
- a declaration of the set of behaviours pertinent to error conditions.

TIPHON enables innovation through allowing specified service capability declarations to be specialised by the addition of attributes or further behavioural

---

[1] Actually there is, of course, an alternative which is to have a much smaller number of network operators and equipment manufacturers who operate on a completely global scale.

declarations. This feature is predicated upon the inheritance principles of object-oriented design and offers the potential to enable service designers to manage the creation of service applications.

16.2.1.2 Service Application Creation

Recognising that TIPHON considers services to be created from service applications which are themselves comprised of sets of service capabilities, the TIPHON approach to developing service applications enables services to be described in terms of their constituent service capabilities. This effectively enables service capabilities to be used as a language for describing the behaviour of service applications at points of interworking between different networks. As an illustration, Fig 16.3 shows how a base service application can be extended through the

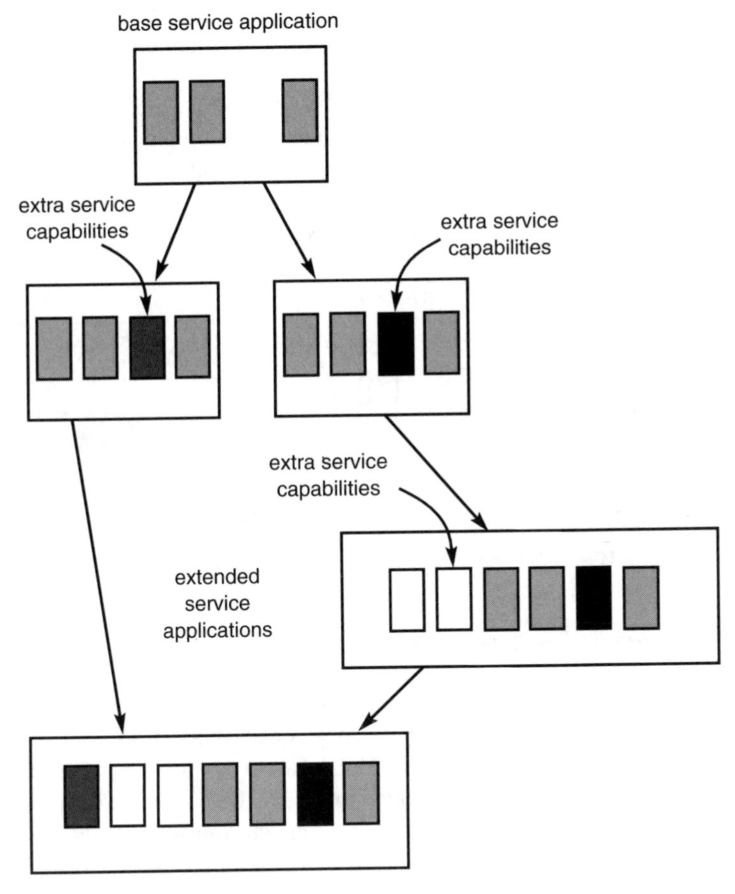

Fig 16.3 Deriving new service applications by adding service capabilities.

inclusion of further service capabilities. Note that as further service capabilities are added, new service applications are created in each case. It is this feature that renders service capabilities rather than service applications the basis for interworking and contrasts this approach with the traditional approach to service-based standardisation.

16.2.1.3 Service Interworking

TIPHON enables service-level interworking by taking a set of service capabilities and defining the set of information flows required to support them. These are then subsequently mapped on to individual network technologies, such as H.323 or SIP. Since the information flows supported by the profile of an individual network technology originates from a common set, any resulting interconnection between such networks should exhibit a high degree of interworking. Service capabilities therefore also emerge as a language for describing the interconnection of disparate network technologies.

Considering the case shown in Figure 16.4, if there are two services, 1 and 2, comprised of service applications A and B, then the common set of functionality that can be delivered when these services interwork should be the common subset of service capabilities between A and B. This common set of capabilities may therefore form the basis of any associated interconnection agreement.

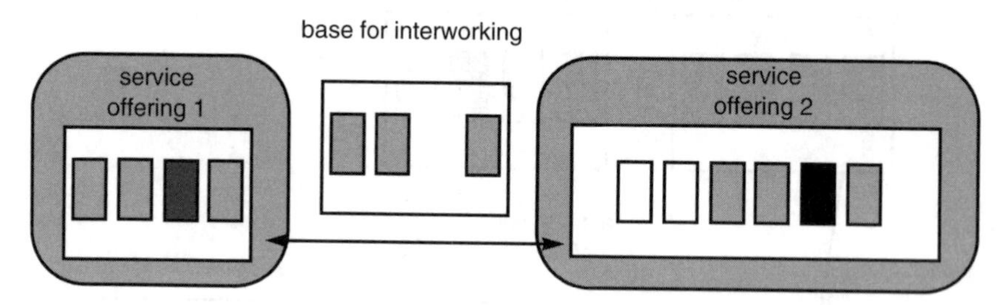

Fig 16.4 Service interworking through service capabilities.

16.2.1.4 Impact on Service Roaming

As described in Chapter 1, it is apparent, from a consideration of how the various network elements register to participate in a VoIP application, that mobility is potentially an integral feature of any VoIP solution. However, the support of differentiated service applications implies that only the home service provider can fully understand and therefore execute a given customised service application. When considering this feature in the context of a multiplicity of network domains, it is apparent that a number of approaches could be taken. In the case where a visited service network does not understand the service application involved, there are three possibilities:

- Internet model — all users contact their home directly;
- HLR-VLR model — entire service is known to the visited network;
- HomeService model — the visited network supports a base set of service capabilities to cope with standard functionality including security, QoS and media events, any additional signalling being relayed to the home network.

In the first case, there are a number of issues concerning security and QoS, both of which are difficult to achieve reliably at scale and this implies the use of the full service standardisation process which has been seen to be deficient. In the second case, exposing the entire service to the visited network tends to reduce the ability for achieving differentiated services at scale and this again implies the use of the full service standardisation process which has been seen to be deficient. The third option, as adopted by TIPHON, therefore aims to provide a meaningful balance between these extremes.

16.2.2 Architectural Design Principles

The TIPHON Release 3 architecture has therefore been developed to embrace TIPHON's approach to service creation while also addressing the shortcomings of the earlier Releases. In particular, it reflects the existence of:

- a multiplicity of networks under different ownership regimes;
- differing network technologies employing a diverse set of protocols;
- diverse service offerings from a multiplicity of network and service providers;
- the provision of quality of service as a service feature;
- appropriate security mechanisms for supporting commercial applications.

16.2.2.1 Planes

As will be appreciated, one of the more challenging aspects is to devise an extensible solution that can allow unforeseen functionality to be embraced without having to continually develop bespoke network architectures. The approach adopted within the TIPHON project has therefore been to start with a model of the problem space and then to clearly separate the concerns of the elements within the model. This was achieved by defining groups of functions, known as planes. As shown in Fig 16.5, there are four planes that comprise the major functional groupings in the TIPHON model.

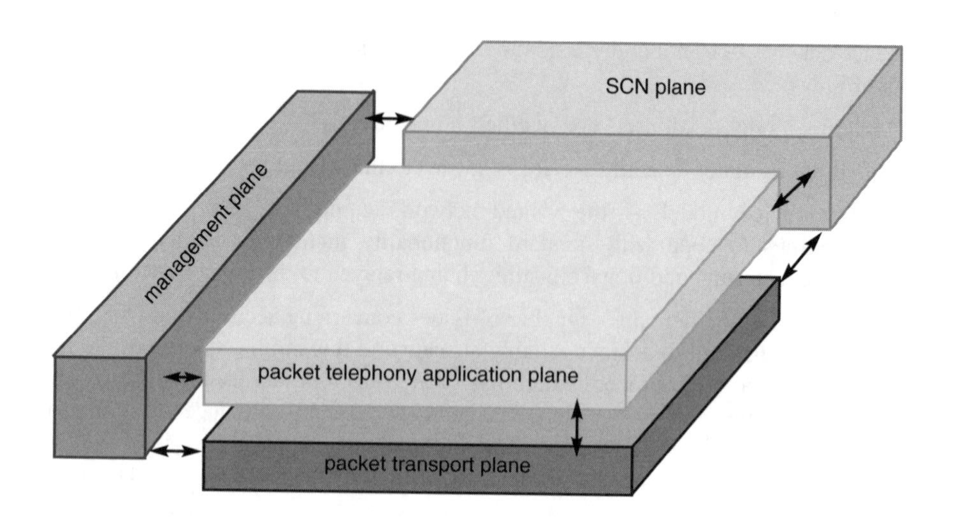

Fig 16.5 TIPHON network model.

- The packet telephony application plane

  The application plane is that part of the architecture specialised to support a given service application. Examples of such applications include voice telephony and videoconferencing.

- The packet transport plane

  The transport plane contains the generic transport functionality required to support and enable telecommunications applications. It hides details of the underlying physical transport technology from the application. The separation of transport functionality into a separate plane is an integral feature of VoIP satisfied by this approach.

- The management plane

  The management plane covers all service and network management facilities. This allows TIPHON to model one management facility across multiple applications.

- The SCN plane

  The switched-circuit network plane consists of the functionality present in legacy-switched telephone networks. Actually this plane is not a functional split *per se,* but rather an instantiation of a particular set of applications on a set of transport network services treated as a monolithic block, the application plane/transport plane not being exposed in this case. Since this functionality is well known, it will not be further described in this chapter.

16.2.2.2 Application Plane

The application plane represents the application-specific part of the architecture. Since voice calls, multimedia calls and conferencing sessions are the main focus of this chapter, this section only deals with functionality in the application plane needed for voice and videotelephony applications, and services related to such applications. Different functionality may therefore be present in this plane for other applications not considered in TIPHON Release 3.

TIPHON subdivides the application plane into five layers. These layers are used to achieve structure and reusability within the resulting architecture. The functional grouping used for each layer is bound to the functions performed and the lifetime of the related information. The layers are defined as follows.

- Services layer

  This layer contains the functionality needed to create services and maintain service-related data, such as databases for user profiles that enable user mobility and roaming. It also includes data relating to routing tables to enable number portability. The information in this layer is perceived to be relatively time invariant compared with the duration of an individual call session. Most importantly, the services layer determines the behaviour when a user is not registered.

- Service control (SC) layer

  This layer contains functionality to control services and provides a distributing interface between the services layer and the call control layer. Some examples of functions provided in this layer include terminal registration and call routing. The information in this layer may exist only as long as the user is registered.

- Call control (CC) layer

  This layer contains functionality for call set-up, call tear-down, and the provision of services during a call. Multi-party sessions are maintained in this layer. The information in this layer lives exactly as long as the call in question. Although call and session are defined differently, they are treated as equivalent in this model.

- Bearer control (BC) layer

  In the bearer control layer, the mapping is performed from the call topology to bearers representing the individual media flows. This layer also negotiates details of the media flows in the bearers, e.g. voice quality and encoding. Information in this layer lives as long as a media flow is desired.

- Media control (MC) layer

  Most calls and sessions require media. This layer is responsible for the realisation of the media flows. The layer negotiates the properties of individual media flows, including media encoding and addresses for the media flow, with the bearer control layer. The media control layer communicates with the transport plane to establish, maintain and destroy a media path across the relevant networks with the appropriate QoS and firewall settings. The information in this layer lives as long as a media flow exists, which may be less than or equal to the lifetime of a bearer.

Despite the description of these layers being rather abstract, the functions in many of today's telecommunications systems can be easily recognised. Using this model of planes and layers, a modular architecture [5] that permits a large amount of reuse can be obtained, for example:

- call routing and user mobility functions may be shared among a number of call-oriented applications;
- a multimedia call may have two or more bearers and as many media flows;
- a multi-party session may have a number of bearers that is equal to the number of participating users multiplied by the number of simultaneous media flows.

16.2.2.3 Transport Plane

The transport plane represents the functionality providing the actual transport of signalling and media packets. Due to the focus on VoIP networks interworking with existing SCNs, in TIPHON Release 3, the transport plane will typically be assumed to be either IP or ATM based. However, the model is equally applicable to other types of transport network including TDM voice networks using C7 signalling and X.25 or frame relay based data networks. Although the transport plane is only required to arrange the bit-pipes necessary to enable the application, there are two important services provided by the transport plane — quality of service and low-level security.

Transport security is an increasingly important topic for packet networks. Users would like to achieve private, end-to-end, uninterrupted and authenticated communication. In addition, the public assume that the network used by the PSTN is not subject to unlawful eavesdropping or theft of service. On packet networks these requirements are more difficult to provide because in the most common form of packet network, IP over Ethernet, everyone broadcasts their messages over a shared access media assuming the intended recipient will see them and everyone else will simply ignore them. This particular issue proved to be a significant problem with early VoIP trials on the equivalent shared access network provided by cable access television networks in Scandinavia, as it was difficult to prevent eavesdropping. From an architectural perspective, TIPHON therefore assumes that appropriate transport security is handled in the transport plane because it is relevant to all applications.

16.2.2.4 Management Plane

This is currently the least developed plane within the architecture. It is intended to offer management procedures that, like the architecture, are not dependent upon the technology used. The intention is to incorporate functions that allow network operators to concentrate on providing management of their services rather than designing to specific technology constraints.

16.2.3 TIPHON Approach to QoS

As discussed in Chapters 2 and 3, quality of service is a recurring issue in deliberations concerning packet transport services and, in particular, those based on IP. It is generally understood that many applications are hampered, if not made impossible, by the lack of QoS in many of today's best-effort IP networks. However, the belief, that QoS-enabled transport will only be deployed if the investments allow the generation of revenue, is an underlying principle of the TIPHON approach. This implies that a QoS-capable network should only offer its services to paying customers. In this context it is important to recognise that TIPHON assumes a scope for the definition of 'customer' that is sufficiently broad to include end users, businesses or other similar bodies, even other service providers and network operators. While it is necessary to have enough bandwidth available for the application, bandwidth alone is not sufficient since, outside a local area network, applications can generally scale to consume all the bandwidth available. Other parameters that are critical to applications that involve real-time media streams are delay, jitter, and packet loss. The ability of today's networks to deliver the desired values for these parameters depends critically on the network load and so no absolute guarantees can be provided for the QoS delivered.

From a commercial point of view, the billing of QoS flows can only be performed if the network can associate particular packet flows with particular users. A QoS-enabled transport network will therefore require some means of access control to protect the network from unauthorised usage and overloading. TIPHON addresses this problem by placing user authentication in the application plane and defines an interface between the transport and application planes that allows the entities in the application plane to request an end-to-end bit-pipe through the transport network. In this way, the application interface can be used to relate packet-flows with customers and application-level billing mechanisms can be used to bill for the QoS received as part of the service offered. How the transport network actually implements the details of its access control and QoS schemes is beyond the scope of TIPHON. This enables innovation in transport network technology, such as diffserv, RSVP and the like [2], to be exploited in a reusable manner by a diverse set of applications. A service provider may wish to exploit such a capability by offering differentiated services. For example, it may be possible to offer a guaranteed service for certain users and applications, and best-effort class of service to other users or applications on the same network. In such cases, it is not unreasonable to expect that multiple grades of service may be charged differently. This differentiation may warrant static provisioning of a certain class of service to a certain user or real-time allocation for users or services.

Naturally, it must also be possible for the service provider to block access to a grade of service for a customer if they are not, or are no longer, entitled to use it. Therefore if quality of service differentiation is to be provided on a network, differentiated access control will be required. The access control mechanism prevents unauthorised users from interfering with the service offered to packet flows with guaranteed QoS. TIPHON has therefore defined control functionality that can oversee QoS, access and security policies. This functionality can address the needs of end-to-end transport policies and may interface with an application domain through a transport interface element to achieve this. However, the requirements for QoS support also imply that the transport network must provide sufficient end-to-end bandwidth capacity to be able to perform within the requirements for absolute delay, loss, and jitter. Fundamentally though, there is, in the end, no 'magic' solution for implementing QoS — it has to be engineered into the solution, as previously identified in Chapter 2.

16.2.3.1 Comparison — the Internet Model

The traditional Internet communication model assumes that each end-point has a unique, fixed, address and that the routers in the network transparently transport the packets these end-points wish to exchange. From a VoIP perspective, Fig 16.6 shows how the end-point communicates with the network and subsequently how the networks communicate with each other using this model.

As described in Chapter 1, application-level signalling may either take place via a call server in a service domain or directly peer-to-peer between the end-points. This approach is a viable option where the transport domains share a common basic set of policies and transport mechanisms. However, addressing, access, QoS mechanisms and policies all have to be identical for this case to work or there has to be an effective one-to-one mapping of heterogeneous mechanisms. An example of this would be the case where two users want to establish a videoconference using Microsoft NetMeeting.

Each user has to know the other party's global IP address and has to establish a QoS-controlled end-to-end path delivering the right quality of the video streams. This means that the users have to be able to:

- reach each other — which implies a global addressing and access policy;
- allow end-to-end QoS set-up — which implies a globally applicable QoS mechanism and policy.

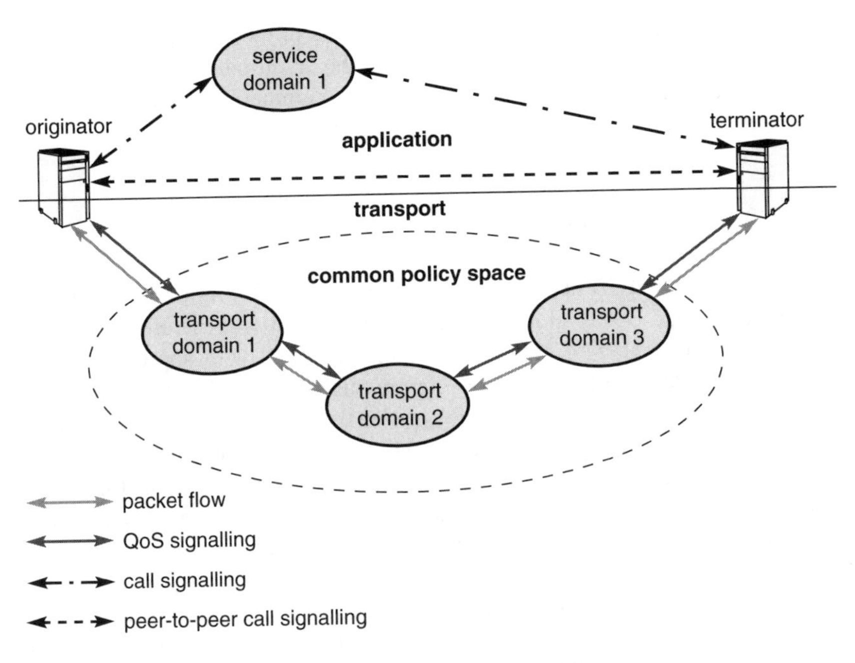

Fig 16.6 The Internet model.

To provide QoS guarantees in such a network implies that interworking between the different mechanisms in use must be standardised, e.g. RSVP to diffserv or MPLS. In fact, not only must the mechanisms and their mappings be standardised, but also charging mechanisms must be in place to account for the QoS delivered. Currently few good mappings are possible because the existing QoS mechanisms lack a consistent set of parameters permitting generalised conversion from one to the other. It is also unreasonable to generally assume that such policies will be homogeneous in a large-scale, commercial deployment. Furthermore, the question of who is responsible for delivering, guaranteeing and billing for end-to-end QoS remains an open question. Problems of lack of homogeneity also become apparent when media streams traverse network address translation (NAT) entities on the edges of enterprise networks. NAT dynamically maps the user's internal IP address and port to an external address and port. However, VoIP and similar session-oriented applications use multiple streams for signalling and media that have to be correlated. For example, the signalling stream has to make reference to the media stream for the remote party to be able to use it. If the address of that stream is changed by an address translator in the middle of the IP network, the signalling stream points to the wrong address and the other party in the call cannot correlate the flows or send back media. Today this issue is fixed on an *ad hoc* basis by the use of application level gateways (ALGs) that are installed in the NAT device. Since ALGs are specific to a given application they have to be updated to support extensions or modifications to the application. Some of the basic premises stated in the previous section are therefore counter to the traditional end-to-end Internet philosophy. Instead they are perhaps closer to the commercial assumptions behind traditional telephone networks.

16.2.3.2 The TIPHON Model

The basis of the TIPHON approach is the commercial offering of packet transport with appropriate QoS. Any global transport deployment will therefore, in general, be made up of a number of separate transport operators with appropriate service level agreements (SLAs) providing the commercial framework for carrying packet flows over their network. The TIPHON model therefore blends elements of both telecommunications and data networks. In this general case, transport is made up of a number of heterogeneous transport domains. The application functionality controls the transport domains and translates the end-to-end transport requirements to domain-by-domain semantics for QoS, addressing, billing, accounting and the like. Each transport operator may govern one or more transport domains. A transport domain is an entity representing a collection of transport resources sharing a common set of policies and QoS mechanisms, and under common administrative control. It comprises transport-related functionality including routing, switching and policy enforcement. Each transport domain will have a consistent set of policies and/or differ from other domains, for example, in terms of administrative control, QoS or transport mechanisms (RSVP, diffserv, MPLS, ATM), access, metering, addressing schemes (global, public, private), and network protocols such as IPv4 or IPv6. Within this model, end-to-end transport involving multiple transport domains has to be achieved through the co-ordination of all the domains involved. If the domains cannot be assumed to be transparent because address translation and firewalls are in place, end-to-end transport can be achieved through application-level control via an application service domain associated with each transport domain. Application-level control is achieved by means of application-level functionality that can ‘see’ beyond the boundary of a single domain, as illustrated in Fig 16.7.

The service domain can therefore request transport resources from each of the transport domains and establish the interconnection in a controlled fashion. Information exchanged between each service domain involved in the session and its associated transport domain(s) will then ensure that the required transport flow parameters are met by each transport domain involved in the session in accordance with the SLAs between the service providers and network operators administering the domains.

Where a service domain has commercial contacts with multiple transport domains it will need to have knowledge of the routing possibilities between the different transport domains and their peers. It is to be expected that a service domain

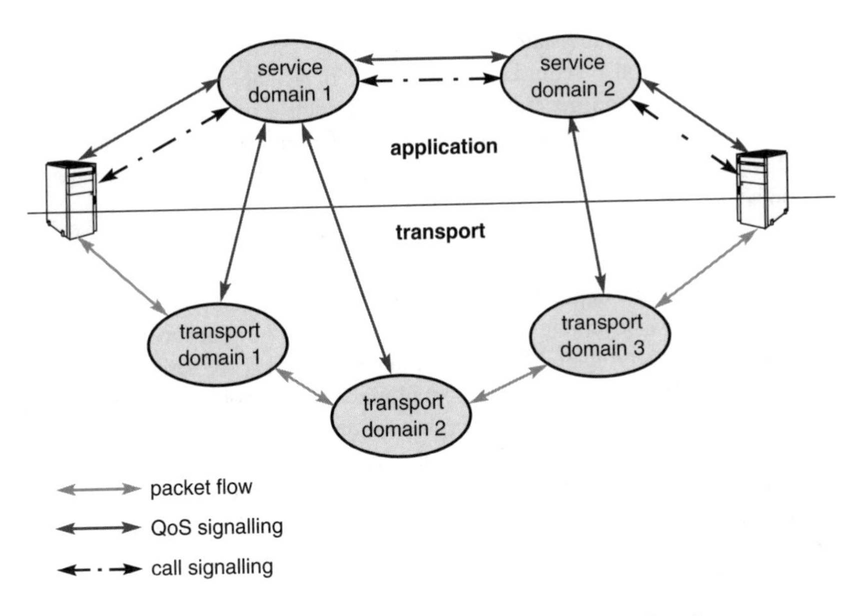

Fig 16.7 Application-level control of transport domains.

faced with this situation will introduce a virtual transport provider in its own domain in order to hide this complexity from its applications.

16.3 TIPHON Release 3 Meta-Protocol

The use of a meta-protocol and its mapping to concrete protocols is an essential feature of the Release 3 approach. This section describes the meta-protocol and illustrates how it can be mapped to some common VoIP protocols.

16.3.1 An Introduction to the TIPHON Meta-Protocol

Traditionally, when signalling protocols are developed, both the sending and receiving parties use an implied state model. It is the messages between the parties that cause them to alter state, based on the semantics conveyed and the previous known state information. The parties share not only coding methods and the semantics associated with the individual encoding, but also rules about how to behave in the event of inconsistencies and methods for overcoming lost, or partially lost, messages.

Generally signalling protocols can be developed upon a number of lines and make different choices concerning transport behaviour, means of coding, and assumptions concerning where state information is held. Therefore, in order to interwork, each signalling protocol must have logic that generates a new state machine that encompasses the states in both signalling systems at the point of interworking. Interworking also requires the definition of a mapping for all the code points in both systems and generates messages in such a way that the messages exchanged between systems cause the desired behaviour in both of the interconnected signalling protocols. Experience has shown that maintaining universally understood semantics and making services transparent are often not achieved.

Since the IETF and the ITU-T, among other groups, are actively engaged with communications protocol development, there are always many protocols in various stages of development and deployment at any one time. In addition, there a number of derived mappings now becoming available. For example, the 'PacketCable' CMSS protocol [6] and the emerging 3GPP SIP signalling protocol [7] are derived from SIP. From an interworking perspective this multiplicity of protocols represents a significant challenge, because each of them has to make assumptions concerning the messages and code points available. This means that the service definition is different on either side of an interworking point, often only in a subtle way, the consequence being that interworking is a complex problem involving a large number of compromises. More quantitatively, if there are n protocols to interwork, then there are $n(n-1)$ interworking approximations that need to be developed.

TIPHON Release 3 addresses this problem by introducing the use of a meta-protocol to manage the overall complexity of achieving multi-protocol interworking. As indicated in Fig 16.8, the process of interworking is defined in terms of rules for encoding the code points, mapping the messages and modifying states. As a consequence, the number of inter-working approximations is reduced to n rather than $n(n-1)$. A service application is therefore defined in terms of the meta-protocol rather than any of the protocols used at the point of interworking. A mechanism is then defined for interworking to each desired or candidate protocol. So a mapping to and from the meta-protocol that is designed to support the services

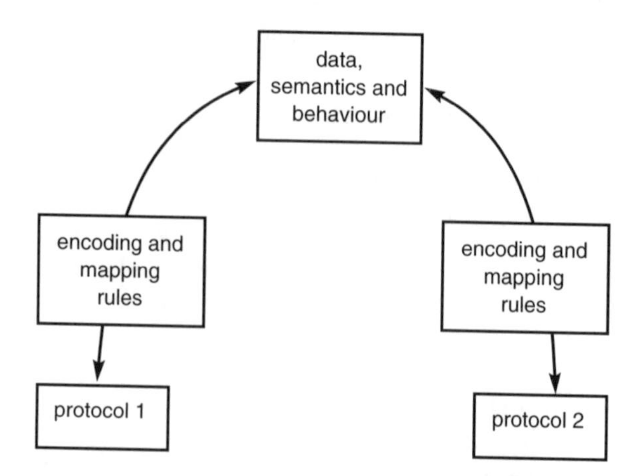

Fig 16.8 Meta-protocol-enabled interworking.

needed is defined for each concrete protocol to be used. This mapping must also take into account the behaviour of underlying transport layers and protect against message loss and so on. The derivation of these mappings is a complex task. However, it is not always possible to apply the meta-protocol to generate a complete mapping in a given protocol. This either results in extensions being discovered for the meta-protocol or deficiencies being identified in the chosen concrete protocol.

16.3.2 Meta-Protocol Derivation

The derivation of the meta-protocol starts with providing a general model of the states in a single media call. For TIPHON Release 3 this is taken to be a voice call, loosely based on the Q.931 and IN call-state models. This is extended for supplementary services and multi-domain services, showing how that model can be refined to provide these services.

The basic call consists of a number of activities that have to happen before the call can be set up.

1. The call is initiated.
2. The call is authorised by the service provider.
3. The call is routed towards the destination.
4. Call associated bearers are set up.
5. Media flows are requested for the bearers.
6. Appropriate transport is arranged for the media flows.
7. The call and bearer requests are forwarded to the next domain (or terminal).
8. The next domain confirms the bearer request.
9. The media flows are established for the bearers.
10. The transport is established for the media.
11. The call is accepted by the recipient.
12. (Optionally) the media is activated (in case the forward channel is disabled to enforce billing policies).
13. (Optionally) the transport is activated.
14. The media flows.
15. The call is torn down, and along with it the bearers, media and transport.

Figure 16.9 shows these steps in a schematic way. If this sequence were not followed unexpected things could happen. For example, the perceived post-pick-up delay, the time from receiver pick-up to media establishment, might be very large or the first few seconds of speech might be choppy until QoS transport is arranged for the media path.

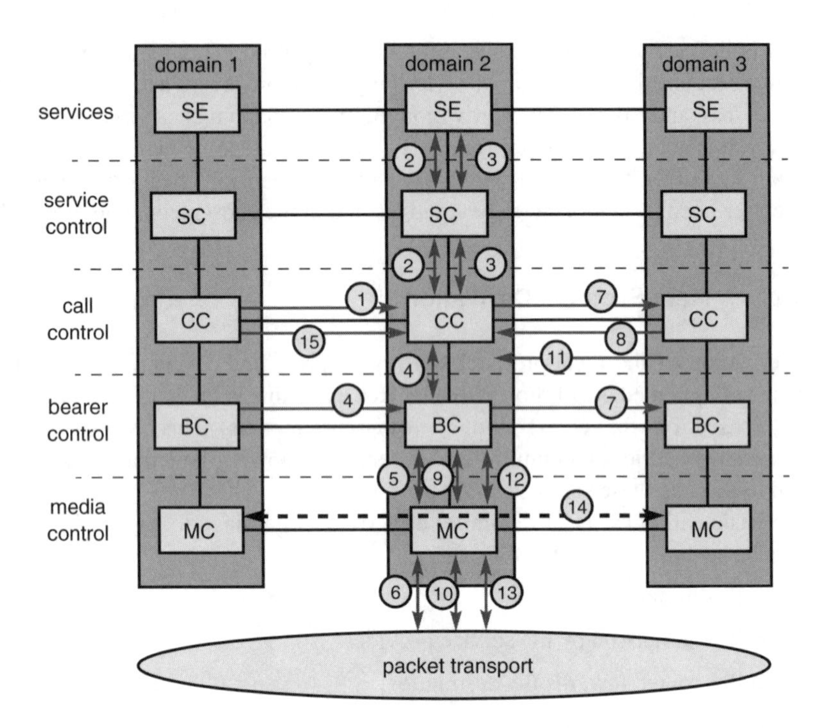

Fig 16.9 Schematic call set-up diagram.

Figure 16.10 shows these fifteen steps in the form of an information flow. However, it does not show special cases such as rejection of requests and overlap sending of called party number. In particular, the routing request flow performed in step 3 hides any enhanced call routing mechanisms such as number portability and intelligent routing schemes. Note that the bearer is set up in three phases:

1 the originator offers their capabilities for sending and receiving and commits to receive what was offered (steps 5, 6);

2 the terminating side responds and offers their capabilities and the appropriate reverse stream can be established (steps 9, 10);

3 when the call is accepted, the reverse speech channel is opened (steps 12, 13).

The service provider can enforce the access policy in the third phase. In particular, information flows down to the transport plane and up to the service control layer can implement strict access and usage control in both the transport plane, where it can block any non-authorised streams, and in the application plane, where it can try to authorise users before committing resources.

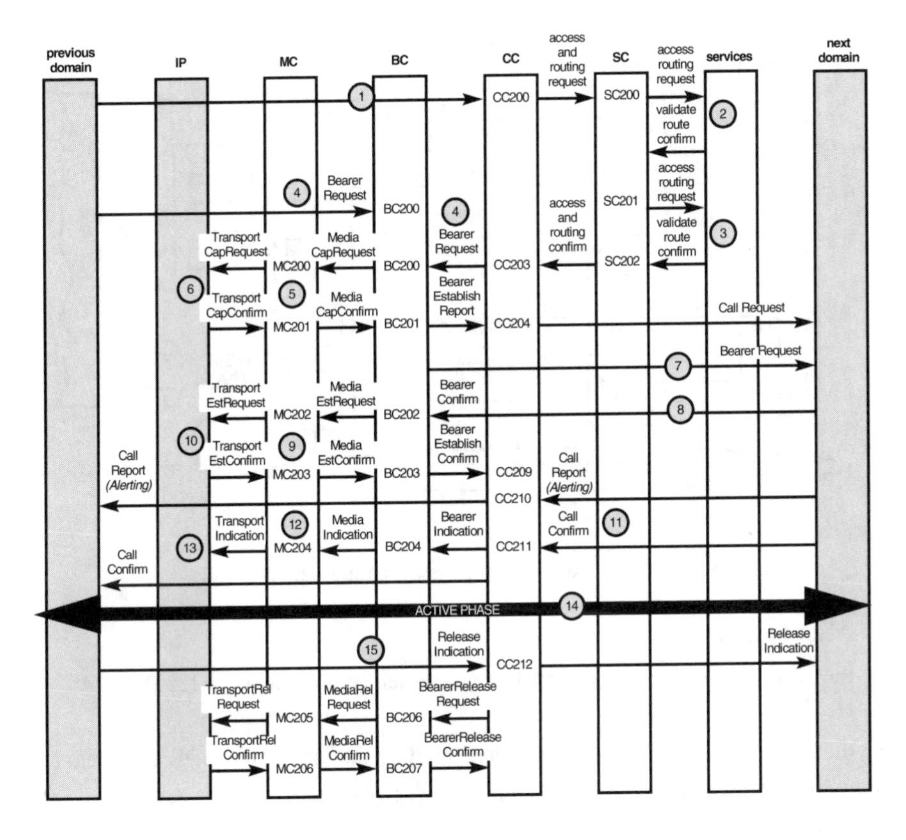

Fig 16.10 High-level normal call set-up information flow.

The next section applies this general model outlined above to enable interworking between the BICC, SIP and H.323 protocols for the case of a basic voice call. It serves as an overview of how the model may be used and how it can be applied in real-world situations. Other protocols may be bound into the model in a similar fashion.

16.3.2.1 Mapping

In this section the information flows are mapped back to real-world specifications. Figure 16.11 gives a high-level overview of H.323, SIP and BICC, which are some of the more common protocols, and their mapping to the architecture. Physical elements, such as a terminal and media gateway, are shown with protocols between them mapped to the layers in the architecture.

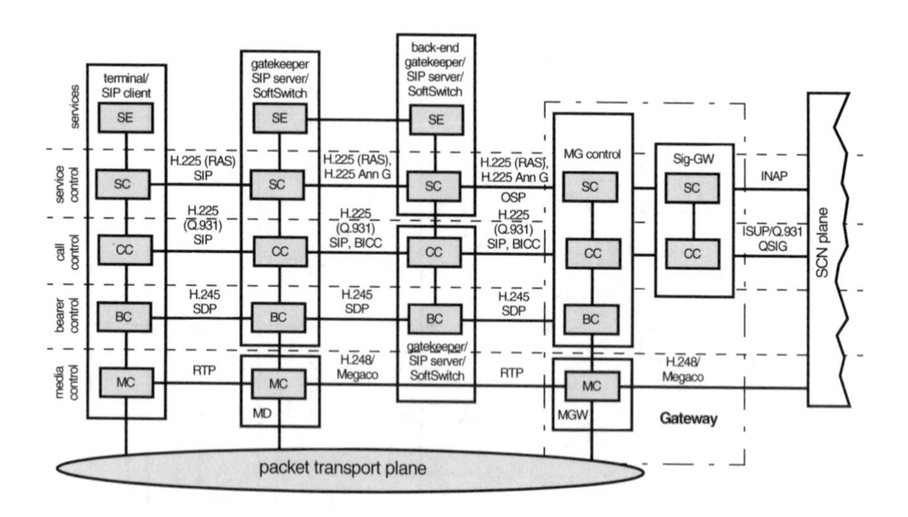

Fig 16.11 Mapping to physical entities and protocols.

Also shown in Fig 16.11 are:

- the H.323 flows (see Chapter 10), represented by the labels H.225.0, H.245 and H.248;
- the SIP flows (see Chapter 11), represented by the labels SIP, SDP and Megaco;
- the BICC flows (see Chapter 8), labelled as BICC (BICC also uses H.248 for the vertical interface);
- media streams, with the label RTP (see Chapter 1).

The mappings to these protocols are described in detail in the remainder of this chapter in the form of examples.

16.3.3 Meta-Protocol to Concrete Protocol Mapping

This section describes the mapping of the TIPHON meta-protocol into three of the VoIP protocols considered most common in the industry today. These examples assume that an IP-based user end-point can communicate with a network service entity, as a front end, while another network entity performs the back-end functionality of call admission control and call routing. Where required an interconnect element is implemented as a controllable firewall or media gateway controlled by the first network element.

16.3.3.1 H.323

Figure 16.12 shows the mapping of the information flows 1-14 (as shown in Fig 16.10) to H.323 messaging flows. In this case, the BC, CC and SC layers are implemented as one gatekeeper. The MC layer is implemented as the media gateway/firewall and the services layer is implemented as the back-end gatekeeper. In this example, the protocol used to communicate with the IP transport is RSVP and the QoS transport is spread across several domains. For simplicity, the domain described here has RSVP between the terminal and the interconnect element; what happens beyond that is outside the scope of these examples.

In the example shown, the call request and bearer request are implemented using the H.323 FastConnect option, as described in Chapter 10. When FastConnect is used, the H.225.0 SETUP message carries H.245 information elements, combining steps 1 and 4 of the information flow of Fig 16.10. If FastConnect were not used, the H.225.0 SETUP message would implement the call request (step 1) and an H.245 open logical channel request would implement the bearer request (step 4). A similar story holds for the subsequent messages (ALERTING and CONNECT) on which the bearer information is piggybacked in the form of H.245 information elements.

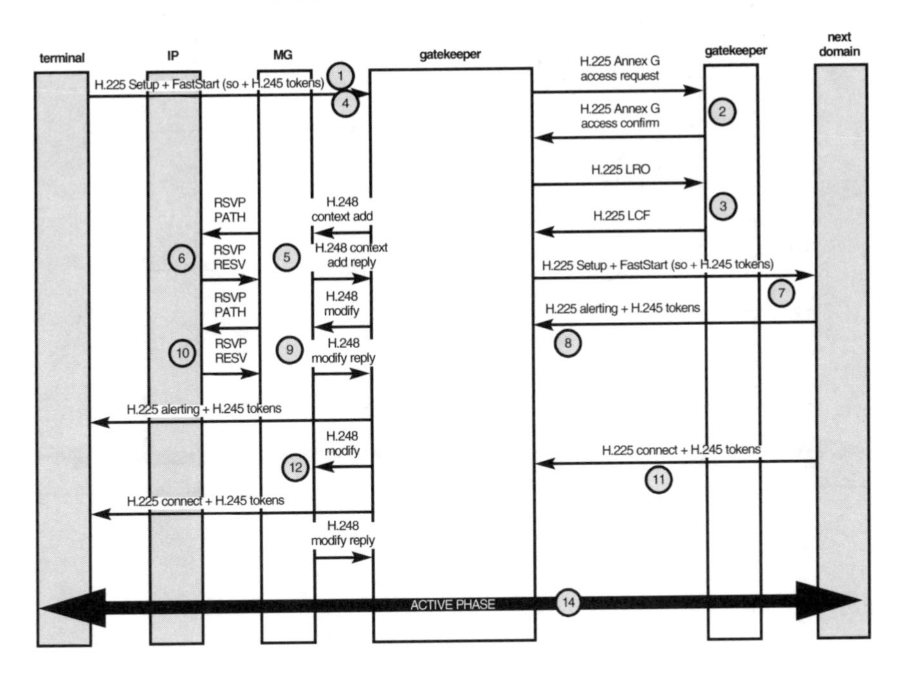

Fig 16.12 H.323 flows.

16.3.3.2 SIP

Figure 16.13 shows the mapping of the information flows 1-14, given in Fig 16.10, to SIP messaging flows. In this example, the BC, CC and SC layers are implemented as a single SIP proxy server. The MC layer is implemented as the media gateway/firewall, and the services layer is implemented as the back-end SIP server. As with the H.323 example considered previously, RSVP is the protocol used to communicate with the IP transport function. Since SIP does not allow separate signalling for bearers but includes SDP session descriptions in the SIP messages, steps 1 and 4 are always performed simultaneously. The same holds for the RINGING and OK messages where the bearer information is piggybacked in these messages in the form of SDP structures.

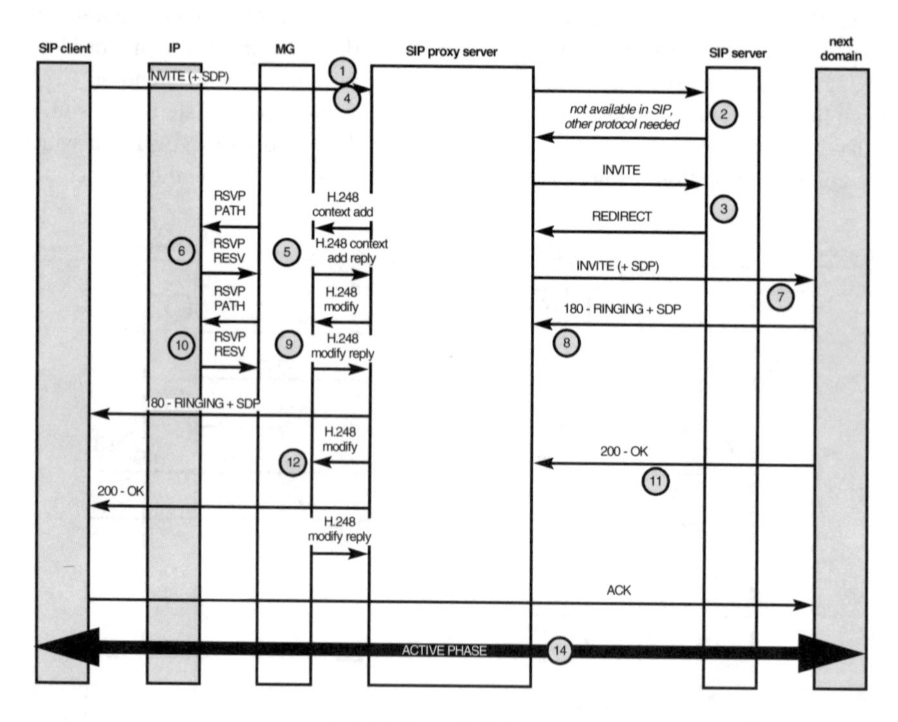

Fig 16.13 SIP flows.

16.3.3.3 BICC

Figure 16.14 shows the mapping of the information flows 1-14 to BICC messaging flows. In this example, the BC, and CC layers are implemented as one BICC call service function (CSF). The MC layer is implemented as the BICC bearer interworking function (BIWF), and the SC and services layers are implemented as a back-end service platform, presumably as an IN service control function. Since the first version of BICC (CS1) supports ATM rather than IP, ATM signalling is used in the communication with the transport network. In this example, an IAM message has been chosen to implement the CallRequest and BearerRequest, so steps 1 and 4 are performed at the same time.

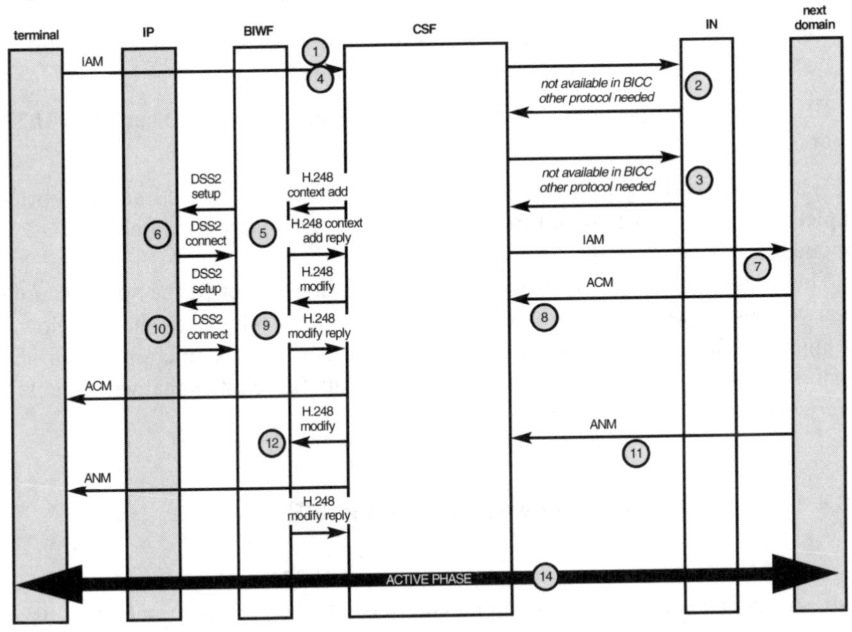

Fig 16.14 BICC flows.

16.4 Interworking through TIPHON

Meta-protocol interworking between SIP, BICC and H.323 is an emerging topic at the time of writing. As a consequence the only available literature is documented in drafts and proposals that have been submitted to standards bodies for consideration. Recognising the relative rate of immaturity of specifications in this area, the following sections aim to demonstrate how the TIPHON Release 3 meta-protocol can be used as the basis for defining interworking between the three example protocols previously considered. From Figs 16.12 to 16.14, it is apparent that H.323, SIP and BICC information flows can be rather similar. The states of the protocols also look similar since they all conform to the Q.931/IN model of call-state behaviour by virtue of the meta-protocol mapping. Interworking may therefore be done at an entity with two faces. For example, on one side it could look like an H.323 gatekeeper while on the other it may look like a SIP proxy server or a BICC

CSF. However, there are issues that may further hamper interworking, depending upon the properties of the protocols in question. In this case these include:

- H.323 allows the separate exchange of call request and bearer request information and SIP does not — in fact, the H.323 base mode of operation is to send these exchanges separately;
- some details of the information exchanged in the H.323 and SIP messages do not necessarily map — this mapping remains for further study;
- BICC for IP transport (BICC CS2) is more limited in its scope than either H.323 or SIP.

While interworking between these protocols is possible, given an appropriate implementation of the specifications, it is not necessarily straightforward in practice.

This is because they may not implement the same subset of the specifications. Interworking between these protocols is therefore not trivial, yet, as shown below, it is apparent that interworking can be greatly aided by the meta-protocol. For example, it is possible to define compliant H.323 and SIP entities that will be able to interwork using the meta-protocol.

16.4.1 Interworking between H.323 and SIP

Figure 16.15 shows the meta-protocol mapping to H.323 to the left and SIP on the right. As can be seen, the individual messages can be mapped relatively directly. In addition, closer inspection of the relevant specifications, shows that the meta-protocol-based behaviour does not seem to contradict the standards.

Extensive study of the individual protocol message parameters, as performed in upcoming TIPHON deliverables for H.323 [8], for SIP [9], and for H.248 [10], show that most of the parameters that are needed for call set-up, QoS and security can be provided by the protocols as they are. However, some small additions are required to be made to allow fully for QoS capabilities at this time.

16.4.2 Interworking between H.323 and BICC

Figure 16.16 shows interworking between H.323 and BICC enabled through use of the TIPHON meta-protocol. As with interworking between H.323 and SIP, the individual messages can be mapped relatively easily. Also the BICC behaviour seems rather well aligned.

Unfortunately BICC CS2 does not support per-call QoS, just like the PSTN, as it assumes that a single call quality is delivered. The TIPHON QoS mechanism is therefore currently being incorporated into BICC CS3, which is an example of how

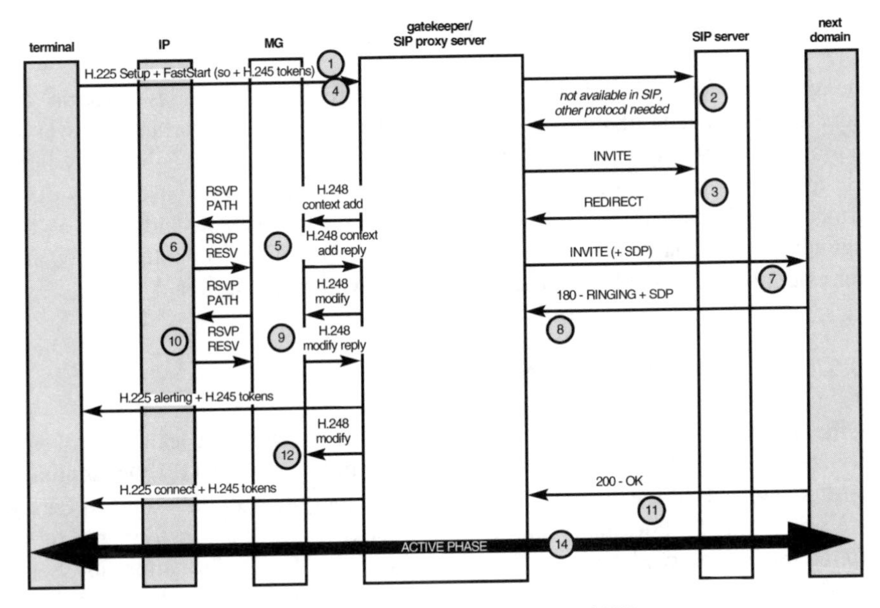

Fig 16.15 Interworking H.323 and SIP.

the TIPHON approach is being used to enable improvements in existing protocol developments elsewhere.

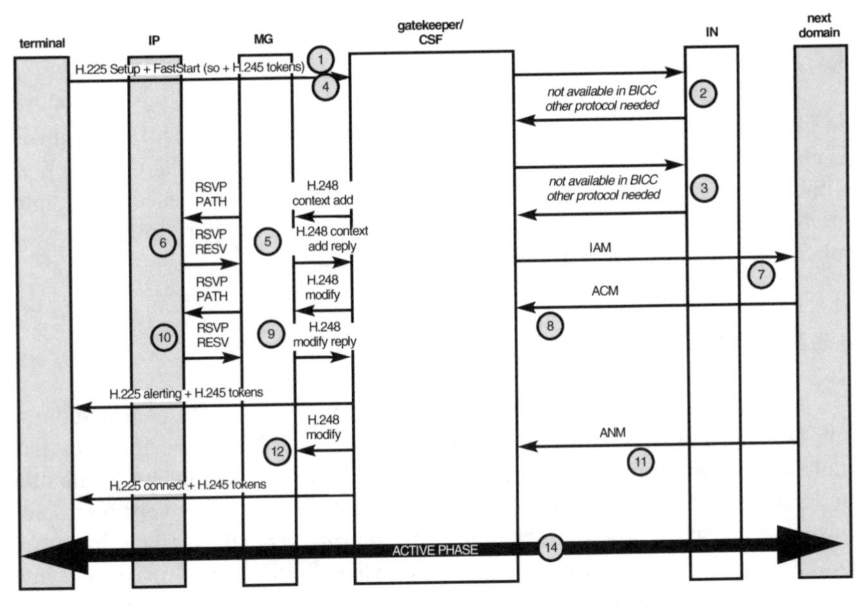

Fig 16.16 Interworking H.323 and BICC

16.5 Possible Future Directions

The current scope of TIPHON Release 3 deliverables provides a basis for addressing the needs of multi-domain, multi-technology interworking. However, due to the complexity involved with the problem space, Release 3 has only been able to address basic telephony services. It is expected that this base-line will be further consolidated in Release 4 of the project. However, as indicated at the beginning of the chapter, the goal is to go much further than simple telephony, and four examples for possible future directions are given in this section.

16.5.1 Third-Party Services

In the TIPHON Release 3 model, communication between the call control and services layers is achieved through the service control layer. The TIPHON approach therefore lends itself ideally to the support of third party service APIs such as Parlay [11] or OSA [12]. Potentially this enables service capabilities in this area to be offered by third party players. As introduced in section 16.3, information flow steps 2 and 3 in the general model provide the basis for developing such mechanisms. TIPHON Release 5 is expected to address further support for APIs by offering in-call media events, such as tones and announcements, and multi-legged call handling. In the same TIPHON Release, alignment with the API choice of 3GPP can be expected.

16.5.2 Presence Applications

The TIPHON model supports a registration session that is not explicitly described in this chapter. This enables TIPHON service providers to know when their users are on-line. This mechanism can easily be employed to support presence-based applications such as instant messaging. Extensions to support presence-aware applications are expected in TIPHON Release 5.

16.5.3 Management

Network and service management are an integral element of any consideration of multi-domain, multi-technology networks. The current TIPHON specifications have sought to develop an appropriate management framework, in conjunction with other standards groups. Since the TIPHON Release 3 activities effectively represent a significant departure from earlier work, the management deliverables have been deferred to a later release. However, one area where TIPHON has made substantial progress is in the service management area of authorisation and settlement. In particular, TIPHON has developed the open settlement protocol (OSP) [13], to address the needs of inter-service provider usage, authorisation and settlement between VoIP networks. Although not a specific focus for the chapter, further details concerning OSP are discussed in Chapter 14.

16.5.4 eCommerce Application Support

The TIPHON registration session may be engineered to form the basis of many electronic and mobile commerce applications. These types of application have the potential to complement those envisaged by the TIPHON Release 3 specifications, by extending the core functionality beyond simple voice communications. It is expected that TIPHON Release 5 will address functionality in this area.

16.6 Summary

This chapter has highlighted the challenges that need to be considered when deploying large public-scale networks involving VoIP. It has reviewed the activities of the TIPHON project that aim to address the problems identified and considered the approach taken by TIPHON Release 3. In particular, this chapter has presented an overview of the TIPHON Release 3 use of service capabilities, the development of an abstract architecture and an associated meta-protocol. It has demonstrated how this approach can be applied to profiling specific network technologies, including H.323, SIP and BICC. It has also demonstrated how the TIPHON approach can be used as the basis for defining interworking functions between network protocols. The ability to extend the TIPHON approach to enable more advanced applications has also been described.

References

1 IETF: '*Hypertext Transfer Protocol/1.1*', RFC 2616 (1999).

2 Willis, P. J. (Ed): '*Carrier-scale IP networks: designing and operating Internet networks*', The Institution of Electrical Engineers, London (2001).

3 ETSI: '*Telecommunications and Internet Protocol Harmonization Over Networks (TIPHON); Requirements Definition Study; Scope and Requirements for a Simple Call*', TR 101 877 (2001).

4 ETSI: '*Telecommunications and Internet Protocol Harmonization Over Networks (TIPHON) Release 3; Service Capability Definition; Service Capabilities for a simple call*', TS 101 878 (2001).

5 ETSI: '*TIPHON Release 3, Information Flow and Reference Point Definition, Network Architecture and Reference Points*', TS 101 314 (2000).

6 PacketCable™: '*Call Management Server Signalling Specification*', pkt-sp-cmss-I01-001128, Cable Television Laboratories Inc (2001).

7 3GPP: '*IP Multimedia Call Control Protocol based on SIP and SDP; stage 3*', TS 24.229 (2000).

8 ETSI: '*TIPHON Release 3, Interface Protocol Requirements Definition, Implementation of TIPHON Architecture using H.323*', TS 101 883 (2001).

9 ETSI: '*TIPHON Release 3, Interface Protocol Requirements Definition, Implementation of TIPHON Architecture using SIP*', TS 101 884 (2001).

10 ETSI: '*TIPHON Release 3, Interface Protocol Requirements Definition, Implementation of TIPHON Architecture using H.248*', TS 101 885 (2001).

11 Parlay — http://www.parlay.org/

12 ETSI: '*Open Service Access*', ES 201 915 (2001).

13 ETSI: '*Telecommunications and Internet Protocol Harmonization Over Networks, (TIPHON), Open Settlement Protocol (OSP) for Inter-Domain pricing, authorization, and usage exchange*', TS 101 321 (December 1998).

17

EVALUATING VoIP TECHNOLOGY — BALANCING TECHNOLOGY PUSH AND MARKET PULL

A J Heron and A H Warren

17.1 Introduction

VoIP is a maturing, yet still emerging, technology, and its products are therefore both relatively new in the market-place and continually evolving. Such a rapidly changing technological environment for service providers, equipment manufacturers and end users is compounded by continually developing regulatory policy. Any potential purchaser of a new technology therefore has the dilemma that equipment may become rapidly dated by new developments or lack of features subsequently identified as essential requirements for commercial applications. As a consequence, establishing the precise capabilities of specific offerings at a given point in time is essential for anyone with a serious interest in the VoIP industry. This chapter describes an approach to evaluating VoIP solutions, the problems associated with the testing of VoIP networks and components, and the tools and techniques employed to do so. It outlines the capabilities of the VoIP technology evaluation test bed, which is administered by BTexact Technologies at Adastral Park, and includes some typical results obtained using perceptual analysis measurement and IP network simulation technologies to assess voice quality.

17.2 Evaluation Stakeholders

There are a number of major stakeholders in the VoIP industry, as discussed in Chapter 1. Each of these groups reflects a specific interest in what the technology currently offers and what it may offer in the future. Exploring the capabilities a particular product offers and predicting what it may be possible to deliver in the future falls within the realm of technology evaluation and assessment. While realising substantial benefits, technology evaluation can represent a significant investment. It therefore needs to be driven by the key stakeholders sponsoring a given assessment study to ensure the potential benefits are fully accrued. The main stakeholders and the relationships that may exist for a particular assessment are reflected in Fig 17.1.

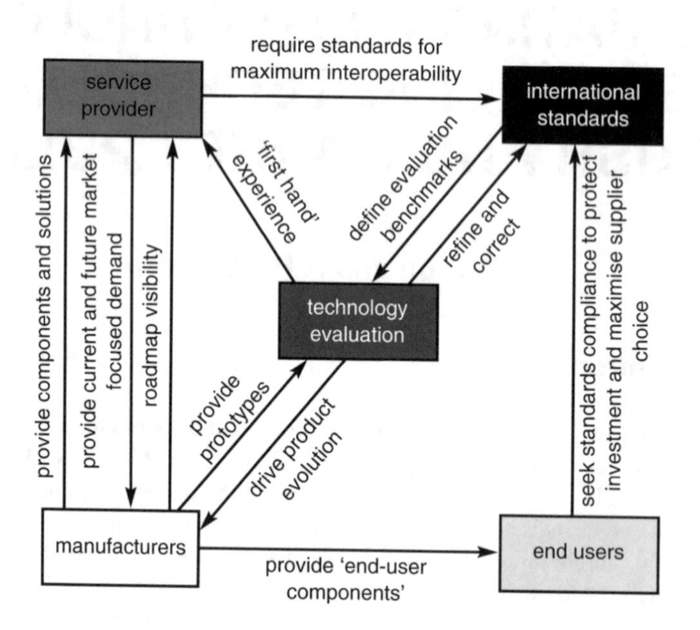

Fig 17.1 Stakeholder relationships and benefits.

The prime mission of the service provider (which for the purposes of this chapter is considered to include network operators) is to economically deploy network solutions for the benefit of end customers. For large-scale network solutions, these should ideally leverage capital investments while requiring minimum recurring or incremental costs. This necessitates that the platform components be resilient, scalable and flexible enough to enable exploitation for multiple services, where possible, over the lifetime of the platform. In seeking to protect its capital investment, a service provider therefore needs to be certain that the equipment market is able to meet current needs, and have a high level of confidence that it will be able to meet future needs where it is pragmatic to do so. This can be achieved by establishing collaborative evaluation relationships with key suppliers.

The continually altering business relationships between service providers is driving change in network infrastructure, and, as a consequence, increasing levels of interoperability are required between individual implementations of a given component. The availability of appropriate, high-quality standards specifications is therefore an essential requirement for achieving interoperability between components. Unfortunately, no matter how well the standards communities do their work, there will always be areas left to proprietary extension or interpretation. These aspects render true multi-vendor interoperability a perpetual challenge, if not an impossibility. As a consequence they drive the need for some degree of solution integration to assure reliable operation between components.

In addition to securing 'first-hand' experience of market offerings through the provision of trial and evaluation systems, there is the potential to address two longer-term objectives for the service provider and the manufacturer. Firstly, the manufacturer obtains market-focused direction from the service provider based on their current or imminent needs in response to specific service opportunities. They can also anticipate future needs derived from their plans for market opportunities. Secondly, and in return, the service provider gains visibility of the supplier's product road-maps, thereby obtaining an understanding of the extent to which the supplier may be able to deliver the capabilities necessary to realise future market opportunities. Through this collaborative and trusted approach, the plans of the two stakeholders may be refined to the benefit of both. Furthermore, the wider industry can ultimately be considered to benefit from this collaborative approach between service providers and suppliers as there is a natural tendency for such work to be reflected in the refinement of specifications from international standards bodies.

International standards organisations (identified earlier in Chapter 5) are important stakeholders in technology evaluation facilities through both contribution and benefit. They contribute by providing fundamental criteria against which components can be evaluated objectively through specification and profiling of protocols. From the service provider perspective, standards compliance can lead to interoperability assurance and greater vendor choice, while avoiding vendor 'lock-in' scenarios that have been associated with traditional telephony solutions. As a consequence, a certain degree of protection is afforded to capital investments. Where a solution comprises components from two or more suppliers, demanding compliance to a recognised standard can prove a useful arbiter when interoperability difficulties are encountered — provided that the standard is specified in sufficient clarity and detail.

The end user also benefits from recognised standards since, in addition to providing wider supplier choice at the time of initial purchase, the interoperability afforded by open standards ensures that purchases are more likely to be independent of service provider. Although the end user probably does not (and should not be expected to) have a noticeable understanding of these standards, most users desire longevity and protection of their purchases. For example, an Internet service provider (ISP) market, in which the modem specification is ISP specific, is clearly not sustainable. Similarly, a VoIP market in which the client end-point is completely equipment manufacturer or service provider specific is unlikely to be a viable proposition in the longer term.

17.3 Commercial Contexts for Technology Evaluation

Technology evaluation is much broader than the simple testing of products and solutions against a test specification — it can have an important, if not vital, contribution to make as part of the commercial model followed in procuring and deploying a network solution. In doing so, it leads to a range of benefits for the stakeholders directly involved. In practice, there are essentially two commercial models that relate to technology evaluation — technology push and market pull.

The technology push model is particularly applicable to any new technology since existing product lines within a service provider's business are less likely to either exist or have highly developed product requirements established. Technology push involves the identification, creation and exploitation of a business opportunity as a direct consequence of technical knowledge. It is driven from the understanding of what the technology can do, rather than what it cannot. In the case of VoIP, this has been the most prevalent and effective model found within the industry to date. However, to be successful in a specific case, the technology push model needs to be based on a sound business proposition — as indeed should the market pull model.

In contrast, the market pull model, shown in Fig 17.2, illustrates the usual commercial model followed in bringing new platforms into a service provider's network.

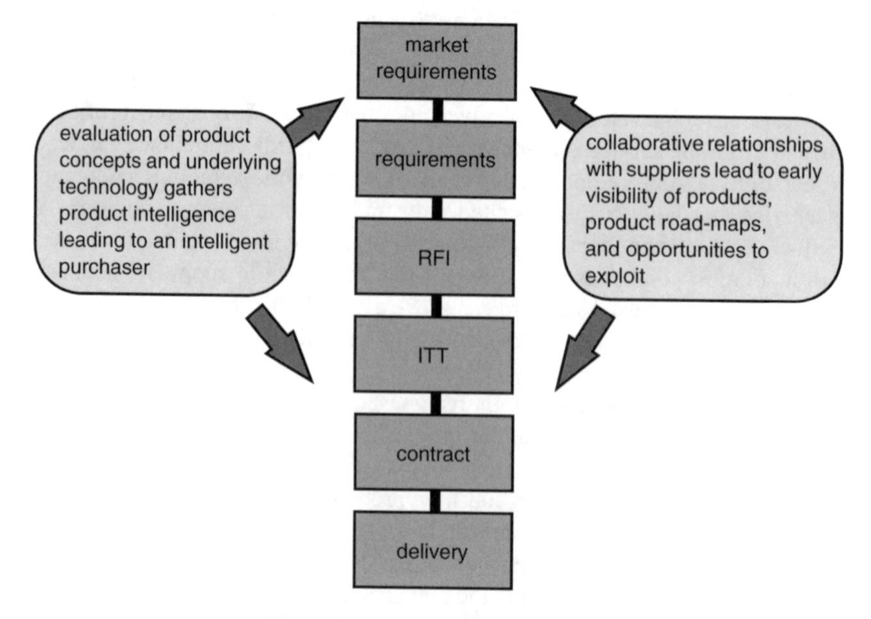

Fig 17.2 The commercial model.

The cycle starts with identifying a market opportunity for a product. This market requirement gives rise to technical requirements for a platform or network capability required to deliver the functionality of the product or service. The usual commercial model takes these requirements through a 'request for information' (RFI) stage to assess, in broad terms, the ability of the market to deliver the requirements and identify the associated costs and time-scales.

If the outcome of this initial RFI phase suggests that the market can deliver the required functionality within acceptable time-scales and at viable cost, potential suppliers are invited to respond to an invitation to tender (ITT). This stage seeks to secure a statement of compliance (SoC) against a more detailed list of requirements, as the basis of a potential contractual agreement. Responses are then analysed through a detailed adjudication process, often including presentations by a short-list of suppliers and visits to reference sites. Finally, a contract is placed with the selected supplier and a delivery project is instigated. Throughout this process the purchaser's direct 'hands-on' contact with the proposed solution is limited and generally restricted to concept demonstrations assembled by the supplier.

The review stages of RFI and ITT often necessitate significant effort in assessing both the basic functionality of the underlying technology and the ability of the supplier to deliver. Clearly, earlier collaborative technology evaluation, independent of any individual direct commercial network requirement, can provide the purchaser with detailed insight, thus enabling stronger focus on the proposed solution rather than the capabilities of component parts. In short, a commitment to technology evaluation equips the network or service provider as an intelligent purchaser, providing scope to influence the development of the market rather than being led by it. It therefore becomes a commercial vehicle for service providers to gain competitive advantage in terms of time to market.

17.4 The Scope of Technology Evaluation

Technology evaluation has many facets. These largely fit within two broad perspectives — component evaluation and solution evaluation.

Component evaluation is concerned with assessing the functionality and performance of a single component. In contrast, solution evaluation is concerned with assessing a complete solution (i.e. engineered from a number of components) against the requirements of a given target application domain.

For both of these evaluation perspectives the activities that ensue may be classified into two broad areas for more detailed assessment. Specifically, these consider the functional aspects and non-functional aspects of the technology being assessed.

17.4.1 Evaluation of Functional Characteristics

Functional evaluation is concerned with assessing the behaviour of a technology — ideally with respect to defined functional requirements for a particular piece of equipment or system. This evaluates what the component or solution does. Establishing an end-to-end call between two VoIP end-points is an example of basic functionality. While it establishes whether a component or solution provides a specific capability, it does not address issues such as scalability of the implementation or the 'quality' of that capability.

17.4.2 Evaluation of Non-Functional Characteristics

Non-functional evaluation identifies 'how well' a capability is provided and establishes whether it is 'usable'. It addresses both quantifiable and subjective aspects such as:

- key performance measures, e.g. voice quality and call set-up time;
- volume characteristics, e.g. number of simultaneous calls, number of simultaneous active users, and scalability;
- management systems provision and usability;
- user documentation;
- security controls;
- auditing provisions;
- resilience.

17.5 Solution Evaluation

When considering the evaluation of a complete solution, the assessment needs to be conducted within a scenario that is representative of how the solution would be used. There are potentially a large variety of possible applications for VoIP technology. However, these can all be considered to fall into a small number of generic scenarios. Assessments are therefore carried out with the equipment configured in one or more scenarios, corresponding to these categories. A candidate set of such scenarios is now considered in detail.

17.5.1 Scenario 1 — PSTN Replacement

The PSTN replacement scenario defines the provision, to end customers, of an end-to-end communications service delivered by IP (i.e. assuming 'always-on' IP access out to the customer premises) and providing the same functionality as that offered by the traditional PSTN. In delivering PSTN equivalence this scenario calls for the call server to deliver the functionality of conventional Class 5 PSTN switches. The scenario can also provide a capability to break out to the conventional circuit-switched PSTN via a gateway.

The generic configuration for this scenario is illustrated in Fig 17.3.

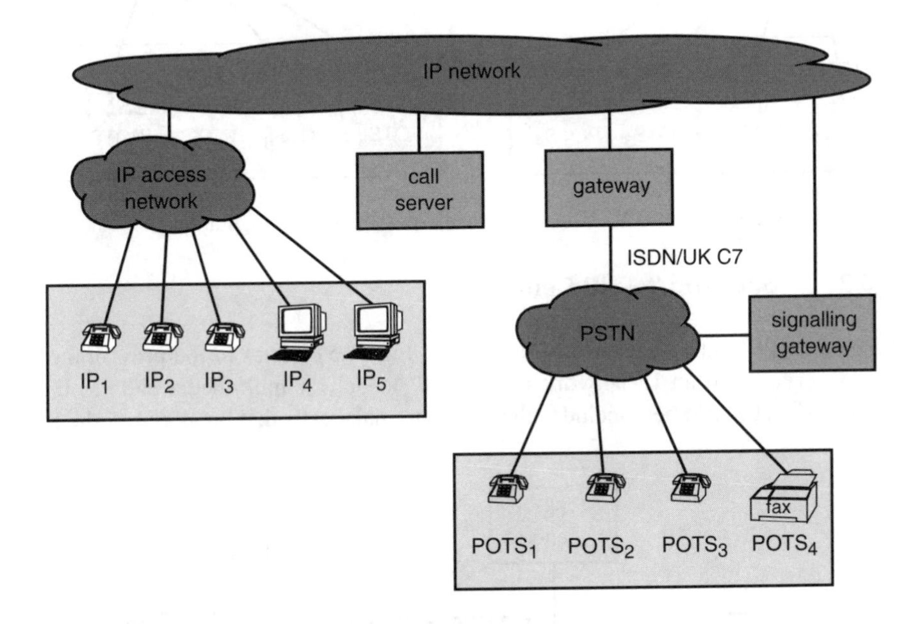

Fig 17.3 PSTN replacement — generic configuration.

17.5.2 Scenario 2 — PSTN Off-Load and Toll By-Pass

This scenario can serve two market niches:

- firstly, to off-load volumes of conventional PSTN access traffic from the trunk level network between access switches (i.e. calling between POTS1 and POTS5) — therefore, the gateway and call server combination are required to deliver the functionality conventionally delivered by a Class 4 switch in a traditional, switched PSTN;
- secondly, as a toll by-pass application allowing PSTN customers to use an IP network for part of their call.

The generic configuration for this scenario is illustrated in Fig 17.4.

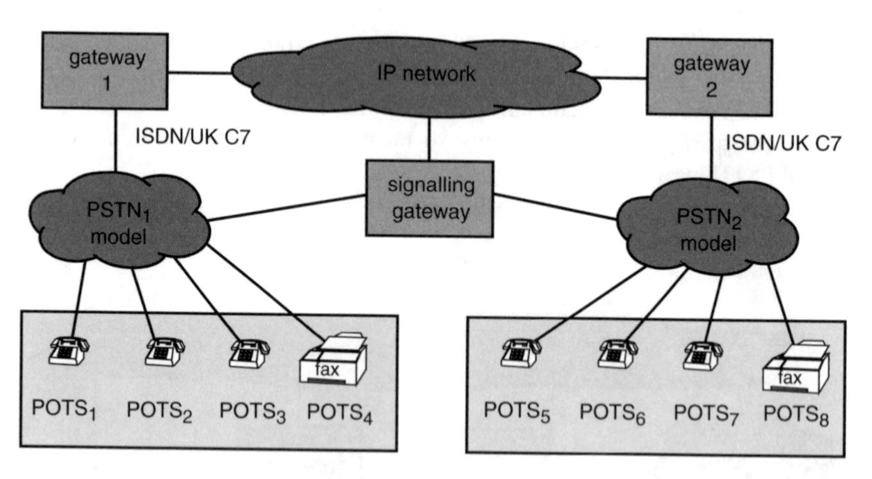

Fig 17.4 PSTN off-load and toll by-pass generic configuration.

17.5.3 Scenario 3 — IP Centrex

The generic IP Centrex scenario (illustrated in Fig 17.5) relates to the provision of centrex services on an IP network using VoIP to deliver multimedia calls to the desktop. The scenario here includes the ability to make calls that break out of the IP

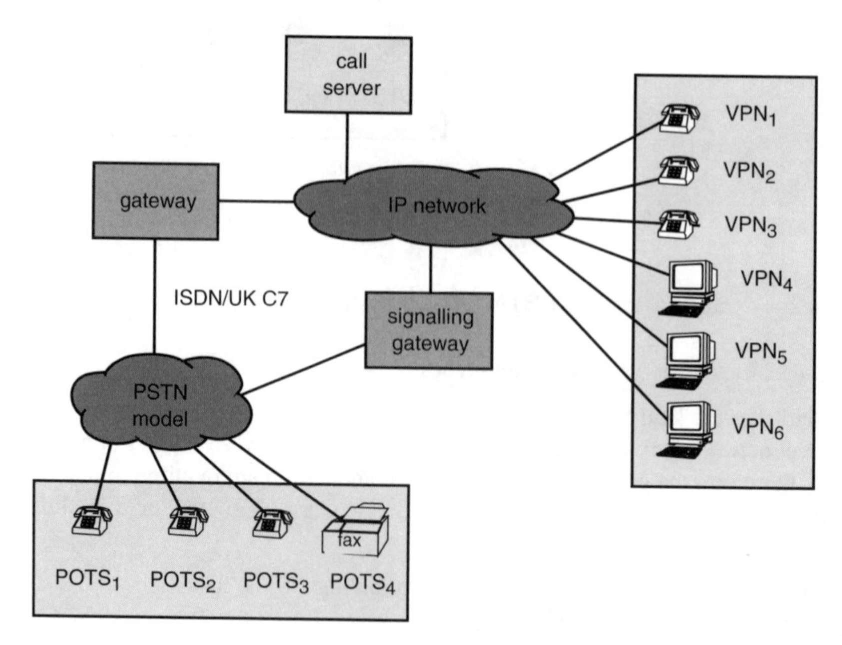

Fig 17.5 IP VPN evaluation — generic configuration.

Centrex network on to the PSTN and for PSTN calls to break in (or direct dial inward) to the IP Centrex network.

The IP Centrex network features a number of hard client devices (VPN1 to VPN3) to support voice-only calls, and a range of soft client devices (VPN4 to VPN6) to support calls with a mix of voice, video and data media. Hard clients refer to dedicated telephony hardware devices, whereas soft clients refer to software running on a PC or similar.

Access to the features of the PSTN is provided by conventional PSTN CPE (POTS1 to POTS4). At least one of these devices will support CLI display and one will support fax transmission.

17.5.4 Scenario 4 — IP Centrex with VPN and PBX Interworking

This scenario is similar to the IP Centrex one, in that it also relates to the provision of multimedia centrex services on an IP network to the desktop. The scenario here includes the ability to make calls that break out of the IP Centrex network into voice VPNs and PBXs and for calls to break in (or direct dial inward) to the IP Centrex network, in addition to PSTN access.

The IP Centrex network features the same client devices (VPN1 to VPN6) but with the addition of voice VPN and PBX clients.

The generic configuration for an IP Centrex with VPN and PBX interworking solution is illustrated in Fig 17.6.

17.5.5 Functional Assessments

The majority of the tests described below can be applied to each of the scenarios, with modifications where appropriate, although some are specific to one or more of them. In all cases, the tests are repeated with different combinations of end-points and with each end-point taking on each of the different roles in the test. Thus, calls are originated and terminated by both PSTN and IP end-points, including hard and soft clients, and public and private networks.

17.5.5.1 Basic Calling

The fundamental test for all scenarios is placing calls between end-points. The basic criteria for success are the ability to set up a call, acceptable speech and the ability to clear the call (by the calling and called party where appropriate). In addition, the tests cover unsuccessful call scenarios, including busy and unobtainable conditions.

Where both end-points have multimedia capabilities, such as video, these are also tested.

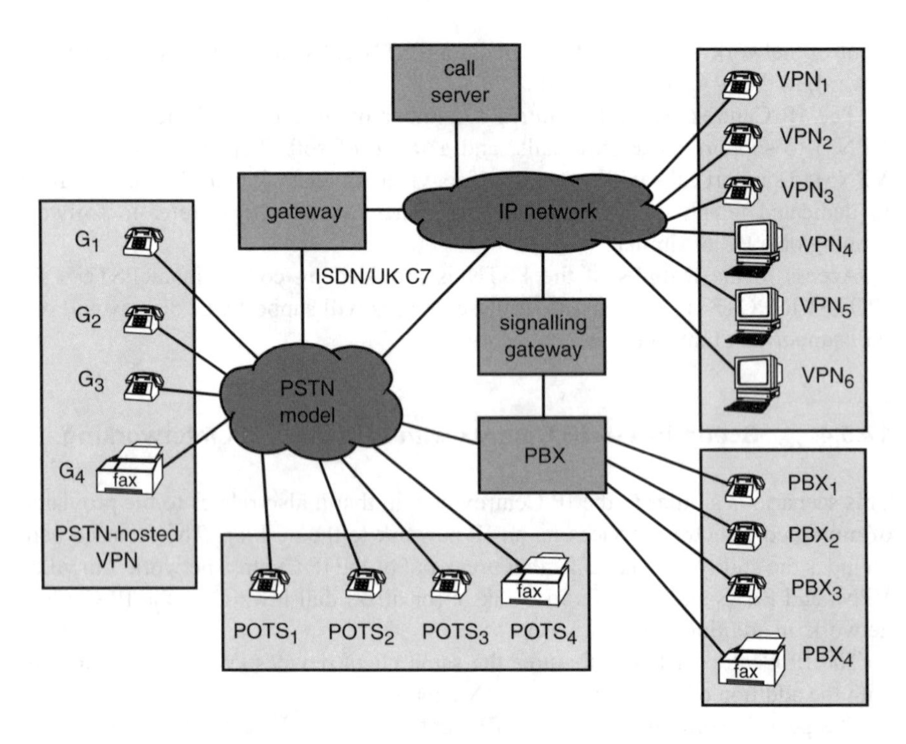

Fig 17.6 IP VPN (with interworking) evaluation — generic configuration.

17.5.5.2 Numbering Plans

Tests are performed to ensure that numbering plans can be configured to allow all end-points to be accessed. This includes allowing IP clients to break out to switched networks, and switched network telephones to break in to IP clients, including the use of direct dialling inwards. The tests involve taking a numbering plan, implementing it on the system under test, using appropriate configuration tools, and then confirming that the plan has been correctly implemented by making calls between end-points.

17.5.5.3 Fax Capabilities

The fax capabilities tested include the basic ability to connect to other fax devices and full interworking through to a variety of devices. The basic tests are performed by making fax calls between normal fax machines; the interworking tests are performed using dedicated fax-testing equipment that tests that each of the protocols involved is correctly carried across the VoIP network. Further tests are carried out that emulate over 100 vendors' fax machines.

17.5.5.4 CLI Features

Calling line identity features centre around the ability of the network to transport the CLI of the calling party to the called party. They include the ability for the CLI to be presented to the called party and to be withheld on request by the calling party and for calls to be rejected by the called party when the CLI is withheld. Presentation of CLI can range from simple presentation of the number, through name look-up, to caller return.

17.5.5.5 Multiparty Features

Multiparty features include call diversion facilities (divert all calls, divert on busy and divert on no reply), call transfers (including call hold/transfer, blind transfer, call park/retrieve), break-in to existing calls (call waiting and executive intrusion), and conference calls.

17.5.5.6 Call-Barring Features

Call barring can be applied to incoming or outgoing calls, and to all calls, or those from/to particular ranges of numbers. 'Do not disturb' and 'call reject' features can also be considered to fall under this category.

17.5.5.7 Voice Messaging Features

Voice messaging covers diversion of calls to voice mail and retrieval of messages, including notification of receipt (message waiting indication). It also incorporates user configuration of voice mail — both local and remote operation.

17.5.5.8 Emergency Service Provision

Emergency service is not yet fully defined for the IP environment, but in the PSTN it includes free access (999/112), caller location, onward connection to appropriate emergency authority, return calls, and last party clear.

17.5.5.9 Directory Service Features

Directories can be specific to the user (personal), cover a wider group (corporate, closed user group), or be global.

17.5.5.10 PSTN Replacement Features

These features only apply to the PSTN replacement scenario — ring back when free, call sign, reminder calls, duration and charge advice, short code dialling, and repeat last call.

17.5.5.11 Toll By-Pass

Toll by-pass requires facilities to interact with the caller to allow account authentication and determination of call destination.

17.5.5.12 Basic Call Centre Capabilities

Call centre capabilities cover a large range of functionalities, but for the current purpose the testing is restricted to the following basic functions — agent registration, automatic call distribution, customer call queuing, answer in order of arrival, agent routing based on most idle, calling party information, and interactive voice response.

17.5.6 Non-Functional Assessment

Non-functional evaluation is that testing undertaken to provide an objective measure on the quantitative and qualitative specifications of a given solution.

17.5.6.1 Call Capacity

The prime consideration is the call capacity of the system. This can be measured in terms of busy hour call attempts and total simultaneous calls. Where a fixed minimum call duration is considered, the former gives a measure of call set-up time. In addition, the capacity of a system is determined by the maximum number of users. These include both the maximum number of registered users and the maximum number of users simultaneously using the system.

17.5.6.2 Voice Quality

Another consideration is voice quality, typically quoted in terms of a mean opinion score (MOS). This can be measured subjectively, using listening tests, or objectively, using equipment that aims to reproduce the results of subjective tests. Objective testing has the advantages of lower cost and the ability to perform a large number of tests in a short period of time.

17.6 Component-Based Assessment

Components are the parts of the solutions described above that can be put together with other components to provide new solutions. Component evaluation tends to concentrate more on the non-functional aspects, rather than the functional aspects, which tend to rely on having a complete solution.

Thus, the evaluation includes interworking and protocol compliance testing, which ensures that the equipment can work successfully with equipment from other suppliers, and voice-quality testing, that measures the basic characteristics of the equipment.

Components include end-points, such as IP telephones and soft clients that run on PCs, and gateways, for connecting VoIP networks to the PSTN and other voice networks. In some cases gateways are implemented as separate media and signalling gateways, together controlled by a media gateway controller. Components also include intermediate equipment, such as H.323 gatekeepers, SIP proxy servers and soft-switches.

17.6.1 Functional Evaluation

Functional evaluation of components typically addresses a subset of the functions tested in solution evaluation, because some functions are only implemented by some of the components of a solution. Unlike solution evaluation, component evaluation tests the ability of the component to interwork with any of a number of different components and thus focuses on the correct implementation of protocols. Components can either be tested in isolation or by connecting them to other reference components. Using reference components allows components to be tested using the same techniques as for solution evaluation. Substituting different reference components gives some degree of confidence that protocols have been implemented correctly.

As with solution evaluation, the main functions tested are those concerned with setting up and clearing calls. However, the focus is now on the correct operation of the protocol in addition to the ability to connect calls. This includes sending correct messages and responding appropriately to received messages. Protocol analysis tools can be used to examine the messages passing over the interfaces between components, and protocol simulation tools can be used to simulate the protocols to check the operation of the component more rigorously.

This allows operation of the equipment to be tested under both normal and exceptional conditions by generating messages with both correct and incorrect parameters. This applies particularly to protocol interworking devices (see Chapter 12) where features in one protocol may be implemented differently, or not at all, in the other.

17.6.2 Non-Functional Evaluation

Non-functional evaluation of components is the same as that of solutions except that it is the performance of the individual component that is measured, rather than the combined performance of a number of components. Thus the evaluation again covers call handling and voice quality.

17.7 Test Bed Infrastructure and Tools

The BTexact Technologies facility at Adastral Park is well placed to offer a world-class VoIP technology evaluation service to the industry. The facilities and capabilities offered at Adastral Park can be characterised by two key differentiators.

Firstly, it has access to a world-class supporting infrastructure, enabling evaluation to be undertaken in the broadest sense. For example, the evaluation facility interconnects with the integration facility (IF), offering access to a range of managed models of all the platforms found in the BT Group's networks including BT's PSTN, intelligent network and data networks. The IF is the only capability of its kind in the UK and represents a world-class capability in the wider context of the European communications industry.

Secondly, it is positioned to offer far more than a basic evaluation service in that it characterises, and possibly identifies problems with, discrete products and complete solutions. The evaluation process can be concluded by calling on the wealth and breadth of technical excellence within BTexact Technologies to provide technical and design consultancy around the results and conclusions of the evaluation. This is illustrated in Fig 17.7.

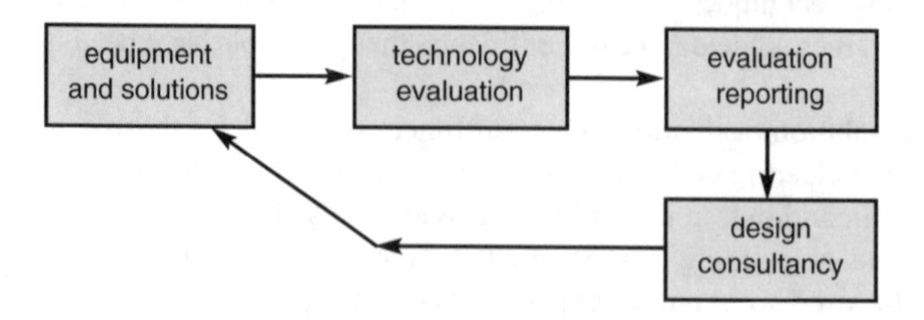

Fig 17.7 Evaluation cycle.

The following sections outline the technical infrastructure and supporting tools available to the VoIP technology evaluation facility.

17.7.1 Call Origination and Termination

The majority of the tests involve placing calls between end-points, whether VoIP gateways or IP client devices, for which a means of generating and terminating calls on a range of terminations is required. For PSTN terminations, the system under test is typically connected to a PSTN network. The BTexact Technologies facility exploits the model PSTN environment within the integration facility at Adastral Park to provide a range of PSTN connections. This provides the ability to originate and terminate calls on normal PSTN telephones. Call-capacity tests require the use of bulk call sending and termination test equipment. These can be connected directly to the equipment under test or via the integration facility.

17.7.2 Protocol Analysis

In theory, different systems developed to conform to common standards should interwork with each other without any difficulty. In practice, differences in the interpretation of standards can result in problems when such systems are interconnected. These problems can be identified using protocol analysers, connected to the appropriate interfaces, be they PSTN or IP. This allows the signalling passing between components to be monitored and decoded, allowing discrepancies from standards to be detected.

Figure 17.8 illustrates the use of call origination and termination, and protocol analysis equipment for functional testing of a gateway, as an example.

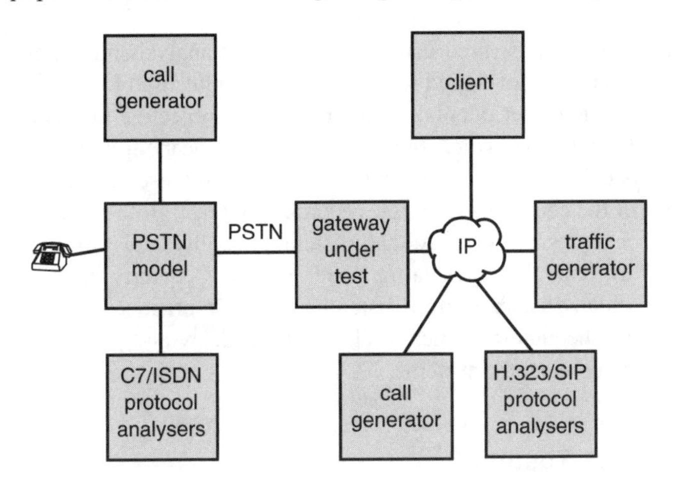

Fig 17.8 Generic functional testing.

17.7.3 IP Network

The functional tests require only simple interconnection of devices at the IP level, which can be achieved with basic IP hubs or switches. However, performance testing needs to take into account the effects of using a real IP network [1], where the level of other traffic can result in packet delay and loss. Packet delay through a network can be variable, leading to jitter, where packets transmitted at a constant rate can be received with varying inter-packet intervals, as discussed earlier, in Chapter 2. Within the end-points, packet jitter is compensated by jitter buffers. These delay packets to allow them to be decoded into a constant rate stream. In contrast, constant delay simply causes the speech to be delayed by a fixed amount. However, the jitter buffer size is limited in order to reduce the amount of delay added. If the level of jitter causes the buffer to over- or under-flow, the resulting speech will be distorted, depending on the codec used.

The effects of a real IP network on the systems under test can be reproduced by simulating the packet delay and loss or by constructing a network of routers and injecting simulated traffic. Simulating packet delay and loss allows the performance to be tested for a given network characteristic, whereas constructing a network allows the limits of performance to be determined. The BTexact Technologies facility is connected to the futures test bed at Adastral Park to provide a routed IP network.

17.7.4 Voice Quality

Voice quality can be measured using the perceptual analysis measurement system (PAMS), a technology developed within BT but now licensed to outside companies (see Chapter 2 for further details). This provides an objective means of predicting the results of subjective voice quality tests. Two mean opinion scores can be obtained — listening quality and listening effort — and these can be used to rate the performance of the codecs used by the systems under test. Both scores lie between 1 and 5, with 5 representing 'excellent' listening quality and 'no effort required' listening effort, and 1 representing 'bad' listening quality and 'no meaning understood with any feasible effort' listening effort. In practice, the voice quality is also affected by the characteristics, such as packet delay and packet loss, of the IP network used to connect the systems.

17.8 Test Results

Equipment from a number of suppliers has been evaluated in the BTexact Technologies VoIP technology evaluation test bed, using infrastructure and test equipment such as that described earlier. As an example, Fig 17.9 shows the configuration for testing the speech quality of gateways, operating over a simulated IP network. The speech quality analyser initiates a PSTN call to gateway 1, which makes a VoIP call to gateway 2 (via the IP network simulation), which in turn initiates a new PSTN call back to the speech quality analyser. At this point the speech quality analyser performs a series of speech quality tests.

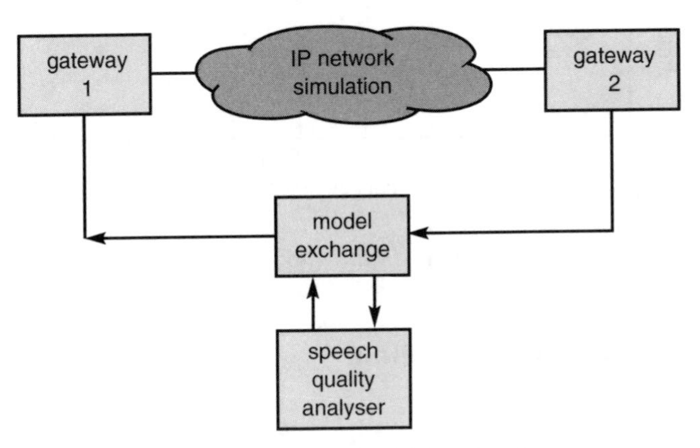

Fig 17.9 Generic PAMS-based testing.

Typical results of such tests are shown in Figs 17.10 and 17.11. In both cases listening effort (LE) and listening quality (LQ) are plotted against a parameter of the IP network simulation. Figure 17.10 shows the effect of jitter, simulated by delaying each packet by a random amount of between 0 and the time plotted.

Figure 17.11 shows the effect of packet loss, simulated by randomly dropping the percentage of packets plotted. A more detailed discussion of the effects of jitter and delay can be found in Chapter 2.

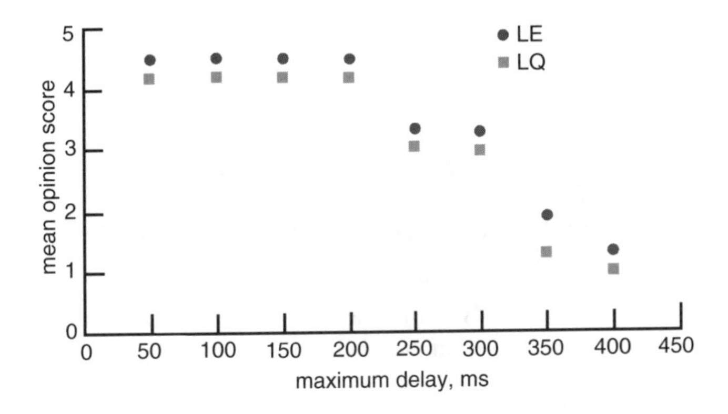

Fig 17.10 Effect of packet jitter.

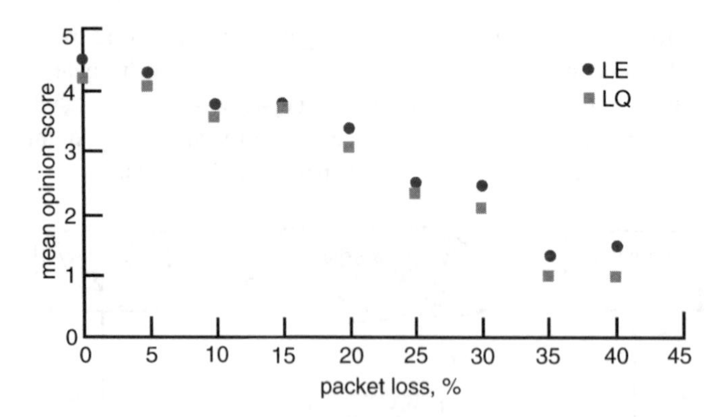

Fig 17.11 Effect of packet loss.

17.9 Benefits of Evaluation

This chapter has described an overall approach to conducting technology assessments on VoIP technologies. Technology evaluation can form an effective business tool for both service providers and equipment manufacturers, by reducing the time to market. From a wider perspective, each of the stakeholders associated with a technology can be considered to benefit as follows.

- Service providers

  — obtain first-hand experience of the products available on the market;

  — gain visibility of products before they reach the market;

  — gain visibility of road-maps and an appreciation of suppliers' 'future thinking';

  — establish the possibility for future requirements to be anticipated by the market.

- Manufacturers

  — obtain feedback from a range of potential customers as a result of exercising components in typical and real network environments, thereby identifying functional suitability and performance for a wider range of applications;

  — establish a collaborative approach with service providers, through which visibility of the service trends in the market can be identified, allowing product road-maps to be more closely aligned with future market demand;

  — the potential to benefit from independent claims relating to the performance and interoperability of components and solutions.

- International standards

  — technology evaluation can provide useful feedback into the standards-forming arena enabling the industry experience to be used to refine and enhance future versions of a standard.

- End users

  — a prime driver behind technology evaluation is to bring better quality products and solutions to the market as quickly as possible and at lowest possible cost — indeed, all three of the benefits described above are ultimately of advantage to the end user.

17.10 Summary

There are two models that form the basis for the introduction of any modern communications technology. This chapter has described how a process of technology evaluation is applicable to both the technology push and market pull models for service development. In particular, technology evaluation has been shown to form an essential role in both models by reducing technical risks of failure in the early stages of the business life cycle. However, irrespective of the approach taken for service development, the technical solution must be complemented with a viable business model. The 'dotcom' industry is now littered with companies that have failed to recognise the importance of this balance in their formative stages — either by poor business propositions or lack of technical competence. While technology assessments cannot resolve poor business planning, a balanced approach to service development that encompasses risk mitigation through technology assessment can help avoid difficulties in the later stages of the project life cycle that may determine the difference between failure and success.

References

1 Willis, P. J. (Ed): '*Carrier-scale IP networks: designing and operating Internet networks*', The Institution of Electrical Engineers, London (2001).

ACRONYMNS

3GPP	Third Generation Partnership Project
AAA	authentication, authorisation and accounting
AAL	ATM adaptation layer
AAP	alternative approval process (ITU-T)
ABNF	augmented Backus-Naur form
ACD	automatic call distributor
ACELP	algebraic CELP
ACF	admission confirm
ACK	acknowledgement
ACM	address complete message
ADPCM	adaptive differential PCM
ADSL	asymmetric digital subscriber line
AINI	ATM internetwork interface
ALC	active level control
ALG	application level gateway
ANM	answer message
API	application programming interface
APM	application transport mechanism/message
APP	application transport parameter
ARJ	admission reject
ARQ	admission request
AS	Applicability Statement (IETF)
ASL	active speech level
ASN.1	abstract syntax notation number 1
ASP	application server process

ASR	automatic speech recognition
ATM	asynchronous transfer mode
ATM-F	ATM Forum
AVT	Audio/Video Transport (IETF)
AXE	automatic exchange equipment
BC	bearer control (layer)
BCF	bearer confirm
BCF	bearer control function
BCP	Best Current Practice (IETF)
BDT	Telecommunication Development Bureau
BE	border element
BER	basic encoding rule
BES	back-end services
BESIWF	BES interworking function
BGP	border gateway protocol
BHCA	busy-hour call attempt
BICC	bearer-independent call control
BIWF	bearer interworking function
BLES	broadband loop emulation service
BR	Radiocommunication Bureau
BRAIN	broadband radio access for IP-based networks
BRJ	bearer reject
BRN	bearer relay node
BRQ	bearer request
C7	CCITT Signalling System No 7
CA	certificate authority
CAC	connection admission control
CBC	call bearer connection
CBR	constant bit rate
CC	call control (layer)
CD	call deflection
CDR	call detail record

CELP	code-excited linear prediction
CGI	common gateway interface
CIC	circuit identification code
CIDR	classless inter-domain routing
CLI	calling line identity
CMN	call mediation node
CMN	configuration management network
CMSS	call management server signalling
CoS	class of service
COT	continuity (message)
CPE	customer premises equipment
CPL	call processing language
CPU	central processing unit
CRV	call reference value
CS	call server
CS	capability set
CSCF	call server control function
CSF	call serving function
CSV	comma separated value
CTI	computer/telephony integration
CUG	closed user group
CVIPS	Concert Voice on IP Solutions (BT project)
DARPAnet	Defense Advanced Research Projects Agency Network
DCF	disengage confirm
DDI	direct dialling in
DECT	digital European cordless telephony
DHCP	dynamic host configuration protocol
DNS	domain name server
DoD	Department of Defense (USA)
DPNSS	digital private network signalling system
DRQ	disengage request
DSL	digital subscriber line

DSS1	digital subscriber signalling system no 1
DTMF	dual tone multifrequency
DTX	discontinuous transmission
DPC	destination point code
EC	European Commission
ECMA	European Computer Manufacturers Association
EDS	ETSI Documentation Service
EFR	enhanced full rate
EIR	equipment information register
ENUM	E.164 number mapping
EoIP	everything over IP
EP	end-point
EP	ETSI Project
ETSI	European Telecommunications Standards Institute
FDDI	fibre distributed data interface
FDM	frequency division multiplex
FoIP	fax over IP
FRF	Frame Relay Forum
GCF	gateway confirm
GCP	gateway control protocol
GEF	general extensible framework
GFP	generic functional protocol
GGSN	gateway GPRS support node
GII	global information infrastructure
GK	gatekeeper
GMSC	gateway MSC
GPRS	general packet radio service
GRQ	gateway request
GSM	global system for mobile communications
GSM-EFR	GSM enhanced full rate
GSN	gateway serving node
GSTN	global switched telephone network

GTT	global title translation
GUI	graphical user interface
GVNS	global virtual network service
GW	gateway
HDLC	high level data link control
HLR	home location register
HPNA	Home Phoneline Networking Alliance
HSS	home subscriber server
HTML	hypertext mark-up language
HTTP	hypertext transfer protocol
IA	interoperability agreement
IAB	Internet Architecture Board
IAM	initial address message
IANA	Internet Assigned Numbers Authority
ICANN	Internet Corporation for Assigned Names and Numbers
ICW	Internet call waiting
ID	identity
IDD	international direct dialling
IETF	Internet Engineering Task Force
IESG	Internet Engineering Steering Group
IFS	Initial Feasibility Study (BT VoIP project)
IGMP	Internet group membership protocol
ILS	Internet locator service
IM	instant messaging
IMPP	instant messaging and presence protocol
IMS	intelligent media server
IMTC	International Multimedia Teleconferencing Consortium
IN	intelligent network
INAP	intelligent network application part
IP	intelligent peripheral
IP	Internet protocol
IPDC	IP device control

IPDR	IP Detail Record
IP-PBX	IP-based PBX
IPS	narrowband switch
IPS7	Internet protocol signalling system no 7
IPsec	Internet protocol security
IPSP	IP signalling point
IPTel	IP Telephony (IETF)
IPTN	international public telecommunications number
IRR	information request response
IRTF	Internet Research Task Force
ISC	International SoftSwitch Consortium
ISDN	integrated services digital network
ISN	interface serving node
ISOC	Internet Society
ISP	Internet service provider
ISSP	inter-softswitch sinalling protocol
ISUP	integrated services user part
IT	information technology
ITSP	Internet telephony service provider
ITT	invitation to tender
ITU	International Telecommunication Union
IUA	Q.921 user adaptation layer
IVR	interactive voice response
IWF	interworking function
IWU	interworking unit
LAN	local area network
LCD	liquid crystal display
LCF	location confirm
LCR	least-cost routing
LDAP	lightweight directory access protocol
LD-CELP	low-delay CELP
LDP	label distribution protocol

LE	listening effort
LQ	listening quality
LRJ	location reject
LRQ	location request
M2PA	MPT2 user peer-to-peer adaptation layer
M2UA	MTP2 user adaptation layer
M3UA	MTP3 user adaption layer
MAC	media access control
MAE	metropolitan area exchange
MAP	Membership Approval Procedure (ETSI)
MAP	mobile application part
MB	middlebox
MC	media control (layer)
MCDN	Meridian Customer-Defined Network
MCU	media control unit
Megaco	media gateway control
MDCP	media device control protocol
MEGACOP	media gateway control protocol
MEM	mediation element manager
MG	media gateway
MGC	media gateway controller
MGCF	media gateway control function
MGCP	media gateway control protocol
MIB	management information base
MIDCOM	Middlebox Communications (IETF)
MIME	multipurpose Internet mail extension
MMSF	media mapping and switching function
MMUSIC	multiparty multimedia session control
MoIP	multimedia over IP
MOS	mean opinion score
MPEG	Moving Picture Experts Group
MPLS	multi-protocol label switching

MP-MLQ	multi-pulse max likelihood quantisation
MS	master socket
MSC	mobile switching centre
MSD	master slave determination
MSF	Multiservice Switching Forum
MSN	multiple subscriber number
MTP	message transfer part
MTU	maximum transmission unit
MWIF	Mobile Wireless Internet Forum
NAPTR	naming authority pointer
NAS	network access server
NAT	network address translation
NDC	national destination code
NDM-U	network data management usage
NGS	next generation switch
NIF	nodal interworking function
N-ISDN	narrowband ISDN
NNI	network-to-network interface
NSN	national significant number
NSO	National Standards Organisation (ETSI)
ODBC	open database connectivity
OLC	open logical channel
OLO	other licensed operator
OPRA	Opinion Poll Registration Activity
OSA	open service access
OSI	open systems interconnection
OSP	open settlement protocol
P(A)BX	private (automatic) branch exchange
PAMS	perceptual analysis measurement system
PAS	publicly available specification (ETSI)
PAT	port address translation
PC	personal computer

PCI	peripheral component interconnect
PCM	pulse code modulation
PDD	post-dial delay
PDU	protocol data unit
PER	packed encoding rule
PESQ	perceptual evaluation of speech quality
PGM	pragmatic general multicast
PICS	protocol implementation conformance specification
PIG	PSTN-to-IP gateway
PIN	personal identification number
PINT	PSTN and Internet Internetworking (IETF)
PIUR	platform-independent usage record
PIWF	protocol interworking function
PIXIT	protocol implementation extra information for testing
PNNI	private network-to-network interface
PNO	public network operator
POTS	plain old telephony system
PPP	point-to-point protocol
PRI	primary rate interface
PSQM	perceptual speech quality measure
PSTN	public switched telephone network
PSU	power supply unit
PUA	personal user agent
PVC	permanent virtual circuit
QoS	quality of service
QSIG	signalling at the Q reference point (ECMA 133 standard)
RADIUS	remote authentication dial-in user service
RAID	redundant array of inexpensive disks
RAM	random access memory
RAS	registration admission and status
RCF	registration confirm
RDBMS	relational database management system

RFC	request for comment
RFI	request for information
RIDE	Recorded Information Distribution Equipment
RNC	radio network controller
ROSE	remote operations service
RPE-LTP	Regular pulse-excited long-term predictor
RRJ	registration reject
RRQ	registration request
RSVP	resource reservation protocol
RTCP	real time control protocol
RTP	real time protocol
RTSP	real time streaming protocol
SAAL	signalling ATM adaptation layer
SAB	start at beginning
SAP	service access point
SC	service control (layer)
SCCP	signalling connection control part
SCE	service creation environment
SCN	switched-circuit network
SCP	service control point
SCTP	stream control transmission protocol
SCU	service/system control unit
SDP	session description protocol
SDPng	SDP next generation
SDSL	symmetric digital subscriber line
SEP	signalling end-point
SET	simple end-point type
SG	signalling gateway
SG	Study Group (ITU-T)
SGCP	simple gateway control protocol
SGSN	serving GPRS support node
SigTran	Signalling Transport (IETF)

SIP	session initiation protocol
SLA	service level agreement
SLG	service level guarantee
SLS	signalling link selection
S/MIME	secure MIME
SMS	service management system
SMS	short message service
SMTP	simple mail transfer protocol
SMU	service management unit
SN	serving node
SNMP	simple network management protocol
SNR	signal-to-noise ratio
SoC	statement of compliance
SOHO	small office, home office
SP	service provider
SPAN	Services and Protocols for Advanced Networks (ETSI)
SPIRITS	Services in the PSTN/IN Requesting Internet Services (ETSI)
SS	signalling system
SS7	Signalling System No 7
SSCF	service specific co-ordination function
SSCOP	service-specific connection-oriented protocol
SSCOPMCE	service-specific connection-oriented protocol in a multi-link and connectionless environment
SSG	Special Study Group (ITU-T)
SSL	secure sockets layer
SSN	SCCP subsystem number
SSP	service switch point
STC	signalling transport converter
STF	Specialist Task Force (ETSI)
SUA	SCCP user adaptation layer
TAP	traditional approval process (ITU-T)
TAR	temporary alternative routing

TC	transaction capabilities
TCP	transmission control protocol
TCS	terminal capability set
TDM	time division multiplex
THD	total harmonic distortion
TIA	Telecommunications Industry Association (USA)
TIPHON	Telecommunications and Internet Protocol Harmonisation Over Networks (ETSI Project)
TLS	transport layer security
ToS	type of service
TMF	Telecommunications Management Forum
TMN	Telecommunications Management Network
TRIP	telephony routing over IP
TS	Technical Specification (IETF)
TSAG	Telecommunication Standardisation Advisory Group
TSAP	transport layer service access point
TSB	Telecommunication Standardisation Bureau
TSN	transit serving node
TUP	telephone user part
TXE	telephone exchange equipment
UA	user agent
UAC	user agent client
UAS	user agent server
UCF	unregister confirm
UCI	universal communications identifier
UDP	user datagram protocol
UI	usage indication
UMTS	universal mobile telecommunications system
UNI	user network interface
URI	uniform resource identifier
URL	uniform resource locator
URQ	unregister request

USB	universal serial bus
USRF	user-selectable routing function
UTRAN	UMTS terrestrial radio access network
VAD	voice activity detector
VHE	virtual home environment
VLAN	virtual LAN
VLR	visitor location register
VoD	voice over data
VoIP	voice over IP
VPN	virtual private network
VSA	vendor-specific attribute
VXML	voice extensible mark-up language
W3C	World Wide Web Consortium
WAN	wide area network
WAP	wireless application protocol
WG	working group
WINA	WAP interim naming authority
WML	wireless markup language
WOS	wait on start
WTSA	World Telecommunication Standardisation Assembly
WWW	World Wide Web
XML	extensible markup language

INDEX